Salt-Affected Soils

Author

István Szabolcs, Ph.D., C. Sc.

Professor, Research Institute for Soil Science
and Agricultural Chemistry
Hungarian Academy of Sciences
Budapest, Hungary

CRC Press, Inc.
Boca Raton, Florida

Library of Congress Cataloging-in-Publication Data

Szabolcs, I. (István)
Salt-affected soils / author, István Szabolcs.
p. cm.
Bibliography: p.
Includes index.
ISBN 0-8493-4818-8
1. Soils, Salts in. 2. Alkali lands. I. Title
S595.S899 1989 88-6492
631.4'16--dc19 CIP

Direct all inquiries to CRC Press, Inc., 2000 Corporate Blvd., N.W., Boca Raton, Florida, 33431.

International Standard Book Number 0-8493-4818-8

Library of Congress Card Number 88-6492
Printed in the United States

PREFACE

Day after day, technical and social developments throw light on new problems, some crucial for the future of mankind, and reveal new aspects of many old problems which have been known for a long time. A fair proportion of such problems is closely related to predictions and prognoses for the next millenium and influences the production and availability of food and other commodities for the growing population of the world. There is an abundance of prognoses of different scales and kinds, but at the same time there is a shortage of reliable analyses of the predictable prospects facing, for instance, fertile soils, sustainable agricultural production, and the carrying capacity of the surface layers of the earth in the face of contamination by harmful elements and destruction by chemical, physical, and biological agents.

One of the hazards, not only for soil fertility but also for water purity and for plant, animal, and human life, is the high salinity of soils and waters. A high concentration of salts may adversely affect waters, food, and the cells of living organisms, but the main medium for their accumulation is the soil. Salinity influences the environment mainly through soil material, and the buffering capacity of the soil plays a dominant role in diminishing or abolishing its effects. This is one reason why soil salinity, which has been known for thousands of years, became a major problem both for theoretical science and practice, particularly during the period of rapid development in science and technology in the last century. Several hundreds of books and many thousands of papers have been published, relating directly or indirectly to the problem of soil salinity in many countries and regions. They generally study the salt-affected soils of various regions and possible methods for their reclamation and utilization. A great part of the technical literature on soil physical, chemical, physico-chemical, geochemical and biological processes also relates to salinity. The interpretation of most of these papers, which investigate particular processes leading to the salinization or desalinization of soils, requires a thorough knowledge of modern physics, chemistry, computer science, etc.

In consequence of the great diversity of salt-affected soils, of the differences in their morphology, appearance, and possibilities of utilization, it is difficult to formulate general rules which apply to all salt-affected soils. It is even more difficult to adapt experiences and methods gained and used in one place to the conditions of another, even if considerable experience has been gained in the other region, as there are quite a few different classification systems in use for salt-affected soils. The situation is no better as regards analytical methods, threshold values, adverse effects, or harmful substances in respect to salt-affected soils or the materials suitable and available for their reclamation.

Thanks to the execution of new surveys, the discovery of new productive areas, the introduction of new methods (such as aerial photography and remote sensing), and the application of sophisticated equipment, salt-affected soils have been identified in many areas where they were unknown before. Today not only is the total recorded area of salt-affected soils much larger than just a few decades ago, but also the variability in their properties is much greater, and the processes of their formation are much more diverse than was previously realized.

Parallel with these developments, over the last decade a great number of publications have appeared on salt-affected soils, discussing different problems concerned with the behavior, main processes of formation, possibilities of utilization, and practical value of these soils.

The aim of this book is to make a state-of-the-art review of the whole problem of soil salinity, including the nature of this phenomenon and the extent of salt-affected soils in different parts of the world in relation to the environmental conditions under different climatic, geochemical, and hydrological circumstances. It is also important to characterize the main areas where salt-affected soils appear and the physical and chemical properties that have a bearing on the possibilities for agricultural production, horticulture and sylviculture, and on environmental problems.

It is also necessary to interpret the pedological characteristics of various types of salt-affected soils as well as their morphology, structure, physics, and chemistry. The processes that result in saline and alkali soils are very diverse, but they have one thing in common: they all develop due to the direct effect of water-soluble salts. This is a common occurrence, so there is a great need to describe and interpret the soil-forming processes leading to the development of different types of salt-affected soils. The physics and chemistry, and in part the physico-chemistry and geochemistry of such phenomena are widely described and discussed in the literature, but a comprehensive compilation is still needed. The same goes for the pedological and biological processes resulting in salt-affected soils, which have been given less attention than the chemistry, physico-chemistry, and even the hydrology of these formations. The elaboration of effective systems for the improvement and utilization of such soils is only possible in the possession of a thorough knowledge of these processes.

There are vast territories on all continents covered by different types of salt-affected soils, but besides those formed by natural, primary soil-forming processes, human activity (particularly irrigation), has also created large tracts of so-called secondary salt-affected soils. The extent of such soils is similar to that of soils which have been formed without human interference; however, the importance and hazard of this phenomenon are much greater. It is therefore vital to characterize man-made soil salinity and to outline, or in some cases to elaborate in detail the methods for combating it.

More than 100 countries of the globe are confronted with problems of soil salinity, and these problems are highly diverse, not only in respect to the extent but also as regards the properties, geochemistry, and value of the soils. There are countries where only a few percent of the total area is saline, whereas in others 30, 40, 60%, or in some cases almost 100% of the arable land can be termed as such. There are salt-affected soils with a coarse texture and others with a heavy clay texture. Some salt-affected soils are strongly acidic, others have an extremely alkaline pH. Since many other diversities also exist, it is evidently not easy to cover all varieties and specifications.

The author of this book is fully aware that to give a comprehensive picture of the problem in a limited volume is only possible through compromises. Mention must be made of nearly all the important problems of soil salinity without going into detail, and sometimes even without going into depth. In consequence, a lot of information and references must be left out in the course of compilation. Nevertheless, if this book represents a step forward toward a better recognition of the global problem of salt-affected soils, of their importance, and of possibilities for their study and utilization, the efforts of the author will not have been in vain.

THE AUTHOR

István Szabolcs, Ph.D., C.Sc., D.Sc., was born in Turkeve, Hungary, in 1924. He graduated from the University of Debrecen, Hungary, and received a Ph.D. degree in chemistry in 1948. In 1953 he obtained the Candidate of Agricultural Sciences degree in Moscow, U.S.S.R., and in 1959 the Doctor of Agricultural Sciences degree at the Hungarian Academy of Sciences in Budapest. He was appointed Professor of Soil Science at the L. Eötvös University, Budapest, in 1968. He was director of the Research Institute for Soil Science and Agricultural Chemistry of the Hungarian Academy of Sciences, Budapest, until 1980.

Dr. Szabolcs is the President of the Hungarian Soil Science Society. He acted as the Chairman of the Subcommission on Salt Affected Soils of the International Society of Soil Science from 1965 to 1982; he later became Deputy Secretary General of the Society. He was a participant and officer at many international gatherings. In 1977 he was the Vice President of the United Nations Conference on Desertification held in Kenya. He is Editor-in-Chief of the journal *Agrokémia és Talajtan* (Agrochemistry and Soil Science), and the *International Scientific Centre of Fertilizers* (CIEC), and a member of the editorial board of eight international journals.

His field of research is soil classification, soil mapping, and particularly, the problems of the genesis, characteristics, amelioration and utilization of salt-affected soils. He has thoroughly studied the effect of irrigation and drainage on soils and their properties.

Dr. Szabolcs has been an invited lecturer and consultant of the Food and Agriculture Organization (FAO), United Nations Educational, Scientific and Cultural Organization (UNESCO), and United Nations Environmental Programme (UNEP), in different developing countries of Africa, Asia, and Latin America. He was the director of several international postgraduate courses and seminars (among them were the International Postgraduate Course on Soil Salinity and Alkalinity organized conjointly by the Hungarian Academy of Sciences and UNESCO in 1973; the Hungaro-Indian Seminar on Management of Salt Affected Soils in 1981; etc.).

Dr. Szabolcs has to his credit more than 500 scientific publications and was author or editor of 10 books published in Hungary and abroad. Among these, mention should be made of the *Handbook of Large-Scale Genetic Soil Mapping* (OMMI, Budapest, 1966), which is the manual for soil survey and mapping in Hungary; *European Solonetz Soils and Their Reclamation* (Akadémiai Kiadó, Budapest, 1971); *Salt Affected Soils in Europe* (Martinus Nijhoff, The Hague; and the Research Institute for Soil Science and Agricultural Chemistry of the Hungarian Academy of Sciences, Budapest, 1974); *Modelling of Soil Salinization and Alkalization* (Agrokémia és Talajtan, Vol. 28, Suppl., 1979); *Review of Research on Salt Affected Soils* (UNESCO Natural Resources Research, XV, 1979); and *Fight Against Hunger Through Improved Plant Nutrition* (Proc. 9th World Fertilizer Congress; CIEC-Hungarian Academy of Sciences, Goltze-Druck, Goettingen, 1985).

Dr. Szabolcs is a foreign member of the Academy of Agricultural Sciences of the G.D.R., an honorary member of the All-Union Society of Soil Science, Moscow, and fellow of the Indian Society of Agricultural Chemistry.

For his achievements, he has received two Governmental Awards, the "Tessedik" Gold Medal of the Hungarian Agronomical Society, the Award of the Hungarian Union of Scientific and Technological Societies, and the "Treitz Medal" of the Hungarian Soil Science Society.

CONTRIBUTOR

Katalin Darab, Ph.D.
Scientific Advisor
University of Horticulture
Budapest, Hungary

TABLE OF CONTENTS

Chapter 1

SALT-AFFECTED SOILS AS A WORLD PROBLEM

I. THE MAIN ASPECTS OF GROUPING SALT-AFFECTED SOILS

Salt-affected soils occur under different environmental conditions and have diverse morphological, physical, chemical, physico-chemical, and biological properties, but one common feature, the dominating influence of electrolytes on the soil-forming processes, joins them into one family.[1] A certain concentration of electrolytes determines not only the morphology of the soil profile, but also those physical, chemical, and biological properties which result, as a rule, in low fertility and limited agricultural value on land affected by salinity.[2] Where the above-mentioned adverse effect of electrolytes exists, their influence dominates the soil-forming processes and soil properties to such a degree that it cannot be balanced by common agricultural methods such as the use of fertilizers, agrotechnics, etc., unless the excess of electrolytes is removed or neutralized.[3,4]

It is evident that all soils contain a certain quantity of water-soluble compounds but they should only be classified as salt-affected when the concentration of the soil solution surpasses a certain threshold value.[5,6] The numerical value of this depends on the geochemical and environmental conditions and on the physical and physico-chemical properties of the soils, as well as on the chemical composition of the salts causing salinity. (A detailed description of these circumstances will be given later in this book.) What must be recognized at the very beginning, however, is that different levels of salinity may result in the formation of saline or alkaline soils in different places and under different circumstances.[7,8]

Water-soluble compounds other than electrolytes (e.g., organic matter in water-soluble forms or nondissociating inorganic compounds) also often occur in various soils. As a rule they do not cause salinity directly. However, by indirect processes these compounds may sometimes lead to the alkalinization or salinization of soils.[9] This will also be discussed later.

While the decisive influence of electrolytes is common in all salt-affected soils, their chemical composition, as well as their concentration hazard, varies depending on environmental, pedological, and even economic conditions.[10,11]

In many soil classification systems the term "salt-affected soils" is limited to those saline and alkali soils, where the neutral or alkaline hydrolyzing sodium salts dominate.[12-18] It is true that most salt-affected soils belong to these two groups, but in a system aimed at casting light on the global importance of the problem, all soils which have developed under the dominating influence of electrolytes should be included.

If it is agreed that certain amounts and kinds of salts produce salt-affected soils, this family of soils must include all the formations which have developed under the influence of electrolytes. If such a process advances, in principle, all kinds of soils in any place and condition will be salinized whatever their original physical, chemical or biological properties, or their original place in the hierarchy of soil classification systems.[19]

Salt-affected soils can be defined as formations under the dominating influence of different electrolytes in solid or liquid phases which alter the physical, chemical, and biological properties, and eventually the fertility of the soil. Salinization has a dominating role in soil formation because whenever it occurs it produces a specific formation — the physical, chemical, and biological properties of which will be determined by it. Its effect will overshadow former soil-forming processes and radically change the properties of the soil. This is why salt-affected soils are to be found so widely in all parts of the world, and why they are so diverse in their morphological, physical, chemical, and biological properties, but so similar in their low fertility and in the difficulties facing rational utilization.[1]

Table 1
TENTATIVE CORRELATION BETWEEN THE MOST WIDELY USED CLASSIFICATION SYSTEMS OF SALT-AFFECTED SOILS

Table 1A — Subcommission on Salt-Affected Soils Classification

	In the legend of	
Basic Grouping, 1967	**Map of European Salt-Affected Soils, 1968**	**Saline and Sodic Soils Map of Australia, 1971**
Saline soils	Saline soils	Saline soils
Alkali soils without a structural B horizon	Alkali soils without a structural B horizon	Alkaline sodic soils AS1 AS2 AS3
Alkali soils with a structural B horizon	Alkali soils with a structural B horizon Solonchak-solonetz and calcareous solonetz Noncalcareous solonetz with A horizon <15 cm Solodized and/or deeply leached solonetz Solonized and slightly salt-affected soils with minor structure formation Deeply leached solodized soils and solods	Nonalkaline sodic soils NS1 NS2

There are many classification systems for salt-affected soils. Some of these are incorporated into the generally accepted world soil classification systems, such as Soil Taxonomy, the classification system of the USSR, the soil classification system of the FAO/UNESCO Soil Map of the World and Soil Map of Europe, and may others.[1] An even larger number of classification systems exists in individual countries, particularly in those where salt-affected soils are common.[19] There are many local names and terms for the different kinds of salt-affected soils in such countries and it is difficult to find correlations, if any, between them.[20] There are also numerous systems which group salt-affected soils for reclamation purposes.[21] Most of these systems are based on one or another adverse property of these soils which should be altered during the reclamation process. It is difficult to elaborate a perfect method of conversion between the different classification systems and this cannot be expected in the near future either, as, in the taxonomical hierarchy of soil classification systems, the salinity of soils is indicated at different levels: in some classification systems at a high taxonomical level, and in others at a low level.[22,23]

In the following chapters different groupings and classifications of salt-affected soils will be applied and discussed, depending on the aim of the treatise. One of the general principles of soil classification, i.e., that there cannot exist a universal system for all places, all scales, and all purposes, is probably also valid for salt-affected soils. As soils have a specific character and a specific place and role in nature, a compromise must be made, accepting the existence of different groupings and classification systems depending on the purpose and kind of soil study. Different systems should be applied when drawing up a map of the distribution of salt-affected soils over a whole continent and when compiling a detailed soil map for a single farm. Specific classification systems must be elaborated for combating secondary salinization on irrigated areas. Evidently there must be a general principle underlying all the different groupings and classification systems, but this cannot be more than the main rule for the definition of salt-affected soils.[4]

In Table 1 a tentative correlation is given, showing the convertibility of the most widely used world classification systems for salt-affected soils.[24]

Table 1B

Soil Map of the World (Soil Map of Europe) FAO/UNESCO Project (ECA Working Party) Classification	Australian Classification 1968	Canadian Classification 1965	French Classification 1967	USDA Classification 1967	USSR Classification 1967
Solonchak Orthic solonchak Mollic solonchak Takyric solonchak Gleyic solonchak	Solonchak	Saline subgroups (part)	Saline soils (except for acidic subgroup) Saline soils with alcalins Gley salt soils	Salorthids Salorthidic calciustolls Salorthidic haplustolls Halaquepts (part)	Fluffy solonchak (non steppic) Crust solonchak Soda solonchak (non steppic) Fluffy solonchak (steppic) Soda solonchak (steppic) Takyra Meadow solonchak
Solonetz	Solonetz and solodized solonetz		Sodic soils with a B horizon and solodized solonetz	Nadurargida Natrargida Natriboralfs Natrudalfs Natrustalfs Natrixeralfs	Desert-steppe and desert solonetz
Orthic solonetz		Brown solonetz (part) and alkali solonetz			
Mollic solonetz		Black, gray, and brown solonetz (part)		Natralbolls Natriborolls Natrustolls Natrixerolls	Steppe solonetz
Gleyic solonetz		Gleyed solonetz		Natraquolls Natraqualfs	Meadow solonetz
Solodic planosols	Solod and solodic groups (part)	Solod and solodic subgroups (part)	Solods	Argiabolls	Solod

Table 1 demonstrates the complexity of the correlation between the classification of salt-affected soils in different systems. In spite of its complexity, the table does not cover all kinds of salt-affected soils. It deals with two major groups only, namely with saline and alkali soils, which are certainly the most frequent and important; other groups of salt-affected soils are not included.

In Table 2 a practical grouping of all salt-affected soils is shown, including all kinds and types developing under the influence of electrolytes and including all practical possibilities for the chemical composition of the salts.[4]

As Table 2 clearly shows, at least five groups of salt-affected soils are influenced by different chemical types of electrolytes. The five electrolyte compositions selected not only represent different salts, but their effect on the formation of soils is also significantly different.

Saline soils developing mainly under the influence of sodium, chloride, and sulfate are closely associated with deserts and semideserts and seldom occur in subhumid or humid regions.[25] Saline soils are the most widespread type of salt-affected soils on the continents, and, with the exception of Europe, they dominate within this soil family.[26]

Table 2
GROUPING OF SALT-AFFECTED SOILS

Electrolyte(s) causing salinity and/or alkalinity	Type of salt-affected soil	Environment	Main adverse effect on production	Method for reclamation
Sodium chloride and sulfate (in extreme cases nitrate)	Saline soils	Arid Semiarid	High osmotic pressure of soil solution (toxic effect)	Removal of excess salt (leaching)
Sodium ions capable of alkaline hydrolysis	Alkali soils	Semiarid Semihumid Humid	Alkali pH Effect on water physical soil properties	Lowering or neutralizing the high pH by chemical amendments
Magnesium ions	Magnesium soils	Semiarid Semihumid	Toxic effect High osmotic pressure	Chemical amendments Leaching
Calcium ions (mainly $CaSO_4$)	Gypsiferous soils	Semiarid Arid	Acidic pH Toxic effect	Alkaliine amendments
Ferric and aluminum ions (mainly sulfates)	Acid sulfate soils	Sea shores, lagoons with heavy sulfur containing sediments	Strongly acidic pH Toxic effect	Liming

Alkali soils formed under the influence of sodium ions capable of alkaline hydrolysis occur mainly in semiarid, semihumid, and humid regions, but they can also be found under practically any environmental conditions, from the polar circle to the equator. They also occur at different altitudes from sea-level to high mountains.[24,27]

A specific group of salt-affected soils develops under the strong influence of magnesium ions. This group has been studied very little, but it has both theoretical and practical importance in semiarid and semihumid areas. The soils of this group are closely related to various heavy-textured formations such as vertisols, smonitzas, etc. Apart from magnesium ions, sodium ions also often play a significant role in the formation of magnesium soils. This is one more argument for the inclusion of this group in the family of salt-affected soils.

The soils affected by substantial amounts of soluble calcium salts, mainly calcium sulfate, are the so-called gypsiferous soils, which appear widely in semiarid and semihumid regions. They are common in North Africa, the Middle East, in the arid zones of the U.S., in China, in the Central Asian Soviet Republics, and in many other places. They constitute another specific group of salt-affected soils.[28]

The acid sulfate soils which are specific soil formations on seashores and coastal lowlands under waterlogged conditions which have developed on sulfur-containing sediments also belong to this family and represent a specific group of salt-affected soils. This type of soil also may be found under various climatic conditions from the polar circle to the equator on the seashores of all continents, and is widespread in the salt marshes of coastal areas.[29]

Formerly it was considered that salt-affected soils were always related to arid conditions, and statements to this effect may still be found in many places in the literature. Keeping in mind the diversity of the different types of salt-affected soils, it becomes obvious that they not only occur in desert and semidesert areas but may develop in practically all the climatic zones of the world. It is true that aridity leads to intensive salt accumulation in rocks, waters and soils, as will be discussed later in this book, and this means that salt-affected soils are more extensive in these areas than in nonarid regions.[4] Nevertheless, the nonarid types of salt-affected soils must also be taken into consideration, particularly because of their significance in connection with irrigation and agricultural utilization.

It is also true that in river basins and bottomlands where, due to geomorphological

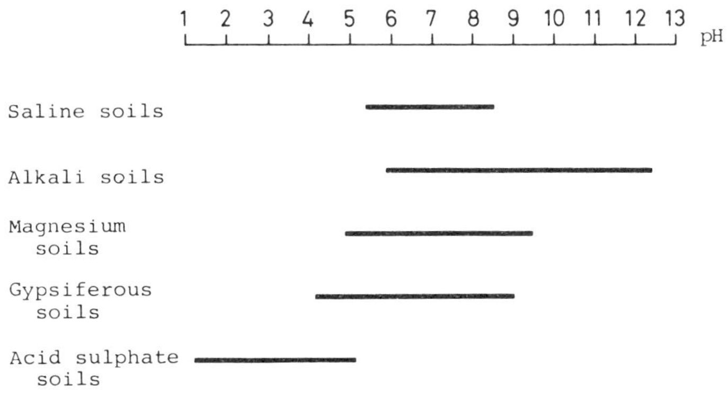

FIGURE 1. pH spectrum of different salt-affected soils.

circumstances, salt accumulation is intensive, conditions are favorable for the salinization of soils. At the same time, in mountainous regions of high and medium heights, salt-affected soils (most of them alkaline) are also more frequent than was previously realized.[30]

Regarding the pH value of salt-affected soils, it can be stated that there is no other soil family where soil reactions may be so diverse and where such extremes can be met with. In Figure 1 the pH spectrum of different salt-affected soils is demonstrated.

Figure 1 clearly shows that the most extreme pH values, both acid and alkaline, occurring in soils in nature are associated with various groups of salt-affected soils. Nearly the whole pH spectrum is covered by the various types of salt-affected soils, ranging from a pH as low as 1.4 to one as high as 12 or more. Evidently, the five groups of salt-affected soils listed in Table 2 occupy different positions in this spectrum. Saline, magnesium, and gypsiferous soils tend to group in the middle, whereas alkali soils have the highest pH. The other extreme end is represented by the acid sulfate soils in the low pH range.[4]

The five groups in Table 2 also differ with respect to the effects of various electrolytes both on soils in particular and on agri-, sylvi-, and horticulture in general. The best known effect of high salinity, the high osmotic pressure of the soil solution which hinders the uptake of water by plants, is associated mainly with saline soils and partly with magnesium and gypsiferous soils.

The toxic effect of salts is also associated mainly with saline and partly with magnesium soils. In alkali soils the high pH value represents the most important adverse effect on both soil properties and plant life. At the other end of the scale, it is the low pH of acid sulfate soils and partly of gypsiferous soils which has the main adverse effect.[19]

Toxic effects, other than high sodium chloride and sodium sulfate concentration, may occur in soils in different groups, e.g., the toxic effect of boron in alkali soils or of aluminum in acid sulfate soils.[31,32]

Table 2 also distinguishes the five groups of salt-affected soils in relation to their utilization and reclamation. As a chief principal it can be accepted that in order to improve salt-affected soils, the causes of their low fertility must be eliminated, i.e., if the salt concentration is too high in relation to the tolerance of the plants the excess salts must be removed. In the case of the extreme pH values of alkali or acid sulfate soils, the soil reaction should be altered by the application of acidic or alkaline amendments, respectively.

Evidently, if toxic compounds are present in the soils they should also be removed or their adverse effects offset. In soil reclamation and irrigation practice, many failures have been due to the disregard of such simple principles.

The five groups of salt-affected soils differ not only in the properties and effects demonstrated in the columns of the table but also in the physical, chemical, and biological processes by which they were formed and in their geochemical and environmental significance;[33] some of these factors will be discussed in later chapters.

Table 3
SALT-AFFECTED SOILS ON CONTINENTS AND SUBCONTINENTS

Continent and/or subcontinent	Area (thousand hectares)
North America	15,755
Mexico and Central America	1,965
South America	129,163
Africa	80,538
South Asia	87,608
North and Central Asia	211,686
South East Asia	19,983
Australasia	357,330
Europe	50,804
Total:	**954,832**

The diversity of salt-affected soils partly accounts for the difficulties in classification mentioned previously. It also explains why different classifications and group patterns are necessary for the study of salt-affected soils, depending on the target. On the following pages numerous systems will be mentioned or applied, but the grouping demonstrated in Table 2 — though it evidently cannot replace the numerous classification systems for salt-affected soils — may nevertheless provide a broad, complete picture of the diversity of their properties, the environmental conditions under which they occur, and the practical possibilities for their utilization, and thus constitute a general pattern not only for the interpretation of salt-affected soils in the forthcoming chapters but also for many other studies on their theoretical and practical treatment.[34-38]

II. THE GLOBAL IMPORTANCE OF SALT-AFFECTED SOILS

The global importance of salt-affected soils, the significance of which is growing continuously at present, can be explained primarily by their wide distribution on all continents, as is demonstrated in Table 3.

Table 3 shows that salt-affected soils are to be found on all continents, covering about 10% of the total surface of dry land. Their extent is different on various continents and subcontinents: e.g., in Australasia they occur on a much higher percentage of the total surface than anywhere else. In Europe, North America, and Central America their extent is comparatively smaller than on other continents.[24,39]

It should be noted that the figures in Table 3 do not always register exactly the areas of existing salt-affected soils, because new territories of salt-affected soils are found extremely often, either by revising former soil surveys or because of secondary salinization due to the extension of irrigation.

In recent years salt-affected soils have been described in many countries where formerly they were unknown, and probably in the future more information will be available on their appearance in different areas.

At present different types of salt-affected soils are known to occur in about 100 countries. A list of countries where salinization and/or alkalization of soils not only appears, but represents a serious problem, is given below (magnesium salinity and acid sulfate soils not included).

- *Europe*: Austria;[40] Bulgaria;[41] Czechoslovakia;[42] Cyprus, France;[43] Greece;[24] Hungary;[15] Italy;[24] Portugal;[24] Romania;[44-47] Spain;[48-50] U.S.S.R.;[51-53] and Yugoslavia[54-57]
- *North America*: Canada[58,59] and the U.S.[1,2,60]

FIGURE 2. Global distribution of salt-affected soils.

- *Mexico and Central America*: Cuba, Haiti, and Mexico[1,61]
- *South America*: Argentina,[62] Bolivia,[63] Brazil, Chile, Columbia, Ecuador, Paraguay, Peru,[64] and Venezuela[63]
- *Africa*: [1,65-67]Afars and Issas Territory, Algeria;[68] Angola, Benin, Botswana, Burkina Faso, Burundi;[69,70] Chad;[71,72] Cameroon;[73] Egypt;[74] Ethiopia, Gambia, Ghana, Guinea, Kenya, Liberia, Libya, Malgash Republic, Mali, Mauritania, Morocco;[75] Niger, Nigeria, Senegal;[76] Sierra Leone, Somalia, South West Africa, Sudan;[77-79] Tanzania, Tunisia, Zaire, Zambia, Zimbabwe[80]
- *South Asia and the Middle East*: Afghanistan, Bangladesh, Burma, India;[21,81-84] Iran;[85] Iraq;[86-88] Israel;[89] Jordan, Lebanon, Kuwait, Muscat and Oman, Pakistan;[90] Quatar, Sarawak, Saudi Arabia, Sri Lanka, Syria;[91] the Trucial States, Turkey;[92] Yemen, the People's Democratic Republic of Yemen[1]
- *North and Central Asia*: China;[93,94] Korea;[95] Mongolia;[96,97] Solomon Islands, the USSR[98]
- *South East Asia*: Indonesia;[99] Cambodia, Malaysia, Thailand;[29] Vietnam
- *Australasia*: Australia,[100-102] Fiji[1]

In Figure 2 a general map of the global distribution of salt-affected soils has been compiled based on available information on the subject. The aim of the map is to give a general survey of the places where salt-affected soils are met with, without exactly indicating the scale of their extent and without specifying the different groups of this soil family. In consequence of the diversity of the technical sources (in some places detailed information is available on the extent, properties and utilization of salt-affected soils, while in many other places little or no information can be found) the map can only be considered as a comprehensive picture without applicability to local problems. Nevertheless, it is obvious from this map that salt-affected soils really constitute a global problem because they occur practically everywhere on the globe, and their extension is quite considerable. The map also shows that:

- The size of salt-affected soils ranges from small spots to large areas
- No continent is free from salt-affected soils
- No climatic zone is free from salt-affected soils, though their appearance is focused either in deserts and semideserts or in bottomlands and river valleys

The map covers all five groups of salt-affected soils in accordance with Table 2.

In consequence of the scale of the map, the spots are not fully proportional, especially the small ones. This is particularly true of the extension of acid sulfate soils, which sometimes only occupy narrow strips on seashores, a kilometer or less in width.

The distribution of salt-affected soils in the different countries is more precisely registered in Table 4. The data for saline soils are included under the headings "Solonchak" and "Saline phase", while the figures indicating the areas of alkali soils are listed in the columns "Solonetz" and "Alkali phase". This kind of grouping is a necessary consequence of the limited availability of information, which could only be obtained for these two groups and not for magnesium, gypsiferous, and acid sulfate soils. But, as has been mentioned before, these two groups represent the majority of the salt-affected soils in the world.[103] In Table 4 the only countries included are those where numerical data are at our disposal either from national or from international sources.[19,22]

In Table 4, according to the type of information available, either the territory of different salt-affected soils is given together with their total amount, or, where information was limited, only their total value is indicated. Table 4 covers all continents and subcontinents except Europe. For Europe more detailed surveys and records on salt-affected soils are available than for any other continent.[24]

The data set out in Table 5 are based on a coordinated project in which 12 European countries participated for the study of salt-affected soils in Europe. In Table 5, not only the saline and alkali soils are distinguished but also the subgroups of alkali soils. An explanation of the significance of these subgroups will be given later. Besides the existing salt-affected soils in Europe, Table 5 also contains data on so-called potentially salt-affected soils which may be salinized due to the incorrect methods of irrigation which will be discussed in detail in the forthcoming chapters.[104]

Those European countries which are primarily interested in the salinity and alkalinity of soils are, with few exceptions, those with a semiarid climate. Most of the salt-affected soils in Europe are located on the territory of the U.S.S.R., which can be explained by the vast dimensions of this country as well as by the extensive dry areas.[105,106]

This is followed by Hungary, where the main factors contributing to the extension of salt-affected soils are not climatic, but related to the geochemical and hydrological conditions of the country.[24] It should be noted that typical Mediterranean countries such as Greece, Italy, Portugal, and Spain, also have considerable areas of salt-affected soils.[24]

III. CHARACTERIZATION OF SALT-AFFECTED SOILS IN DIFFERENT CONTINENTS AND COUNTRIES

As a consequence of the dissimilarity in the extension of salt-affected soils on the various continents, and due to the diversity in the maps and materials available, it is impossible to obtain a comprehensive view of salt-affected soils in all the interested countries. The exact location and distribution of salt-affected soils have been studied in varying degrees of detail, different systems of classification and grouping are employed in the individual countries, and, in addition, the maps have not been prepared on a uniform scale. Efforts have been made to group and classify soils according to uniform principles to make it possible to compare them. Although an FAO/UNESCO Soil Map of the World has been compiled,[22] due to its 1:5,000,000 scale and to the characteristics of its classification system, it cannot act as a substitute for detailed maps of the salt-affected soils of given territories.

Below, based on the FAO/UNESCO Soil Map of the World and many other available maps, data, and material, information is given on the distribution of salinity and alkalinity on the continents, areas, and countries most extensively affected.

Table 4
DISTRIBUTION OF SALINITY AND ALKALINITY IN COUNTRIES MAINLY AFFECTED

Country	Solonchaks	Saline phase	Solonetz	Alkaline phase	Total
North America					
Canada		264	6,974		7,238
U.S.		5,927	2,590		8,517
Mexico and Central America					
Cuba		316			316
Mexico	242	1,407			1,649
South America					
Argentina	1,905	30,568	11,818	41,821	85,612
Bolivia		5,233	716		5,949
Brazil	4,141		362		4,503
Chile	1,860	3,140		3,642	8,642
Columbia	907				907
Equador	387				387
Paraguay		20,008	1,894		21,902
Peru	21				21
Venezuela	1,240				1,240
Africa					
Afars and Issas Territory	59	1,682			1,741
Algeria	1,132	1,889		129	3,150
Angola	126	314	86		526
Botswana	1,131	3,878		670	5,679
Chad	2,417		3,728	2,122	8,267
Cameroons				671	671
Egypt	3,283	4,077			7,360
Ethiopia	319	10,289		425	11,033
Gambia		150			150
Ghana	200			118	318
Guinea		525			525
Kenya	3,501	909		448	4,858
Liberia		362	44		406
Libya	905	1,552			2,457
Madagascar	37			1,287	1,324
Mali		2,770			2,770
Mauritania	150	490			640
Morocco	42	1,106			1,148
Niger			111	1,378	1,489
Nigeria	455	210		5,837	6,502
Portuguese Guinea		194			194
Rhodesia				26	26
Senegal	141	624			765
Sierra Leone		307			307
Somalia	1,043	526	3,754	279	5,602
South West Africa	562		1,751		2,313
Sudan		2,138		2,736	4,874
Tanzania		2,954		583	3,537
Tunisia	990				990
Zaire		53			53
Zambia				863	863
South Asia					
Afghanistan	2,924	177			3,101
Bangladesh		2,479		583	3,017
Burma	634				634
India	2,979	20,243		574	23,796
Iran	24,317	1,582		686	27,085

Table 4 (continued)
DISTRIBUTION OF SALINITY AND ALKALINITY IN COUNTRIES MAINLY AFFECTED

Country	Solonchaks	Saline phase	Solonetz	Alkaline phase	Total
Iraq	6,679	47			6,726
Israel	28				28
Jordan	74	106			180
Kuwait	209				209
Muscat and Oman	290				290
Pakistan	1,103	9,353			10,456
Quatar	225				225
Sarawak		1,538			1,538
Saudi Arabia	6,002				6,002
Sri Lanka	180	20			200
Syria		532			532
Trucial States	1,089				1,089
Turkey					2,500
North and Central Asia					
China	7,307	28,914		437	36,658
Mongolia	3,728	342			4,070
Solomon Island		238			238
U.S.S.R.	11,430	39,662	30,062	89,566	170,720
Southeast Asia					
Indonesia		13,213			13,213
Cambodia		1,291			1,291
Malaysia		3,040			3,040
Thailand		1,456			1,456
Viet Nam		983			983
Australasia					
Australia	16,567	702	38,111	301,860	357,240
Fiji		90			90

Note: Areas in 1000 ha.

A. Europe

In at least 12 European countries, the salinity and alkalinity of the soils hinder the satisfactory utilization of lands in many regions and cause problems in the planning of agricultural production and in environmental protection. Obviously, the importance of the problem varies from area to area; in at least half the countries it is a matter of grave concern, while in other cases it bears little significance.

In Figure 3 a schematic map is presented of the distribution of salt-affected soils in Europe.[24] Four mapping units have been distinguished on the map:

1. Alkali soils with a structural B horizon
2. Alkali soils without a structural B horizon. These are the dominating groups of salt-affected soils in Europe, in contrast with the situation on other continents, where saline soils prevail
3. Saline soils
4. The so-called potentially salt-affected soils are also indicated.

Table 5
DISTRIBUTION AND EXTENT OF SALT-AFFECTED SOILS IN EUROPE[24]

Countries and originators	Mapping units[a]					Total area (in 1000 ha)
		Alkali soil				
			With a structural B horizon			
	Saline soil	Without a structural B horizon	(non calc.)	(calc.)	Potential salt-affected soil	
Austria						
J. Fink	0.5	—	—	—	2.5	3.0
O. Nestory	(25.0)	—	—	—	(75.0)	
Bulgaria						
L.P. Raikov	5.0	—	20.0	—	—	25.0
H. Trashliev	(20.0)	—	(80.0)	—	—	
Czechoslovakia						
J. Hrasko	6.2	7.5	2.7	4.3	85.0	105.7
	(5.8)	(7.1)	(2.5)	(4.1)	(80.5)	
France						
E. Servat	175.0	—	75.0	—	—	250.0
J. Servant	(70.0)	—	(30.0)	—	—	
Greece						
E.P. Papanicolaou						
Hungary						
I. Szabolcs	1.6	58.6	294.0	31.9	885.5	1271.6
G. Várallyay	(0.1)	(4.7)	(23.1)	(2.5)	(69.6)	
J. Mélyvölgyi						
Italy						
O.T. Rotini	50.0	—	—	—	400.0	450.0
L. Carloni	(20.0)	—	—	—	(80.0)	
Portugal						
J. Carvalho Cardoso	—	—	—	—	—	25.0
M. Branco Marado	—	—	—	—	—	
Romania						
N. Florea	40.0	100.0		110.0	—	250.0
J. Munteanu	(16.0)	(40.0)		(44.0)	—	
Spain						
P. Grajera Torrez	—	—	—	—	—	840.0
R. Grande Covian						
J. Bardaji Cando						
J. M. Ontanon						
U.S.S.R.						
V.V. Egorov	7546.0	1616.0	20382.0	—	17781.0	47325.0
N.I. Bazilevich	(16.0)	(3.4)	(43.1)	—	(37.5)	
E.I. Pankova						
Yugoslavia						
N. Miljkovic	20.0	50.0	110.0	75.0	—	255.0
G. Filipovski	(7.8)	(19.6)	(43.1)	(29.5)	—	
N. Plamenac						
P. Blaskovic						

[a] Area in 1000 ha per area in percentage of the total salt-affected area. The first number refers to the actual areas of salt-affected soils; parentheses indicate percentage of the total area of such soils.

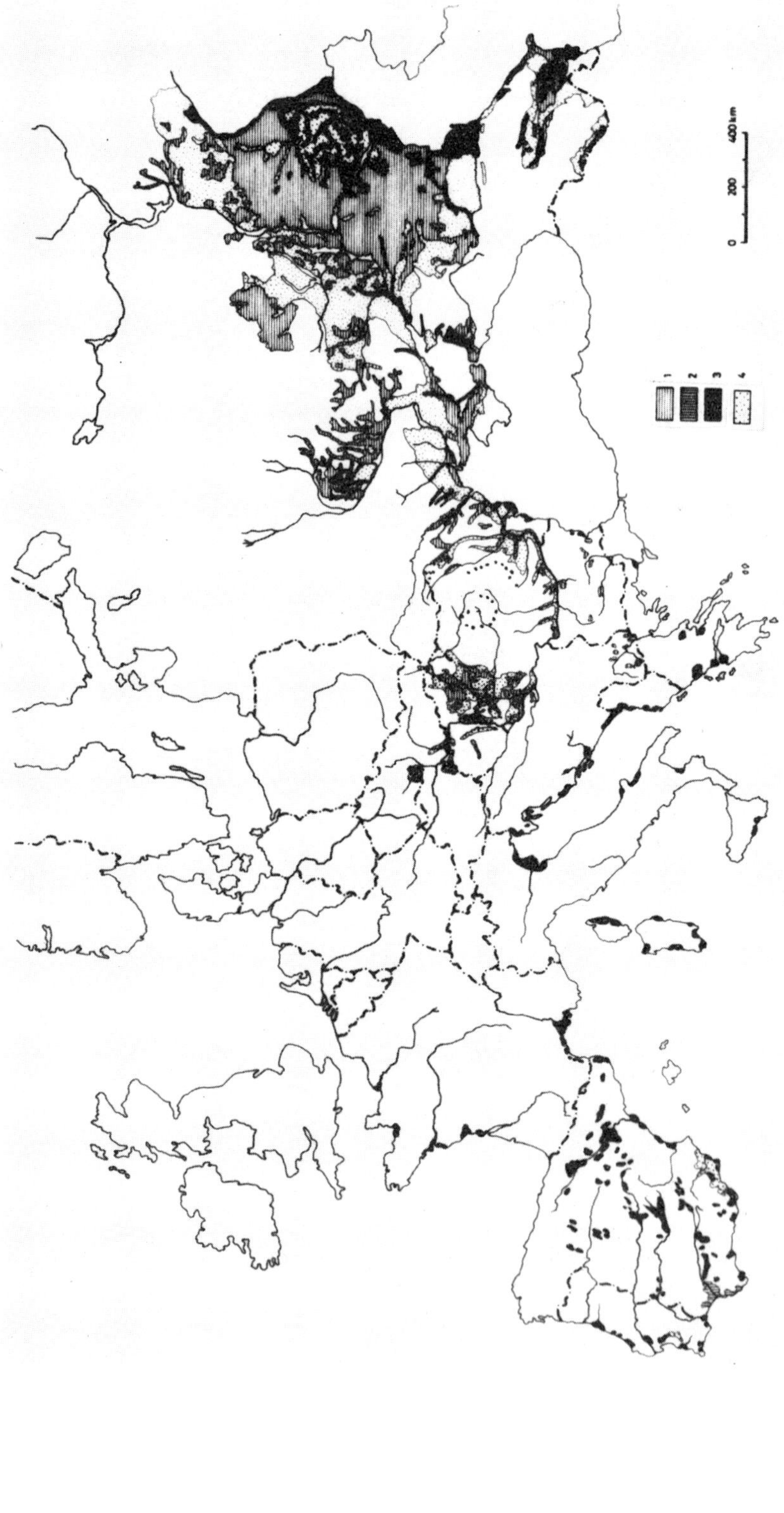

FIGURE 3. Distribution of salt-affected soils in Europe. (1) Alkali soils with a structural B horizon; (2) alkali soils without a structural B horizon; (3) saline soils; (4) potentially salt-affected soils. (After Szabolcs.[1])

These soils are not (or only to a very low degree) saline and/or alkali for the time being, but they may become salinized and/or alkalinized due to human intervention, especially due to incorrect irrigation.

Gypsiferous soils are practically unknown in Europe and the extension of magnesium soils is also so limited that it is almost impossible to indicate them on such a scale. Due to the shortage of data, acid sulfate soils, which appear on some seashores in Europe, are not indicated on Figure 3 either, but can be found on the map giving the global distribution of salt-affected soils (Figure 2).

1. Austria

In Austria the area of existing salt-affected soils is rather limited, making up only 25,000 hectares. The only part of Austria where salt-affected soils are extensive is the southwestern part of the country adjoining the alluvial territories bounding Lake Fertö (Neusiedler See). Here, river, swamp and other sediments are the parent materials for the formation of different types of salt-affected soils, mainly alkali soils.[107] Both the lake and groundwater contain a considerable amount of water-soluble salts, which together with the high groundwater table, result in the formation of stratified soda-saline soils. These soils are very similar to those located on the other shore of Lake Fertö on the territory of Hungary. This situation can be observed in Figure 3.

Irrigation is also part of Austrian agricultural practice, with the consequence that secondary salinization occurs. The estimated area of potential salt-affected soils in this country is three times greater than the extent of current salt-affected soils, as was reported by Finck and Nestroy.[24]

2. Bulgaria

Salt-affected soils (solonetz, solonchak, and solonetz-like soils) are of limited occurrence in Bulgaria (See Figures 3 and 4). The main reason for this is the predominant relief which is cut, undulated, and well-drained. Climatic conditions, in combination with hydrological and soil-formation conditions do, however, cause soil salinization in individual places, chiefly on young terraces. The total area of salt-affected soils is about 25,000 hectares. The major part of this area consists of meadow (and dry meadow) solonetzes, and meadow solonchak-solonetzes. The area occupied by solonchaks is very small: in the Karaboaz Lowland, along the Danube, and in small patches in the districts of Bourgas and Varna, along the Black Sea coast. Salt-affected soils are usually found in small patches measuring a few hundred hectares adjacent to solonetz-like soils and nonsaline soils typical of the area (meadow cinnamonic, meadow chernozems, meadow smonitzas, leached cinnamonic, alluvial meadow, and other soils). Although areas occupied by salt-affected soils are not large, they represent an important problem for Bulgarian agriculture. They are located in areas under intensive irrigated farming, where land is very valuable. Arable land is limited in Bulgaria (less than 50% of the total territory) and there is almost no possibility of expansion. Since more than 20% of the arable land is under irrigation, there is a danger that areas adjoining the salt-affected soils may be affected by secondary salinization, unless adequate measures are taken to reclaim the land already salt-affected. The latter, with its adverse properties, low fertility, and costly tillage requirements, causes considerable losses to agriculture.

The solonetzes and othere salt-affected soils have been mapped and surveyed in detail over the past 25 years in connection with irrigation studies. The sketch map (Figure 4) shows that the area of salt-affected soils is greatest in South Bulgaria, particularly in the Thracian Lowland, along the valley of the River Maritsa (Plovdiv District), the valley of the River Toundja and some of its tributaries (districts of Sliven and Yambol) and in the district of Bourgas. The salinization of the soils in the interior of the country is usually sodic-sulfate

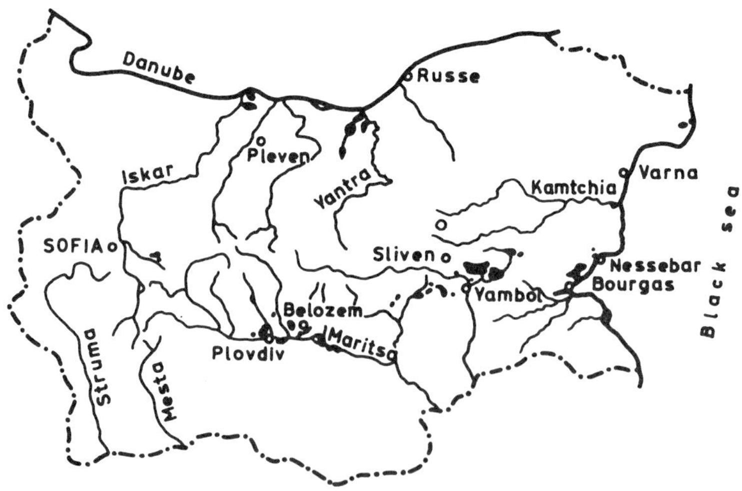

FIGURE 4. Salt-affected soils in Bulgaria.

or occassionally chloride-sulfate. Solonchak salinization in the Karaboaz Lowland near the Danube is chloride, and in the Nessebar and Staro Oryahovo plain near the Black Sea, chloride-sulfate.[108]

Adverse tectonic movements preceded salinization and played an important part in the formation of the terrain of the Thracian Lowland, an area which is very difficult to drain. This part of the country is still affected by tectonic movements. The 1928 earthquake is an example of this. It affected the areas of Chirpan, Belozem and Pervomai, where the ground sank by about one meter and deep fissures opened. As a result of this land sinkage, the level of the groundwater rose, particularly of that originating from the region of the Sredna Gora Mountains. The drainage of these depressions deteriorated and salinization appeared in previously cinnamonic, meadow cinnamonic, and other nonsaline soils. This phenomenon also caused the salinization of soils near Nova Zagora and Bourgas. It should be noted that, on the southern bank of the River Maritsa, groundwaters coming from the Rhodope pass through coarser and better drained material. Here, salinization processes are rarely observed.

Continuous rice cropping in Bulgaria, after the arrival of the Turks in the 14th century, has likewise contributed to salinization in areas where irrigation was not efficient and there was no drainage. Widespread salinization near the villages of Belozem, Tsaratsovo, and others can be attributed to rice cropping.

3. Czechoslovakia

The distribution of salt-affected soils in Czechoslovakia is also shown on Figure 3.

Three geographical areas can be distinguished for the occurrence of saline and alkali soils in Czechoslovakia:[109]

- The South Moravia region
- The Danubian Lowland (part of the more extensive basin of Komárno or of the Lesser Hungarian Plain)
- The East-Slovakian Plain (part of the more extensive Tisza Lowland, or the Great Hungarian Plain)

The saline soils of South Moravia have, in principle, the character of solonchaks. The substrata of these soils are formed by calcareous alluvial sediments, by marlaceous rocks of the Ždánice Flysch, by micaceous deposits of the marine neogene, or by loess. This dissimilarity of the substrata is caused by the peculiarity of the geological and geomorphological development.

The saline soils are found at lower altitudes, seldom exceeding 200 m above sea level. The most typical sulfatic saline soils, primarily in the form of solonchaked chernozem or its modifications, occur in the Hustopečska Pahorkationa region. These soils exhibit a residual sulfatic salinization. Their substratum consists of marls from the Ždánica Flysch. They originated both in plains and in undulate relief, mainly in undrained depressions. They contain large quantities of sulfates, calcium, magnesium, and sodium. Since evaporation considerably prevails over precipitation, the salts were transported by capillary rise, from the strongly mineralized underground water towards the surface, and they sedimented at different depths.

Two further large areas of saline soils lie on the river alluvia between Rokytnica and Jevišovka, and on the Dyje alluvium. Smaller districts likewise are located on the alluvia of other rivers.

In the Danubian Lowland, alkalization is absolutely dominant, and, consequently, the soil reaction is very high (pH 9-10 or more.) A further characteristic feature of these soils is that they are usually calcareous right from the surface, which means that in lower parts of the profile, the content of carbonates exceeds 20%. The amounts of salts determined from the water extract (evaporation at 105°C) in the upper horizons, do not usually exceed 0.4%. In general, the solonetzic meadow and alluvial soils contain 5 to 20% of adsorbed sodium and the meadow solonetz-solonchaks 20 to 67% sodium, or even more in extreme cases.

Soil salinization in the Danubian Lowland is primarily observed on the strongly calcareous alluvia of the Danube and its left-bank affluents, but also on loesses which sedimented in former water reservoirs, or neotectonically sank, and are now influenced by the groundwater.

The East Slovakian Plain forms the northern part of the Great Hungarian Plain. It makes up 3/4 of the Danubian Lowland area, i.e., about 4000 km^2. During the Miocene, it was still connected with the Kosice Basin, but in the Pannonian-Sarmatian Epoch, the volcanic Slánske Pohorie hills arose and separated them.

Salt-affected soils originated here on two geomorphological units, i.e., on the alluvia of some rivers and on loess-like sediments.

In the plain, these conditions are determined by the Ondava, Laborec, Latorica, Bodrog, and Tisza rivers and their affluents, and by the entire geomorphology of this region, which forms a reservoir collecting the waters from the environing mountains. For this reason, particularly in the past but also today to some extent, this region was often flooded, especially during thaws. The rivers are unable to drain away all these waters with the consequence that many gley soils have originated here.

Salt-affected soils are found more or less continuously on the loessial covers and alluvial sediments of the Laborec, Latorica, Ondava, and Bodrog rivers. While solonetzic soils with well-developed structural horizons prevail on loess sediments, on the alluvia of the above-mentioned rivers, saline soils of the solonchak type are most widespread.

4. France

The extent of salt-affected soils in France is fairly limited. They occur in a few places on the Atlantic coast and in the Rhone Delta on the Mediterranean coast. They can also be found on the east coast of Corsica.

The majority of salt-affected soils in France belong to the saline type, only a few having an alkaline reaction. The agricultural areas of the country, except for some areas near the coast, are free from the hazard of soil salinity due to the fortunate hydrogeological and geomorphological conditions of this country.

5. Greece

In coastal areas of Greece small areas are covered with salt-affected soils, practically all

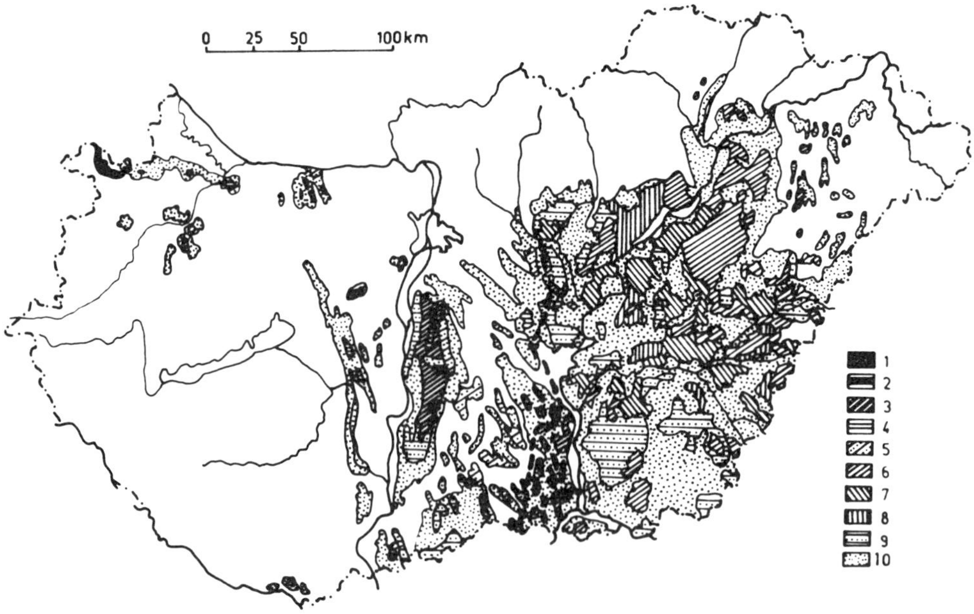

FIGURE 5. Distribution of salt-affected soils in Hungary. (After Szabolcs, Várallyay and Mélyvölgyi[1]).

According to the Hungarian soil classification:	According to the classification accepted by the Sub-commission on Salt-Affected Soils of the ISSS:
1. Chloride and/or sulfate solonchak	Saline soils
2. Sodic solonchak 3. Sodic solonchak-solonetz	Alkali soils without a structural B horizon
4. Calcareous meadow solonetz 5. Calcareous solonetzic meadow soil	Alkali soils with a structural B horizon
6. Meadow solonetz 7. Meadow solonetz turning into steppe formation 8. Solonetzic meadow soil	Alkali soils with a structural B horizon, non-calcareous
9. Chernozem and meadow chernozem salty in deeper layers 10. Potentially salt-affected soils	Potentially salt-affected soils

of which are saline, having developed under the influence of seawater. The total territory of salt-affected soils is less than 3,000 ha, so it does not constitute a very serious problem. Nevertheless, it disturbs the intensive utilization of land, particularly in orchards, vineyards, and vegetable production. Due to the geomorphological and hydrological conditions of the country, the hazard of salinity and alkalinity is negligible in most areas.[24]

6. Hungary

The total area of salt-affected soils in Hungary exceeds one million ha, which is more than 10% of the territory of the country. There are few countries in the world, and none in Europe, where the ratio of salt-affected soils is so high. Salt-affected soils include both the existing and the potentially saline and alkali soils which have developed mainly in the Hungarian Plain and only in small spots in other regions of the country. More than 95% of the existing salt-affected soils are situated in the valleys of the Danube and the Tisza. The sediments of these rivers formed and filled up the Hungarian Plain, the alluvial origin of which is closely related to the location of salt-affected soils (szik).

Figure 5 shows the map of salt-affected soils in Hungary. In Hungary the main factor in the salinization and alkalization of soils is not the climate. The average precipitation on the plain varies between 450 to 600 mm and the mean annual temperature is 10°C. Under such conditions chernozem soils, blackmeadow soils, and other types with good fertility and without saline or alkali properties develop in some areas of the Hungarian Plain. It is mainly the hydrogeological properties which result in the formation of salt-affected soils in other areas of the lowland.[110]

In the groundwaters of the Danube and Tisza valleys considerable amounts of dissolved salts often are present, with sodium salts frequently dominating. Whenver the salty groundwater is linked with the soil profile, salt accumulation takes place, leading to the formation of salt-affected soils. This phenomenon never occurs in the first terraces of the rivers, only in the second and third ones. In the valleys of both rivers, salt-affected soils are mainly situated along the left bank and only to a smaller extent on the right.[111]

Mineralized groundwaters represent only one precondition, though a very important one, for the formation of salt-affected soils in the Hungarian Plain. Surface waters collecting during the wet season also contribute to their development. The surface waters accumulate in depressions and often link up with the groundwater table, which rises during the same period. As a result, stagnant water covers the surface of the soil for a considerable stretch of time. The downward movement of the water is prevented by the heavy texture of the soils, as well as by the high groundwater table. The upward movement of the water, accompanied by the salts, dominates and, as a result, salt accumulation occurs in the soil profile. However, the effect of the periodical downward movement of water and salt must not be underestimated. The repeated changes in the direction of their flow leads to the formation of the most common type of Hungarian salt-affected soils, solonetz or, according to the classification of the Subcommission on Salt-Affected Soils of the International Society of Soil Science (ISSS), alkali soils with a structural B horizon.

In the Danube and Tisza valleys great flood control projects were completed in the second half of the 19th century and the first decades of the 20th century. As a result, the river-beds were straightened, the floods practically arrested, and millions of hectares of land, formerly bogs and swamps, desiccated. This huge ameliorative operation created the possibility for the transformation of the soil cover of the territory. However, the leaching of salts from the soils was not consistent in many places and was severely hindered by the poor physical properties of the soils as well as by the high water table, which generally retreated in the wake of the amelioration of the river valleys — but, in many cases, not sufficiently so to ensure good natural drainage. Consequently, in places where former saline swamps and bogs had been desiccated, the migration of slightly salty solutions frequently resulted in the formation of solonetz or solod soils. In other places, where drainage remained even poorer, or in depressions with practically no drainage, solonchak soils developed. In areas where the amelioration radically improved natural drainage, formerly salt-affected soils have probably disappeared as a consequence of desiccation. The climate of the Hungarian Plain is sufficiently humid to leach the salts from the soil profile in the case of good hydraulic conductivity and drainage.[112]

In the salt-affected soils of Hungary, alkalinity, as a rule, plays a more important role than salinity. Except for a few cases, saline soils are not common; nearly all the salt-affected soils belong to the alkali class. Both alkali soils with a structural B horizon (solonetz) and alkali soils without a structural B horizon (soda-solonchak) are met with in the Danube and Tisza valley. In the Danube valley calcareous varieties prevail and in the Tisza valley noncalcareous varieties prevail.[113]

There are similarities and differences between the salt-affected soils of the two river valleys.[114]

The Danube deposits are calcareous while the deposits of the Tisza river are of a slightly acid character. However, in the deeper layers (in some places at a depth of 10 to 20 m, in others at 1 to 2 m only), the deposits are also rich in calcium carbonate in the Tisza valley.

The explanation of this phenomenon is to be found in the geological history of the Carpathian Basin and the Hungarian Plain. During the Holocene, the primeval Danube covered the whole land with calcareous deposits. The Tisza river emerged in a later period of geohistory. It carried acidic materials from the Carpathian Mountains, and these sediments were spread in an uneven layer over the calcareous Danube deposits. Evidently, the borderline between the two kinds of sediments changed in many cases under the influence of soil-forming processes. As a consequence of a rising groundwater table and solution flow, $CaCO_3$ and even Na_2CO_3 are often present recently near the surface of salt-affected soils in the Danube valley. On the other hand, when the downward solution flow dominates, even if only periodically in the Tisza valley, the top layer of the soil may become free of lime and soda. The texture of salt-affected soils is, as a rule, lighter in the Danube valley, and heavier in the Tisza valley due to the different granulometrical composition of the sediments of the two rivers.

In Figure 5 both the Hungarian national classification units and those of the Subcommission on Salt-Affected Soils of the ISSS are indicated. The map shows that a fairly good correlation exists between the two classification systems. In Hungary a wide variety of salt-affected soils occur, including both saline and alkali soils, but the solonetz type prevails.[115]

While the extensive flood control of the river valleys in Hungary definitely lowered the basis of erosion in the lowland, the effects of later irrigation projects have not always been beneficial. Since the end of World War II the area of irrigated lands has increased more than twenty-fold in the Hungarian Plain, particularly in the valley of the Tisza river.[116]

As in many other alluvial plains of the world, the hazard of secondary salinization and alkalization exists in the Hungarian Plain, too. The sharp rise of irrigated agriculture after World War II led to the accumulation of salts in many places. As is well known, there are two main sources of the secondary salinization and alkalization of irrigated soils, namely the salt content of the irrigation water and that of the groundwater. If the groundwater rises, the salinization and alkalization of the overlying soil layers cannot be ruled out.[117]

A further hazard of this harmful phenomenon is very serious due to the geomorphological and hydrogeological conditions of the Hungarian Plain,[118,119] which is the second, nearly closed plain through which the Danube flows and is connected with the southern and southeastern parts of this lowland in the territories of Yugoslavia and Romania.

The factor mainly responsible for this secondary salinization and/or alkalization of soils in the Danube and Tisza valleys is the rise of mineralized groundwaters near the soil surface.

7. Italy

The occurrence of salt-affected soils in Italy is also shown in Figure 3. The present territory of salt-affected soils is fairly small and is located mainly along the coasts of the mainland, and of Sicily and Sardinia. In consequence of the effect of seawater, they are practically all saline soils dominated by the effect of sodium chloride. The largest single extension of salt-affected soils in Italy can be found around the Poe Delta, near Venice.[24]

Alkali soils occur in several places in Italy but in very small spots, which is why they are not indicated on Figure 3. These soils are mostly of deluvial origin in submountainous regions, e.g., in Volterra near Florence. Interestingly enough, Rotini and Carloni[24] estimate the territory of potentially salt-affected soils in Italy to be quite large, amounting to five times the existing territory. Keeping in mind that irrigation is widely practiced in the country and is likely to be increased, the hazard certainly exists.

8. Romania

Figure 6 indicates the distribution of salt-affected soils in Romania.[120] Romania has considerable territories of salt-affected soils and the study and improvement of these areas have long traditions.

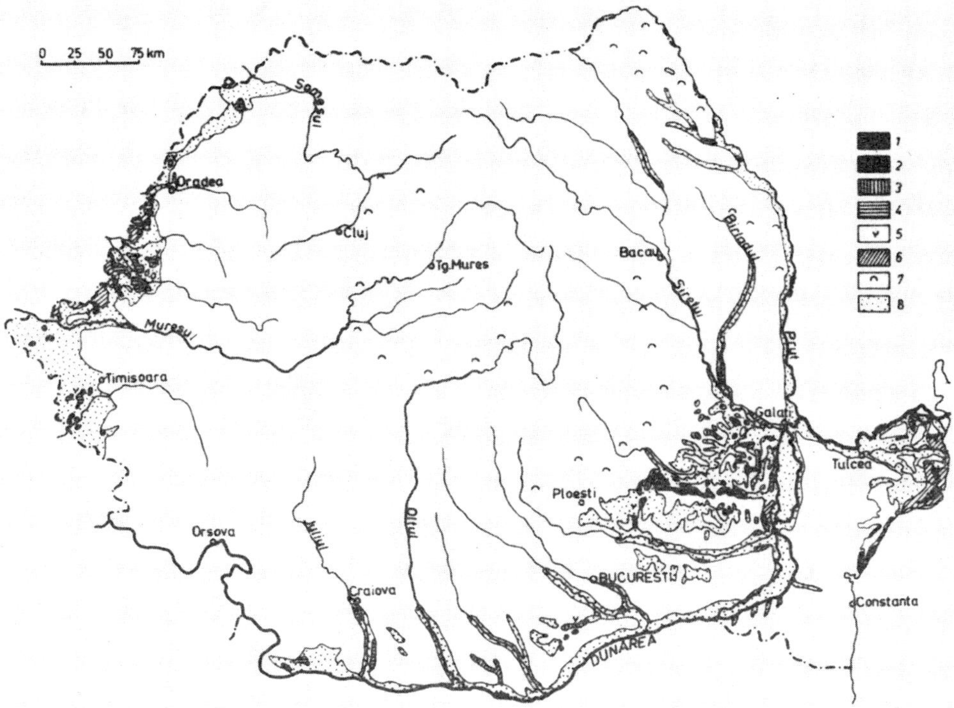

FIGURE 6. Distribution of salt-affected soils in Romania. (1) Saline soils; (2) alkali soils without a structural B horizon; (3) solonchak-solonetz and calcareous solonetz; (4) non-calcareous solonetz with an A horizon not thicker than 15 cm; (5) solodized and/or deeply leached solonetz and solods; (6) solonized and slightly salt-affected soils with minor structure formation; (7) residual and semiresidual salt-affected soils; (8) nonsalinized soils (generally with a groundwater table at small depth) which may be enriched in salts, locally saline and alkali soils (After Florea.[1])

Saline and alkali soils, together with soils which are salinized or solonetzic to various degrees, cover an area of some 400,000 ha in Romania.[121] A large part of this area consists of solonetzes and solonetzic soils. These appear in the steppe and forest steppe zones of territories with a certain type of lithologic, pedologic, and hydrogeologic conditions. These conditions are found in the Romanian Plain, in the Western plain, in the Danube Delta and flood plain, as well as in Moldavia and the Transylvanian Plateau.

The soils mentioned occur either as continuous massifs or as patches within territories of high fertility; they frequently are found in territories linked with a depression having a water accumulation character and unsatisfactory internal and external natural drainage (cones of dispersion, low terraces, flood plains, at the point of contact between two forms of relief). The groundwaters have a varied degree of mineralization.

These soils frequently develop in a complex with chernozem soils having a grassland regime. Crops often develop unevenly and yields are often depressed. The presence of these soils in areas of prospective irrigation is a permanent danger, threatening to degrade the soil in neighboring territories by processes of salinization, solonetzization, and alkalization.

The accumulation of soluble salts in the salinized soils and groundwaters of Moldavia and the Jijia-Bahlui depression is associated with the presence of diapir folds and deposits of sarmatic marls rich in sulfates and relatively poor in chlorides. In the Danube Delta and flood plain the genesis of solonetzized soils is due to the presence of salt sources in high adjacent terraces and fields, and to the unsatisfactory natural drainage of the plain below the terraces. These conditions assure the accumulation of salts. Large oscillations in the level of the groundwaters assure alternate salinizing and desalinizing of the soils, thus favoring the accumulation of sodium in the complex.[121]

As Figure 6 shows, salt-affected soils are extensive mainly in two areas of the country: in the west in the Transylvania and Bánát regions, and in the southeastern part of the country in the lower reaches and delta region of the Danube. The former area belongs to the second lowland, and the latter to the third lowland of the Danube. There are considerable differences between the climates of the two regions: while on the second plain the yearly precipitation averages nearly 500 mm and the mean annual temperature is about 11°C, on the third Danube plain the yearly precipitation average is as little as 300 mm or less, with a mean annual temperature of 13°C or even more. The differences in the aridity index are reflected in the properties of the salt-affected soils. The salt-affected soils of the second Danube Plain belong mainly to the alkali type, while those of the third plain are more saline and consist partly of saline and partly of saline-alkali soils.

In Figure 6 the national soil classification system is applied containing units for the characterization of different salt-affected soils. Saline soils, where salinity is caused mainly by sodium chloride or sodium sulfate, are extensive in the lower reaches and delta of the Danube and are associated with alkali soils without a structural B horizon, which also contain considerable amounts of soluble salts. In the same areas, solonchak, solonetz, and calcareous solonetz soils occur. Noncalcareous solonetz, solodized or deeply leached solonetz, and slightly salt-affected solodized soils are met with only on small areas in the eastern part of the country, while such types are dominant among the salt-affected soils in western regions. This can be explained partly by hydrogeological factors, but mainly by the climatic differences between the two regions. The salt-affected soils in the western part of Romania are closely related to those bordering the eastern districts of Hungary.

On the Tisza Plain the genesis of solonetzes is mainly linked to the presence of a marly-clayish layer containing sodic sludge, a $CaCO_3$ content (resulting from the withdrawal of the Pannonian Sea), and a high capacity of adsorption. It is also associated with the existence of irregular alluvial deposits from the rivers.[122] Carbonates, bicarbonates, and sodium sulfates are predominant in the composition of these soils and their groundwaters.

Besides the two Danube plains there are small spots of salt-affected soils even in mountainous parts of Romania. These soils are indicated as residual and semiresidual salt-affected soils on the map. They are present either in small depressions or on the site of geological salt accumulations. They can sometimes be found near spas and mineral water karsts. In contrast to the chemical composition of salt-affected soils in Transylvania, they contain mainly sulfates of sodium or magnesium, and sometimes other salts, too.

The term "nonsalinized soils" (see Figure 6) is used in practice to indicate potentially salt-affected soils. In Romania irrigation is gaining ground, particularly in dry regions of the country, i.e., in the third plain of the Danube and in the valleys of the Danube, the Prut, the Oltul, the Mures, and other rivers. The extension of potentially salt-affected soils is comparable with the extension of salt-affected soils already developed by natural processes without irrigation.

9. Spain and Portugal

The distribution of salt-affected soils in the Iberian Peninsula (i.e., Spain and Portugal) is shown in Figure 7. According to the figure, both countries of the peninsula have considerable areas of salt-affected soils. In Spain they are distributed over most regions with the exception of the northwest part of the country; the areas most severely affected are the valley of the Guadalquivir and the marshes in the delta region. Other areas affected are the central plateaus and the southern part of the country.[108]

In Portugal saline soils are extensive along the coasts and in the valley of the River Tajo.[123]

The soils of the peninsula are affected both by salinity and alkalinity. Saline soils prevail along the coasts, where the main salt influencing the soil-forming processes is sodium chloride. In some places a considerable amount of sodium sulfate can also be found, together

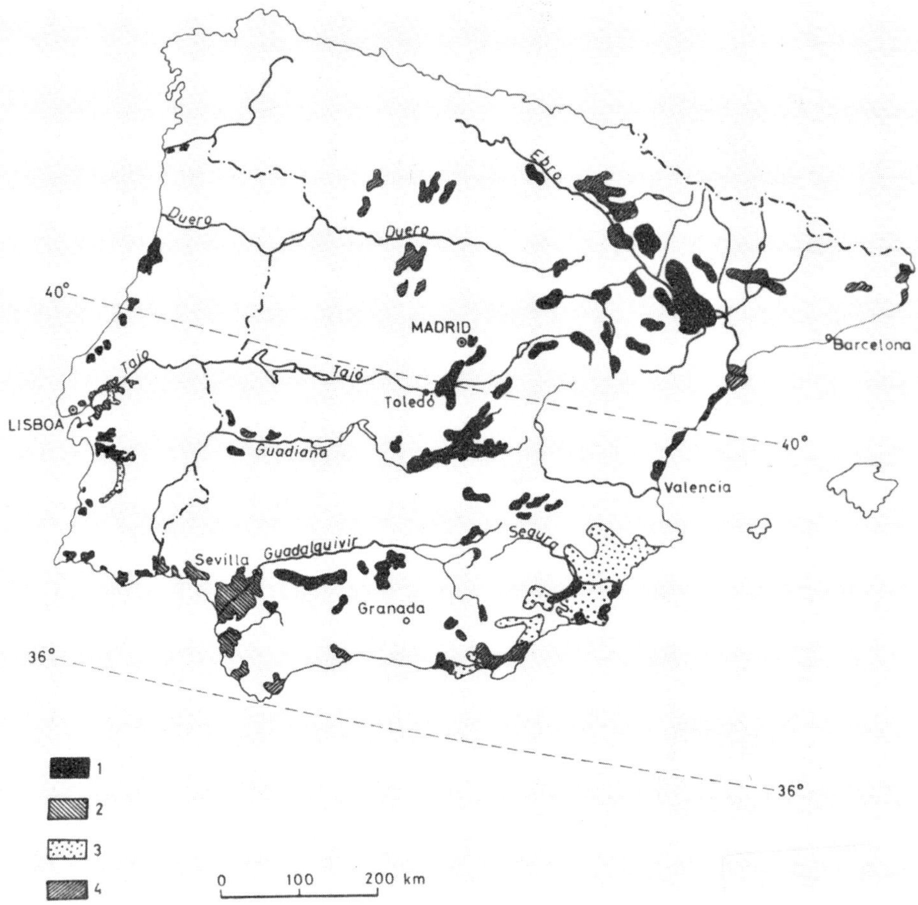

FIGURE 7. Distribution of salt-affected soils in Spain and Portugal. (1) Saline soils (EC >4 mmho/cm); (2) saline-alkaline soils without a structural B horizon (EC >4 mmho/cm; ESP >15); (3) potentially saline soils; (4) complex of soils 1 and 2. (After Grajera Torrez and Carvalho Cardoso.[1])

with a certain quantity of magnesium salts. Alkalinity prevails in the valley of the Guadalquivir and in the "Marismas" to be found in the same area. Alkalinity is also to be found in the Ebro Valley and in some southern parts of the peninsula. Salinity and alkalinity often occur jointly; soils of this type are indicated as saline-alkali soils on the map. In these soils the high salt content is associated with an alkali pH. On the map of the conventional limit values are applied both for salinity and alkalinity, i.e., saline soils have an electrical conductivity of more than 4 mmho/cm, and alkali soils are distinguished if the exchangeable sodium percentage exceeds 15.

In Spain, among the areas of saline soils which are capable of being plowed, the most important are the so-called "Marismas del Guadalquivir", covering the area of more than 140,000 hectares. Due to the origin, evolution, and development of these soils they are considered of interest for reclamation.

A fundamental phenomenon involved in the evolution of these soils is still taking place in the estuary. The Guadalquivir River carries soils down from the surrounding mountains during its winter flood and deposits them where its flow is slowed down by seawater. The first to be deposited is the sand, which needs a higher transportation speed. Then the finer material and finally the clay are deposited, when the water is almost still. In the river north of Seville the suspension is complex and made up of sand, silt, and clay: once it has passed Seville, the river carries only silt and clay, the sand having been deposited earlier on the river bed. If a water sample is taken midway between Seville and the sea, the suspension consists of only the very fine elements (clays) which are deposited there.

The old estuary has been filled up. In the beginning it formed a shallow bottom, but still allowed navigation. Deposits formed deltas, where the water speed was minimum. The river opened a course through those deposits and formed the characteristic arms of a delta formation. Thus, an alluvial formation of a pasty consistence (swamps) appeared at the beginning of the present century. Since then the evolution has been controlled by man. The bed of the river was diverted to protect special areas and true marsh soil appeared. This resulted in the formation of saline clay alluvia of great impermeability.

Morphologically, the deposit is a uniform mass with similar characteristics and great horizontality. The highest level of this horizontality is the maximum tide level. At high tide, the river floods and at low tide it flows along its bed. This action formed the present topography of the area. When the tide rises and the river flows, it loses speed and leaves a large volume of deposits along its edges. On the islets in the Guadalquivir the edges are a few inches higher than the rest and are called ''river mountains'' or levees. In the center of the islets are ''lucios'', or depression zones consisting of deposits from the flood water with no possible exit. During the rainy season the marsh becomes a series of lagoons surrounded by grayish soil.

When the hot summer months arrive and the river has less flow, the water in the ''lucios'' evaporates, the soil becomes dry, and the lagoons are surrounded by beaches. A whitish color develops due to the deposit of salt left by evaporation. While the lucios are covered by water, the marginal areas are covered by salt without any kind of vegetation; and the higher areas are covered by halophile vegetation of varying density, being greatest at the edges.

The marshes were formed by the deposit of small solid elements, precipitated in a saline sodic liquid environment (seawater). This formation gave the main characteristic of the soil. The clay suspended in the river waters fixed sodium on its exchange complex and lost its calcium. Because the soil was formed from expansive sodic clays, it has extremely great impermeability.

In winter the soil is practically covered by rain or overflowing water, which remains until it is evaporated.

The potentially salt-affected soils, located mainly in the southern part of the Mediterranean coast, are also indicated in Figure 7. According to the data on this map, the extension of potential salt-affected soils in Spain is comparable in order of magnitude with the existing area of salt-affected soils. However, the spread of irrigation will probably result in a further increase in the extent of such soils.

The total area of salt-affected soils in the Iberian Peninsula is nearly one million hectares, which is not too high a figure in comparison to the total territory of the country or with the agriculturally utilized area. However, due to the fact that salt-affected soils occur mainly in river valleys, fertile plains and other areas of intensive agriculture, the importance of their study and of their utilization should not be underestimated.

10. The European part of the U.S.S.R.

Approximately 90% of all European salt-affected soils are found in the European part of the U.S.S.R. This may be explained not only by the size of the country but also by the climatic and geological characteristics of this area. In the European part of the U.S.S.R., salt-affected soils can be located in eight regions which differ in their climatic, geological, geochemical, and other conditions. One large area of salt-affected soils is to be found in the Ukraine, partly in the valley of the Dnepr and partly in the southern part of the republic, while another has been formed on the Russian Plain, along the rivers Oka, Don, and Volga, particularly along the lower reaches of the Volga and around its delta. A massive area of salt-affected soils and the largest salt-affected area in Europe occurs in the Pre-Caspian Lowland.[124,125]

Another region of salt-affected soils in the European part of the U.S.S.R. is located in the Trans-Caucasian Republics, partly in Georgia and Armenia, but mainly in Azerbaidzhan.[126,127]

The salt-affected soils of the Ukraine belong mainly to the alkali type and are in close genetic relation to the solonetz soils in Romania, Bulgaria, Hungary, and Czechoslovakia.[128] The extension of these soils is particularly great in the valley of the Dnepr and also to some extent the valley of the Dnestr.[129] In the Dnepr Valley, large areas of solonetz soils can be found south-east of Kiev and, following the direction of the river, they continue as far as Dnetropetrovsk and Zaporozhe.

Another massive area of salt-affected soils can be found in Moldavia around Kishinev, including the Soviet territories of the Danube Delta, and near Odessa. Alkali soils are also extensive in the southern part of the Ukraine, in the Northern part of the Crimean peninsula, and in adjoining territories.

Saline soils also occur in the Ukraine, partly on the coast and partly in association with alkali soils.

The salt-affected soils of the Ukraine adjoin similar soils in the neighboring Russian Federation, and near Karkov and Voronezh such soils occur in association with the fertile chernozem soils of the Russian Plain. The chemical character of the salt-affected soils in these areas is characterized by the occurrence, and in some cases dominance, of sodium carbonate and bicarbonate and by the fact that alkalinity is more pronounced than salinity.

In the Pre-Asuvian Plain, where soil salinity and alkalinity again overlap and are in close relation to temporary waterlogging and swamps, considerable territories are covered with saline and alkali soils.

On the territory of the Russian Federation, in the southern regions of the European part of the U.S.S.R., the extension of salt-affected soils follows the valleys of the large rivers, especially the Don and the Volga. The salinity of the soils increases in a north-west to south-east direction, but the degree of alkalinity shows the opposite tendency. This means that alkali soils are widely distributed along the Don, while southwards on the Volga, soil salinity prevails. It should be noted that in the lower reaches of the Volga and in its delta, soil salinity and alkalinity overlap and both types constitute a serious problem for agriculture and for environmental protection. The Volga Delta, which is the largest salt-affected area in Europe, has very favorable natural conditions for the salinization of both soils and waters. The Pre-Caspian Lowland has a very specific hydrogeological character, because it is influenced not only by the Volga, but also by the Ural, the Emba, and in part by the Terek. Another special feature of the Pre-Caspian Lowland is its very low altitude. It is one of the few maritime lowlands which lies below sea level (at depths of up to −26 m). This can be explained by the specific character of the Caspian Sea which is completely closed and isolated from other seas and from the World oceans.[130]

As a result of the low altitude of the Pre-Caspian Lowland, it is a typical region of salt accumulation both of alluvial and diluvial origin. The poor drainage and limited outflow of salty water contribute to the salinization of the soils and waters of the region. The salt content of the soils on the plain differs from place to place, increasing towards the shores of the Caspian Sea. Among the cations, sodium dominates followed by magnesium and calcium, while the anions are different in different areas. From northwest to southeast there is an increase in the amount of sulfates and even greater incrase in the chlorides, which predominate in coastal areas. Carbonates and bicarbonates are also present in most cases but they are of inferior importance whenever high salinity exists. According to the Soviet soil classification, solonchak soils are dominant in the Pre-Caspian Lowland. They are often in close association with solonetz-type soils, particularly in areas far away from the delta and the coastline.[131]

The Caspian Sea has numerous peculiarities when compared with many others. While the salt content of the Caspian is much lower than that of the Black Sea, the Mediterranean, or the oceans, the salinity of the adjoining territories is higher than that of other coastal regions. This is due to the fact that the salinity of the Pre-Caspian Lowland is the result not of the influence of the salty water of the Caspian Sea, but rather of the specific geological and hydrogeological conditions of the area, which were outlined above. Apart from this, the level of the Caspian Sea has been sinking and this tendency will continue in the future. Consequently, new drylands or flats will appear on the shores, and these are obviously always saline.[132]

More than half of all the salt-affected soils in the European part of the U.S.S.R. can be found along the lower reaches of the Volga and in the Pre-Caspian Lowland.

Some of the Caucasian and Trans-Caucasian salt-affected soils, which also belong to the European part of the U.S.S.R., are adjacent to those of the Pre-Caspian Lowland along the valley of the River Terek. All the characteristics of these soils are similar to those outlined above.[133,134]

The vast territory of salt-affected soils on the left bank of the River Volga is included among the salt-affected soils of the Asian part of the U.S.S.R. in spite of their close geographical and pedological relationship to the salt-affected soils on the right bank of the river.[135]

The three Trans-Caucasian Republics of the U.S.S.R., namely Georgia, Armenia, and Azerbaidzhan, have considerable areas of different types of salt-affected soils. This sequence also indicates the extent of salt-affected soils, which are rare in Georgia and most common in Azerbaidzhan.

On the Trans-Caucasian Plain, salt-affected soils are extensive in the valleys of the rivers Kura and Araks, and run parallel with these rivers from Georgia (starting south of Tbilisi) into Azerbaidzhan, right down to the southern shores of the Caspian Sea. These soils are mainly saline, in some places with a high percentage of water-soluble salts which varies from carbonate to chloride. As in many other places, the dominating role of chloride appears parallel with an increase in aridity and is inversely proportional to the distance from the seashore. Sulfate-solonchak soils, and in some cases gypsiferous soils, are also frequent in the Azerbaidzhan Soviet Republic. Soil alkalinity, which may or may not be associated with salinity, is also widespread, particularly in the Kura-Araks Plain. In Azerbaidzhan the percentage of salt-affected soils is nearly as high as in some Near Eastern countries (Iraq, Iran, etc.). Different types of salt-affected soils are extensive in this Republic, and their significance, particularly in irrigated agricultural areas, is considerable.

Armenia is one of the most southern republics of the Soviet Union, and due to the vertical variability of this country — where high mountains and even lowlands can be found — many latitudinal bioclimatic zones are represented. The Ararat Plain, in the southern part of the Republic, is a mountainous area with an altitude of over 500 m above sea level. The irrigated agriculture in the Ararat Plain originated in the remote past during the period of Chaldian (Urartian) civilization in Asia Minor, and is as ancient as the peoples inhabiting this country. It is necessary to characterize the hydrogeological conditions of this plain in order to explain the wide occurrence of salinity and alkalinity in the soils and waters of this region. In Figure 8 a hydrological section of the Ararat Plain is given, showing the distribution of the different geological layers.[133]

Figure 8 explains the regular features of salt accumulation which lead to the formation of salt-affected soils on the Ararat Plain. In Figure 8 an indication is also given of the groundwater aquifer and other factors which result in the development of recharge and discharge areas in the groundwater of Armenia, as shown in Figure 9.

Hydrocarbonate and normal soda (from 10 to 40 mg/ℓ) have been found in the pressing waters of artesian wells and the pumped water of vertical drainage wells within the confines of the Ararat Plain.[136]

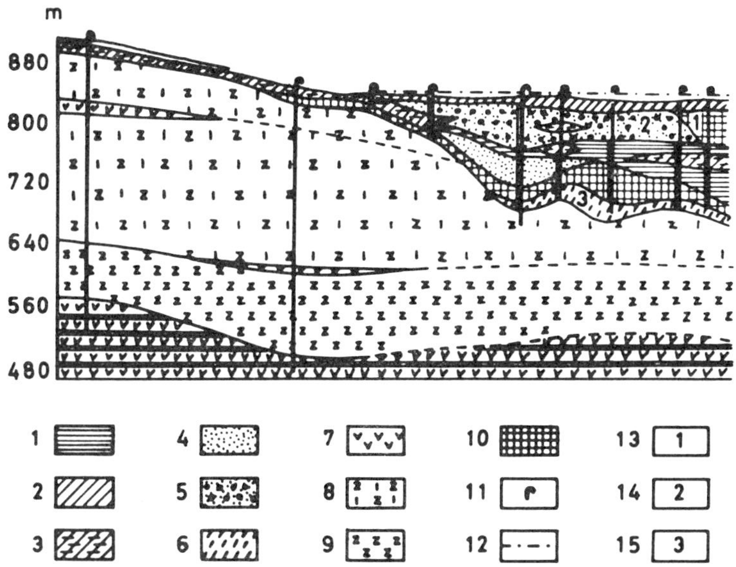

FIGURE 8. Hydrogeological section of the Ararat Plain (along the line: village of Agavnatun — village of Markara). (1) Cinnamonic clays; (2) blue-green clays with inclusions of fauna; (3) loams with interbedded clays and sandy loams; (4) sands; (5) pebble and gravel with sand; (6) boulders and pebble with sand; (7) tuffs; (8) basalts, andesite-basalts andesites and dacites; (9) doleritic basalts (Upper Pliocene); (10) rocks of gypsiferous and saliferous thickness (Middle Miocene); (11) well; (12) piezometric level; (13) groundwater aquifer; (14) first (low-head) aquifer; (15) second (high-head) aquifer.

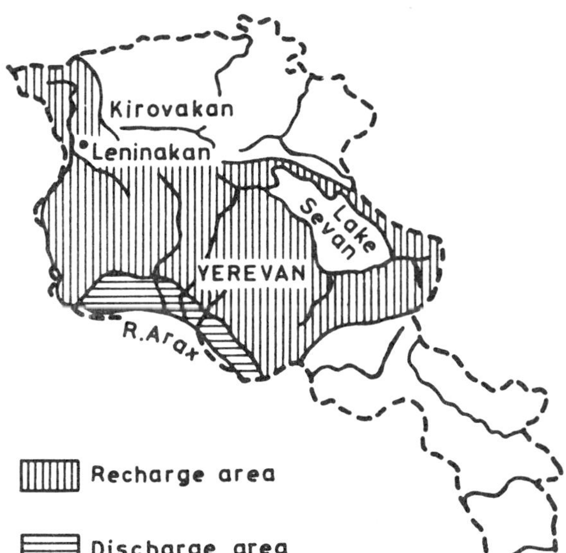

FIGURE 9. Groundwater recharge and discharge areas

The hydrocarbonate and carbonate alkalinity of artesian waters suggests the possible presence of abyssal geohydrobiochemical processes of soda formation, necessitating profound theoretical investigations.

The soils of the Ararat Plain were formed on the alluvia of the River Araks and its left-bank tributaries, as well as on the proluvial deposits of numerous mud blows. Consequently, they are represented lithologically by stratified sediments with rather a motley texture.

As a result of geological and hydrogeological processes, soil salinity is fairly substantial on the Ararat Plain. The dominating salt is sodium carbonate, which results in a strongly alkaline pH and the formation of alkali soils mainly without a structural B horizon. Armenia is one of the few places where soda-saline soils are common at high altitude (where salinity is usually rare) and under extreme conditions where in many other places saline or gypsiferous soils prevail. This fact again proves that soil alkalinity can be found under very varying climatic and geological conditions.

In Figure 10 the distribution of salt-affected soils in the European part of the U.S.S.R. is demonstrated (according to Egorov and Bazilevich et al.). This map includes not only the different groups and types of salt-affected soils discussed above, but also potential salt-affected soils which are very widespread in all the affected regions of the European part of the U.S.S.R. As in many other European countries, the extent of potential salt-affected soils, which are a real hazard due to the increasing use of irrigation, is of the same order of magnitude as the area of existing salt-affected soils. These potentially affected areas are distributed partly in river valleys and delta areas of the Dnestr, the Dnepr, the Don, the Volga, the Kura, the Araks, and the Terek, where salinity and alkalinity occur, and partly on arid or semiarid territories of the country. The map clearly shows that both existing and potential salinity and alkalinity increase towards the southeast. Besides the salt-affected soils of the European part of the U.S.S.R., certain areas affected by salinity and alkalinity on the left bank of the Volga are also indicated on the map, proving the statement made earlier, that these are in close pedological relation with the salt-affected soils on the right bank of the river. It should be noted that some areas on the left bank of the Volga form part of the Pre-Caspian Lowland.[137,138]

11. Yugoslavia

The appearance of salt-affected soils in Yugoslavia is a significant agricultural, economical, and political problem which has not yet been completely solved. Salt-affected soils are found in varying extents in different parts of the country, and total 264,000 ha in all. The greatest part (235,000 ha) of these soils occurs in the province of Voyvodina, representing a loss of 18.6% of its total arable land. In the southern part of the country, in Macedonia, there are about 11,000 ha of salt-affected soils. Besides the continental salt-affected soils in Yugoslavia, maritime saline and alkali soils (18,000 ha) formed by the action of sodium and chloride from seawater are found on the Adriatic coast.[139]

The origin of the salts in salt-affected soils in the Pannonian Plain is connected with the weathering of various minerals which are present in eolian loess (which predominates as a parent material) and sands. Another important contribution to the water-soluble salts in the soils is brought by the Carpathian water streams, which transport to the plains compounds released by the weathering of primary rocks in the Vrshac, the Banat, and the Carpathian Mountains.[140] However, it seems that the main source of salts are maritime (neogene) sediments, which appear closer to the surface along the tectonic breaklines due to direct communication between the lower (salt) and upper groundwaters.

The salt-affected soils in Macedonia are located in various valleys. The salt accumulation in the valleys is closely related to the sources of the salts. Some paleogenic deposits contain 0.8% salts and are rich in lime, Na-sulfate and borate. In some basins the weathering of igneous rocks and thermal springs can also be recognized as the origin of the salts.

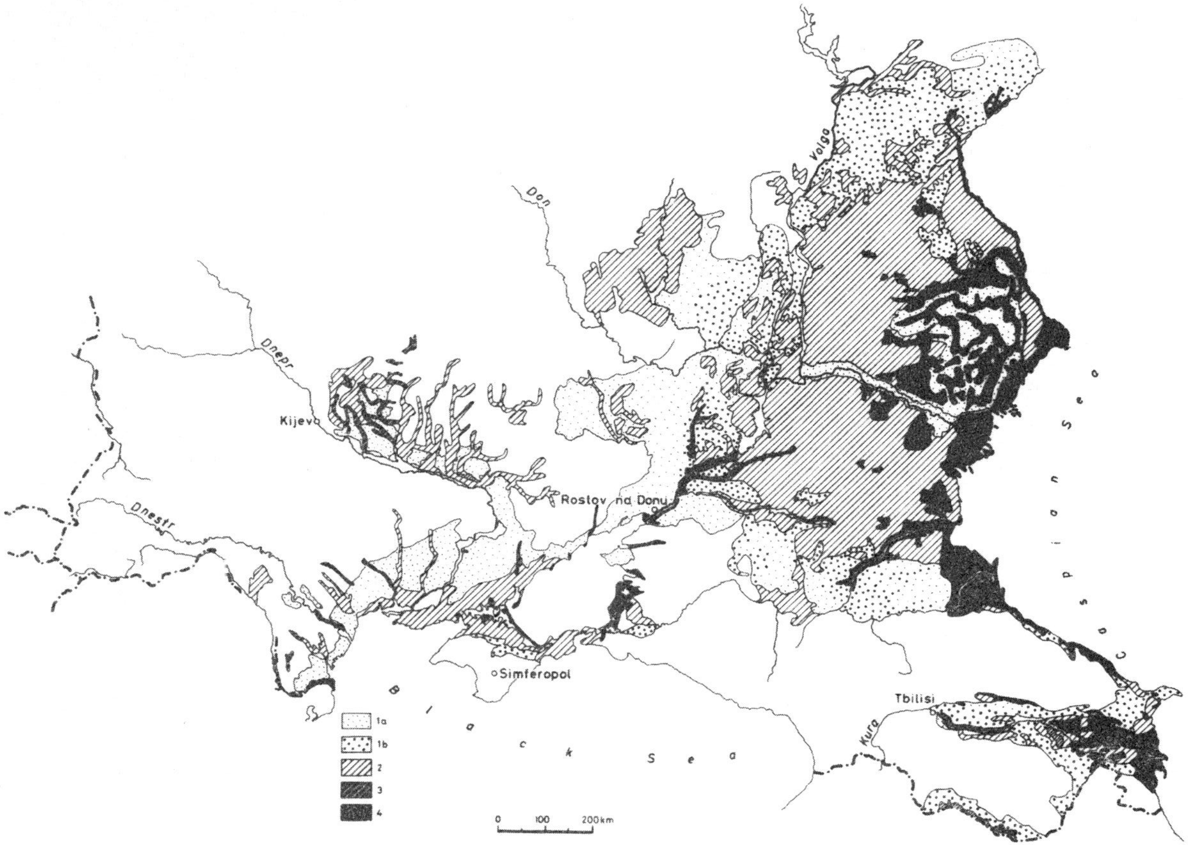

FIGURE 10. Distribution of salt-affected soils in the European part of the U.S.S.R. (1a) Potentially salt-affected soils (with salts at a depth from 2 to 5 m) or nonsaline soils confined to poorly drained plains in the areas of contemporary accumulation of salts; (1b) soils with salt accumulation within 2 m of the surface; (2) alkali soils with a structural B horizon; (3) alkali soils without a structural B horizon; (4) saline soils. (After Egorov, Bazilevich, and Pankova.[1])

Mineralized plant residues from natural vegetation (succulent and dry halphytes) and irrigation waters appear as a source of salt in soils also, but to a lesser degree.

In the genesis and evolution of Yugoslavian salt-affected soils, the warm-dry and cold-wet periods, which exist throughout the course of the year, are of prime importance, followed by alterations in the upward and downward movement of water in the soil. In warm-dry periods, salts accumulate in the soil due to intensive evaporation, while in cold-wet periods the salts are leached. Alternating salinization and desalinization are thus characteristic of the province of Voyvodina, under meteorological conditions which periodically bring excessive moisture into soils having impermeable subsoil strata ("blue clay").

In southern Banat, the alkali soils originated as a result of alluvial (lacustrine) salinization. This salinization was caused by the draining of surface waters into large depressions, where salts accumulated around the borders after evaporation.

In the Strem (Syrmian) region between the Danube and Sava rivers, solod soils are present as a predominant type of salt-affected soils. These formed due to natural processes of desalinization and dealkalization under the influence of flood waters, particularly from the Frushka Gora mountains, as well as under the influence of the forests.

The solonetz soils in Macedonia are formed in certain higher, better drained areas, which have a more pronounced alluvial water flow and a lower erosion basis, with the additional influence of the melting of accumulated snow.[141]

The origin of sulfate-soda solonetz soils in Macedonia, which are in the first stage of their development, can be explained by the effect of soda formed under anaerobic hydromorphic conditions, capillary movement and the saturation of the exchange complex with Na-ions during the process of alternating salinization and desalinization. Under such conditions the A and B_1 horizons are totally leached; therefore, lime and gypsum accumulations appear only in the B_2 or C horizons.

Besides the existing salt-affected soils in Yugoslavia, secondary salt-affected soils also occur, due to irrigation, mainly in the northern part of the country, namely in Voyvodina, Croatia, etc. Voyvodina in particular, which lies on the second Danube Plain, is vulnerable to secondary salinization due to its hydrogeological and geomorphological conditions as well as to the fact that a great part of the irrigated agriculture of Yugoslavia is carried out in this region.[142]

The processes of salinization and desalinization are active during the wet and dry periods in the valleys of all rivers flowing into the Adriatic. Desalinization takes place during the rainy winter months, while salinization takes place during the dry summer months. In the dry period of the year the water only remains in deeper depressions, where saline-alkali soils have developed. On the Adriatic coast there are several places affected by salinity and alkalinity. This phenomenon is characteristic not only of Yugoslavia and the Adriatic coast, but also of several other coastal regions in Southern Europe. This type of salinization was described by Gracanin and Blaskovic on the coastal area of the Wellebit Mountains in Yugoslavia and on the Island of Pag.[108]

A specific character is given to these soils by salted winds, causing a permanent salinization in the coastal area and on the islands. Two main processes took part in developing the coastal salt-affected soils: leaching and salinization. With regard to the genesis of the soils, M. Gracanin divided the soils of the Island of Pag into four groups:[108]

1. Stony soils (brown solonetz)
2. Dry meadow soils (solonetz)
3. Humid meadow soils (solonetz)
4. Muddy soils (muddy solonchak)

According to the kind of salinization, this author described the above classification as follows:

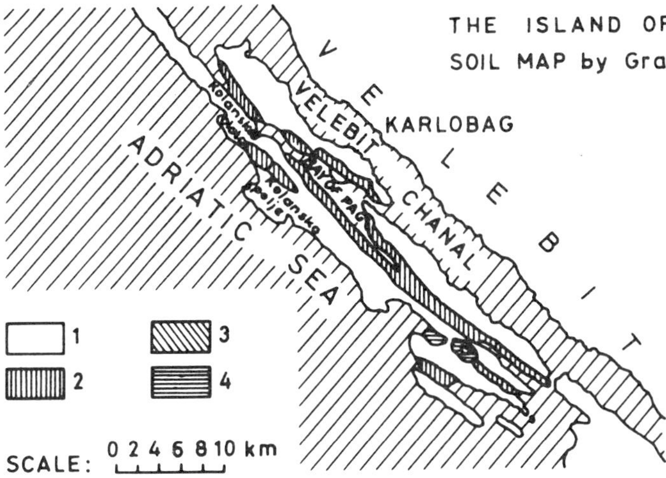

FIGURE 11. The soil map of the Island of Pag. (1) Stony soil with brown solonetz soil; (2) solonetz; (3) mineral solonchak soils; (4) organogeneous swampy solonchak soils.[108]

1. Stony soils have the character of an undeveloped skeletal brown solonetz because of their shallowness and permeability.
2. Dry meadow soils are saturated by ions of alkali and calcium, and they contain a small quantity of readily soluble alkali salts. The author included them in the alkali-calcic solonetz group.
3. Humid meadow soils contain a high concentration of alkali and Ca ions in their solution, so they were placed in the group of alkali-calcic solonchaks.
4. Muddy soils are rich in readily soluble salts and they were placed in the group of organogeneous swamp soils (alkali-calcic solonchaks).

It is not possible to put all these soils on the soil map because of the small scale, so they are grouped with the bare karstic soils, along with some spots of terra rossa and alluvial or hydromorphic soils (Figure 11).

In all cases when saline and/or alkali soils develop, the joint effect of seawater, salt-transporting water, and groundwater conditions result in the formation of salt-affected soils.

In Table 6 the regional occurrence, climatic characteristics, and altitude of European salt-affected soils is demonstrated. The data in this table include practically all the existing European salt-affected soils, even if their territory is limited. The Table clearly shows that a wide range of mean annual temperature, mean annual precipitation, and altitude can be responsible for the development of salt-affected soils. Conditions vary from humid or semihumid to semiarid and the altitude ranges from below sea level to 700 m above it. As will be seen later, similar or even more pronounced variations can also be found on other continents.[143]

In the driest regions of Europe, namely along the Volga and in the Trans-Caucasian Soviet Republics, gypsiferous soils occur, while magnesium soils are extensive in numerous countries, commonly in association with alkali soils. These are particularly frequent in the Soviet Ukraine and in related provinces of the Russian Federation, as well as in Romania, Hungary, Yugoslavia, and Bulgaria.

As on the seashores of all continents, acid sulfate soils occur along the Atlantic coast and partly on the Mediterranean coastline of Europe as well. They are particularly frequent along the seashores of Sweden, Finland, Denmark, Germany, Belgium, and the Netherlands, but are also found on the Atlantic coast of Spain and to some extent on the Mediterranean coast.

Table 6
REGIONAL OCCURRENCE, CLIMATIC CHARACTERISTICS AND ALTITUDE OF EUROPEAN SALT-AFFECTED SOILS

Country, region	Mean annual temperature (°C)	Mean annual precipitation (mm)	Altitude (m)
Austria			
Pulkantal (Illmitz)	9.2	566	187
Seewinkel (Apetlon)	9.8	623	140
Bulgaria			
Danube Valley	11.6	585	50—100
Maritza, Tundja, Strema Valley	12.3	500	100—200
Czechoslovakia			
South Moravia	9.0	500—550	200
Danubian Lowland	9.0—10.0	500—600	100—150
East Slovakian Lowland	9.0	600—650	100
France			
Atlantic sea coast	10.5—11.5	700—800	0—100
Mediterranean sea coast	14.0—15.0	550—650	0—100
Greece			
Ionian sea coast	17.5—18.5	650—750	0—100
Thessaloniki Plain	14.0—15.0	500—600	0—50
Hungary			
Hungarian Plain	10.0—10.5	524—585	80—120
Italy			
Northern Italy	13.7	744	0—100
Southern and island region	17.6	478	0—100
Portugal			
Atlantic sea coast	14.0—18.0	660—1400	0—100
Romania			
Black Sea shore — Danube Delta	11.0—11.3	359—439	0—50
N.E. Romanian Lower Danube Plain	9.6—11.1	400—515	0—100
Western Romanian Lower Danube Plain	10.6—11.5	480—570	100—200
Tisza Plain	10.7—10.8	558—620	80—100
Moldavian Table-Land	9.0—10.5	399—588	200
Spain			
Southern region	17.6—18.3	535—651	30—100
Southwestern region	17.0—17.6	295—419	0—60
Basin depression of the Ebro	14.6—15.1	324—378	118—380
Central plateaus	11.6—14.2	389—403	700
U.S.S.R.			
Ukraine	7.0	500—550	100
Oka-Don plain	7.0—8.0	500—550	100—200
Preazovian plain	9.0—10.0	300—550	100
Prevolgian plateau	6.5—7.5	500—600	200—300
Precaspian lowland	6.0	150—250	0—100
Transcaucasian plains	14.0	200—250	100—200
Armenia	11.3—14.0	200—220	500—700
Transvolga region	4.0—5.0	300—400	100—200
Yugoslavia			
Vojvodina	10.0—10.5	550—600	100
Macedonia	11.2	652	100—200
Adriatic sea coast	14.5—15.5	500—900	0—50

Acid sulfate soils also occur on some coastal areas in Britain and have been described in detail in North Kent, where they cover an area of over 70 km^2 in the marshes between Sheerness, on the Isle of Sheppey, and Cliffe.

It is worth mentioning that apart from acid sulfate soils, saline soils can also be found on some parts of the European coastline. These are affected by the direct influence of

FIGURE 12. Distribution of salt-affected soils in Asia. (After Kovda and Lobova.[1])

seawater, and in consequence, they contain mainly sodium chloride. In certain cases, such soils are transformed into alkali soils affected by moderate leaching, as described in the Netherlands by van der Molen[24] and in North Kent by Hazleden, Loveland, and Sturdy.[144]

B. Asia

Asia is not only the largest of the continents but also has the greatest stretches of salt-affected soils. More than one third of the salt-affected soils of the world can be found here, and if those parts of Australasia which are geologically close to the Asian continent are also taken into consideration, this figure will exceed 50%. Very few countries in Asia are free from salt-affected soils and in many countries in this continent soil salinity and alkalinity have developed to such a degree that they constitute cardinal problems for the national economy. Countries where the soil cover is dominated by salt-affected soils can also be found in the arid and semiarid regions of Asia.[145]

In Figure 12 a schematic map can be seen, showing the distribution of salt-affected regions in Asia based on the Soil Map of Asia compiled by Kovda and Lobova in 1971.[1] The map, which does not include Turkey, clearly indicates that the Arabian Peninsula and several other Asian countries have extensive areas of salt-affected soils, particularly Iraq, Iran, Afghanistan, Pakistan, China, the Indian subcontinent, and the Indo-Chinese Peninsula. Salt-affected soils occur extensively in Indonesia, Vietnam, Thailand, Cambodia, and Burma. Large areas of salt-affected soils can also be found in the Asian part of the U.S.S.R. and in Mongolia.[146]

The salt-affected soils in Asia are dominated by the saline group, including both sulfate and chloride versions. Saline or solonchak soils constitute more than two-thirds of all the salt-affected soils on this continent. They lie mainly in the arid and semiarid zones but also appear spotwise in many other places in the semihumid and humid climatic regions.

Alkali soils develop in many Asian countries, in most cases jointly with salinity (as occurs in some places on other continents). Vast territories of saline alkali soils can be found in India, the Middle Eastern countries, the Soviet Union, Mongolia, Indo-China, etc. Alkali soils with a low degree of salinity are met with in Asia to a comparatively lesser extent in deserts and semideserts and to a wider extent in Siberia, Mongolia, Indo-China, etc. By contrast with the regular features mentioned when describing salt-affected soils in Europe, typical solonetz soils are comparatively less widespread in Asia.

Although Figure 12 is mainly concerned with the saline and alkali soils of Asia, gypsiferous, magnesium and acid sulfate soils are also very common on this continent. Gypsiferous soils are widespread in the Middle East (Syria, Iraq, Iran) and Pakistan, as well as in the Central Asian Soviet Republics and in China; magnesium soils can mainly be found in association with alkali soils in Western Siberia, the Indian subcontinent, the Middle East, and Indo-China. There are several types of magnesium soils in Asia, and these are far from having been sufficiently studied, described, and registered. Some of them are closely related to saline soils, others to alkali soils and some to soil-forming processes in semihumid and humid tropics (e.g., in the Deccan Plateau in India).[147]

Along the sea coasts of Asia, acid sulfate soils are widespread. A large part of the mangrove ecosystems of salt marshes is also covered by acid sulfate soils. These are located most frequently in narrow strips following the tidal coastline of the Arabian Gulf, the Indian subcontinent and particularly Indo-China, covering practically all the southeastern coastal areas of the continent. Their occurrence is most frequent in Malaysia, Indo-China and on the coast of the Yellow Sea, and even in certain coastal areas of Japan and Korea.[95]

Salt-affected soils, the way they are distributed and the processes by which they are formed in the Asian countries of the Middle East are closely related to those in the African countries of the Middle East. This close correlation is demonstrated in Figure 13, in a map compiled by El-Gabaly[1] and representing soil salinity in the Middle East. This map includes the saline, alkali and magnesium soils as well as the potentially saline soils of the area. It should be noted that potential salinity is only shown in places where it constitutes an imminent hazard, as it threatens the soils in the countries indicated on a much wider scale.[148]

On the following pages salt-affected soils in the Asian countries mainly affected will be briefly characterized.

1. Turkey

According to the UNESCO/FAO Soil Map of the World there are altogether 2,500,000 ha of salt-affected soils in Turkey. However, on the basis of many other surveys it can be assumed that the real figure exceeds this estimate and is most probably several times as much.[92,149] This is plausible because the climatic, geochemical, and hydrological conditions of the country often promote salt accumulation in the groundwater and the soil profile. Consequently, salt-affected soils are widely extended, particularly in the central regions of the Asian part of Turkey, especially in semiarid and arid lowlands where the groundwater contains considerable amounts of water-soluble salts and its table is linked with the soil profile.[149]

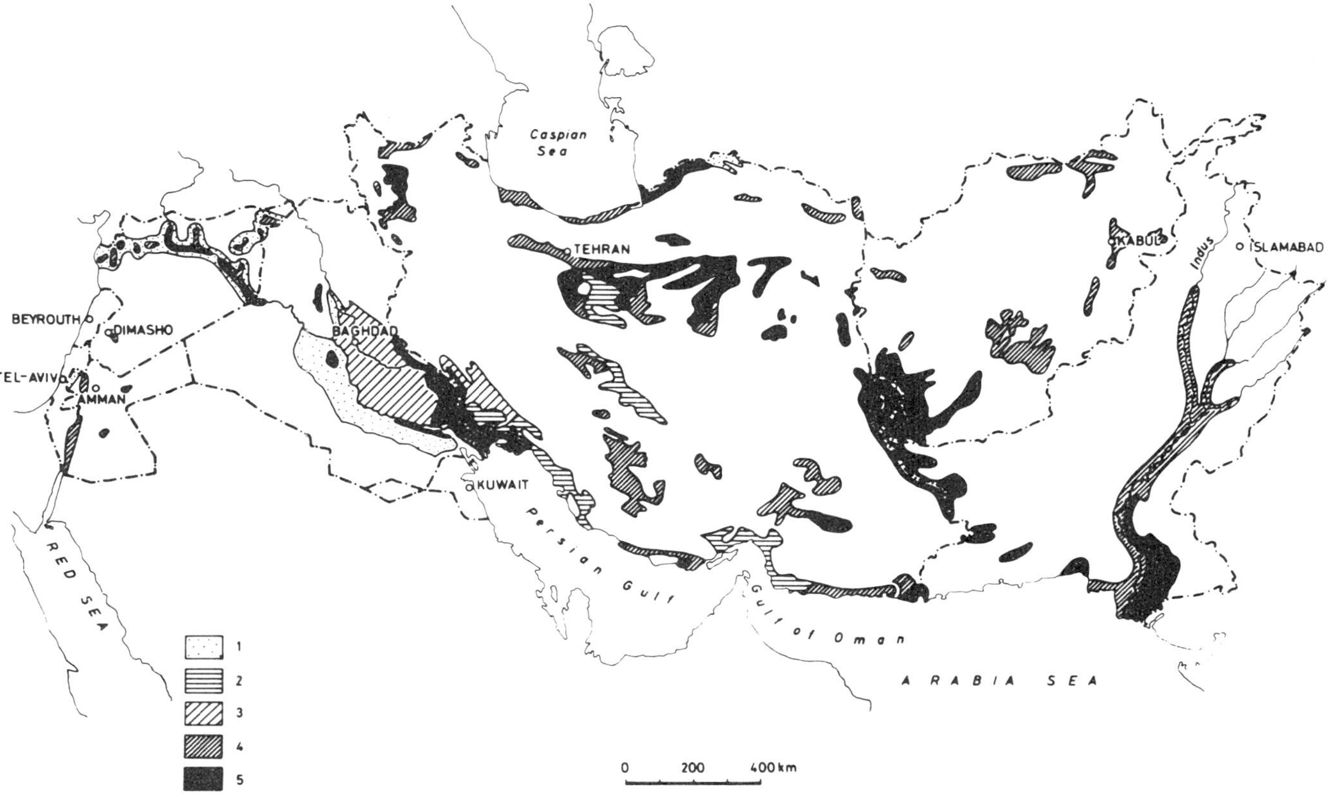

FIGURE 13. Soil-salinity map of the Near and Middle East. (1) Potentially saline soils; (2) saline-alkali soils; (3) spotted saline soils (Ca + Mg salts dominant); (4) spotted saline soils (Na salts dominant); (5) saline soils (Na salts dominant). (After El-Gabaly.[1])

One of the biggest areas of salt-affected soils in Turkey is the Great Konya Basin, where a major part of the soils belongs to various salt-affected types.[92] The most extensive is the saline group which, in most cases, develops under the influence of salty groundwaters. In temporary ponds and along the shores of permanent lakes, sodium chloride and sodium sulfate, accompanied by calcium and magnesium salts, accumulate on the soil surface forming solonchaks. The degree of salt accumulation is so intensive that a white salt crust covers considerable territories. Such flooded solonchaks occur in the low, central part of the basin near Konya and other inhabited areas.

Salt-affected soils with the accumulation of hygroscopic salts (mainly magnesium and calcium chloride) on the surface have the local name of "sabbakh". These soils attract moisture from the air, especially in the early morning when relative humidity is high, and turn the surface into a muddy mixture of salt and soil, often making the sabbakh soils slippery and inaccessible. This type of salinity is characterized by the occurrence of darker patches in the field.

Saline soils may develop in the Konya Basin, even when the groundwater table is not linked with the surface, in consequence of local salt accumulation (partly airborne) and the aridity of the climate. Sometimes salinity does not occur on the very surface of the soil but at a shallow depth along the profile.

Saline soils often develop in the Konya Basin in association with marshes, due to the frequent salinity of the latter. In these and some other cases the accumulation of gypsum is so considerable that gypsiferous soils are also common.

All the saline soils of the Konya Basin contain both chlorides and sulfates, the ratio of which differs subject to local circumstances. The eastern half of the basin, where marine sediments can be found, can be assumed to be rich in chlorides, though locally, sulfates may predominate. The western mountains supply more sulfates, although spots with almost pure chloride salinity have been recorded, in particular in vertisols with deeper water tables (below 1.80 m). These are probably due to the greater mobility of the chloride ions. Deeper in the profile, sulfates become more important.

In general the basin is a catchment area and salts are carried into it in three ways: by rivers, by subsurface seepage, and by surface runoff. The transported soils and those already present in the basin are redistributed and accumulate in very saline areas by lateral and vertical transport. In the former case, besides water, winds also play a considerable role.

There are only minor occurrences of alkali soils in the Great Konya Plain as compared with saline soils. Alkalinity, practically always in association with salinity, can be found in the low central area of the basin. In alluvial soils, which are affected by river water, alkalinity may occur to varying degrees.

There are other districts in Turkey with salt-affected soils similar to those in the Great Konya Basin in respect of the regular features of their occurrence and formation. One of these is the Kücük Menderes Valley south-west of Izmir. In this region salt-affected soils were observed mainly in the river valleys and, fairly extensively, along the coast. It is a feature of the area that all the salt-affected soils are alkali as well. In the Kücük Menderes Valley inland salinization and coastal salinization should be distinguished. In the former case it is always a relatively high water table which is responsible for salt accumulation, and the surface layer is expressly saline.

Coastal salinization is common not only in the Kücük Menderes Valley but also in other regions of western Turkey.

It should be noted that in the salt-affected regions of the country, particularly in the Great Konya Basin, the calcium carbonate contents of the parent material and the soils are considerable and this has a strong influence on the soil-forming processes, particularly on salinization and alkalization.

In addition to the above groups of salt-affected soils, acid sulfate soils also occur in some of the marshes of Turkey.

2. The Asian part of the U.S.S.R.

In the Asian part of the U.S.S.R. the extension of the different groups of salt-affected soils is considerable. More than half of all the salt-affected soils of the Asian continent can be found here. The territory of salt-affected soils is nearly four times as large in the Asian part of the U.S.S.R. as in the European part and altogether they account for about one fourth of all the salt-affected soils in the world.[150]

The distribution of the different groups of salt-affected soils in the Asian part of the U.S.S.R. is closely related to environmental conditions. Saline soils are widely distributed in the Central Asian Soviet Republics, namely in the Kazakh, Uzbek, Turkmen, Tajik, and to some extent in the Kirghiz Soviet Socialist Republics as well as in some parts of the Russian Federation.[151]

Salt-affected soils in the Central Asian Soviet Republics are particularly widespread in deserts like the Kara-Kum and Kizil-Kum and evidently at the Soviet fringe of the Gobi desert, adjoining Mongolia as well as in irrigated areas like Ferghana. They also occur in the valleys of the major rivers of the region, like the Sir Darya and the Amu Darya, near the Aral Sea and on the shores of several lakes with saline and alkali waters, such as Lake Balkhas in Kazakhstan. Many of the salt-affected soils in these areas have been thoroughly studied, and the processes by which they were formed are good models for the general interpretation of the development of such soils under similar environmental conditions.[3] Some of these studies will be cited later when describing the processes of salinization and alkalization.

The development and distribution of gypsiferous soils are also characteristic of the areas which adjoin the related salt-affected soils in the territories of neighboring countries (Turkey, Afghanistan, Iran, Pakistan, China and India). The saline soils of these areas mainly contain sodium sulfate and sodium chloride, the chlorides becoming dominant parallel with an increase in aridity and the nearness of the sea.

It is worth mentioning that in the Asian part of the U.S.S.R. alkalinity is much less widespread than salinity and its geographical distribution is somewhat different. Alkalinity sometimes occurs jointly with salinity in desert and semidesert areas, but soil alkalinity becomes dominant towards the north. One of the largest alkali areas can be found in Western Siberia, along the valleys of the Rivers Ob and Irtis. The alkali soils of Western Siberia belong to a wide spectrum of different types of this soil group. They include alkali soils with and without structural B horizons. Solonetz and solod soils are also widespread and are sometimes associated with magnesium soils. In Western Siberia, the Baraba, and Kulunda lowlands are the areas where alkali soils are the most extensive and where they constitute not only a theoretical problem but also a severe practical hindrance to the development of agriculturc and irrigation.[152]

The vast semidesert areas east of the lower reaches of the River Volga are very rich in saline and alkali soils. Those territories adjoining the Central Asian Soviet Republics contain both alkali and saline soils and the dominance of the latter increases in a south-southeast direction. The same tendency exists with respect to the ratio of saline and alkali soils, where the degree of salinity increases parallel with an increase in aridity. In this area the dry steppe changes into semidesert and desert as we approach the Central Asian Republics.[153]

In several Central Asian Soviet Republics, mainly in Kazakhstan, and partly in Kirghizia and Uzbekistan, the influence of increasing altitudes on the development of salt-affected soils can also be found.[154] (With increasing altitude salinity will evidently decrease).[155]

In Eastern Siberia, alkali soils are widespread in the far north around and beyond the arctic circle in Yakutia. Large areas of alkali soils, both with and without structural B horizons, have developed in the rather extreme conditions of permafrost and constitute one of the northest regions for the occurrence of salt-affected soils. They may develop at high altitudes, and due to their high salt content and strongly alkaline reaction, they hinder the utilization of the soil under conditions where agricultural production is very difficult anyway.

Salt-affected soils appear spotwise in many other regions of Eastern Siberia. On the southern part of the Pacific coast even acid sulfate soils occur.[155,156]

In the Asian part of the U.S.S.R., as in many other places, Takyr formation is also widely extended.[157] This type of salinity, which often overlaps with alkalinity, leads to typical morphological, physical, and physico-chemical properties, including surface cracking, sticking when wet, and extreme difficulties in soil management.[158]

3. Mongolia

Judging by the UNESCO/FAO World Soil Map, only a little more than four million hectares of salt-affected soils can be found in Mongolia, which does not sound too large a quantity in relation to the vast territory of the country; however, their importance is greater than this ratio would suggest, because they occur in areas where the agricultural production of the country is or should be focused. It should be noted that the boundary of permafrost crosses Mongolia and there are salt-affected soils on both sides of this boundary.[96,97]

Bespalov,[96] however, who carried out a more detailed survey in Mongolia, estimates the total area of salt-affected soils to be more than four times that given in the World Soil Map. He states that salt-affected soils account for 162,000 km^2 in Mongolia, which is more than 10% of the total area of the country.

The noteworthy fact that, particularly in developing countries, a detailed survey tends to find much larger expanses of salt-affected soils than global studies indicate, demands further investigation and a new interpretation of salinity problems in many countries of Asia, Africa, and Latin America.

According to Bespalov's data,[96] 90% of all soils in the Gobi desert are salt-affected. In the western districts of the country, in the basins of large lakes, this figure is in the neighborhood of 18% and in the eastern lowland around 12%. Even in highland regions, such as the Mongolian part of the Altai Mountains, the extension of salt-affected soils is nearly 3%. Salt-affected soils are also located in the Hangai Mountains, covering about 3% of the area and sometimes occurring as high as 2500 m above sea level.

Among the saline soils in Mongolia practically all the chemical types can be found, namely chloride, sulfate and carbonate. Gypsiferous soils are also widespread. In saline soils, the neutral and alkaline hydrolyzing sodium soils are present jointly, which result not only in high salinity but also in a high pH.

Alkali soils with comparatively low salt contents are also widespread in Mongolia, for instance in the Gobi desert and southwest of Sani Shand. The alkali soils of this district border on those in China. Alkali soils are extensive in the East Mongolian Lowland, especially in the Suhe Bator District. They are frequent in the western districts of the country near the great lakes, being found somewhat higher in altitude than the bordering saline soils, and occur in many places in the dry steppe regions of Mongolia.

The presence of salt-affected soils in Mongolia is closely related in some regions to saline swamps and marshes.

A. Kowalkowsky and J. Borzyszkowski described a young, not yet fully developed organo-mineral deposit in the area of Gurvan Turuu at a height of 1372 m where turf-sedge solonchaks had developed on basalt 2 1/2 m above the dry bottom of a perennial lake.[97] Permafrost occurs here at a depth of 40 to 60 cm. Salt accumulation is so intensive that both plants and solid surface are covered with crusty salt efflorescences. Regarding their chemical composition, both neutral and alkaline hydrolyzing sodium salts are present.

About half the salt-affected soils in Mongolia lie north of the permafrost boundary, like those mentioned above. These, too, are mainly solonchaks (though salt accumulation is not confined to the upper layers of the soil), but can also be found in deeper horizons and groundwaters (Figure 14). Figure 14 shows the borderline of permafrost in Mongolia and the distribution of salt-affected soils to the north and south of this borderline.[97]

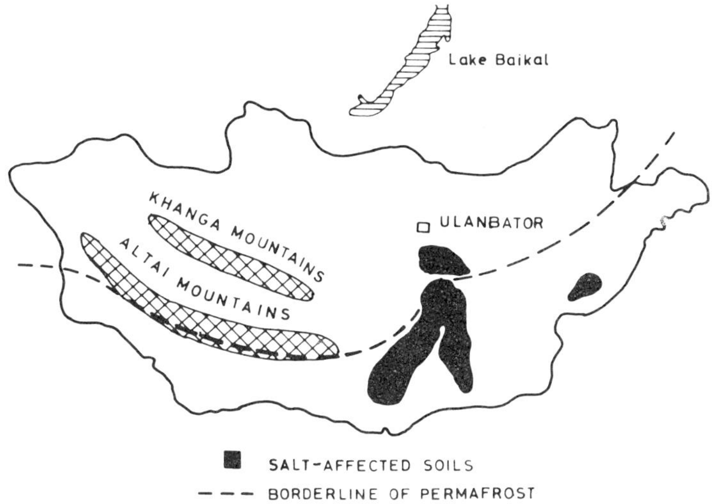

FIGURE 14. Distribution of salt-affected soils in Mongolia. (After Kowalkowsky and Borzyszkowski.[97])

4. China

There are vast territories of both salt-affected soils in China; this country takes second place in Asia after the U.S.S.R., in the extent of such soils. According to Wang-Zunqin,[159] salt-affected soils can be found almost all over China, from the tropics to the frigid zone, from the coastal areas in the east to the deserts in the west, from the lowlands at or below sea level to the high mountain plateaus. Partly as a result of contact with seawater, there are various salt-affected soils in the coastal areas of east and southeast China (including Taiwan and the islands in the South China Sea). In the arid, semiarid and semihumid inland regions of the north (from the line of the Huaihe river to the Qin-ling mountains), all the water converges on lowlying land due to the impeded drainage of the surface runoff and the underground flow, and forms a watershed where both water and salts are accumulated. The large areas of salt-affected soils are therefore generally distributed in the basins, valleys, and alluvial plains in these regions. Salt-affected soils can be found both in the Aidin lake in Xinjiang, 152 m below sea level (the lowest area in the country) and on the Tibetan plateau, 4,000 m above sea level.[93,94,159]

The salt-affected soils in China represent all the groups described earlier. Saline soils occur mainly in the deserts and semideserts of the western and northwestern provinces. Chloride and sulfate dominate among the accumulated salts in the soils. Saline soils can also be met with in nonarid regions, in the valleys of the rivers Hoang-ho and Jangce-Kiang. Gypsiferous soils are extensive mainly in the dry areas adjoining saline soils, while magnesium soils can be found in association with the heavy soils of the lowlands. In the coastal areas saline soils are also frequent, associated of course, mainly with sodium chloride accumulation.

Alkali soils occur in many regions of China. They are particularly widespread in the northeast and in the central provinces of the country. Acid sulfate soils develop along the seashores, mainly on the eastern part of the south coast of the Yellow Sea and the Pacific.

Kovda et al. compiled a schematic map of the distribution of salt-affected soils in China, on which saline and alkali groups of salt-affected soils and potentially salt-affected soils are demonstrated (Figure 15).

Recently Wang-Zunqin gave the following, more detailed description and regionalization of soil salinization and alkalinization in China:[159]

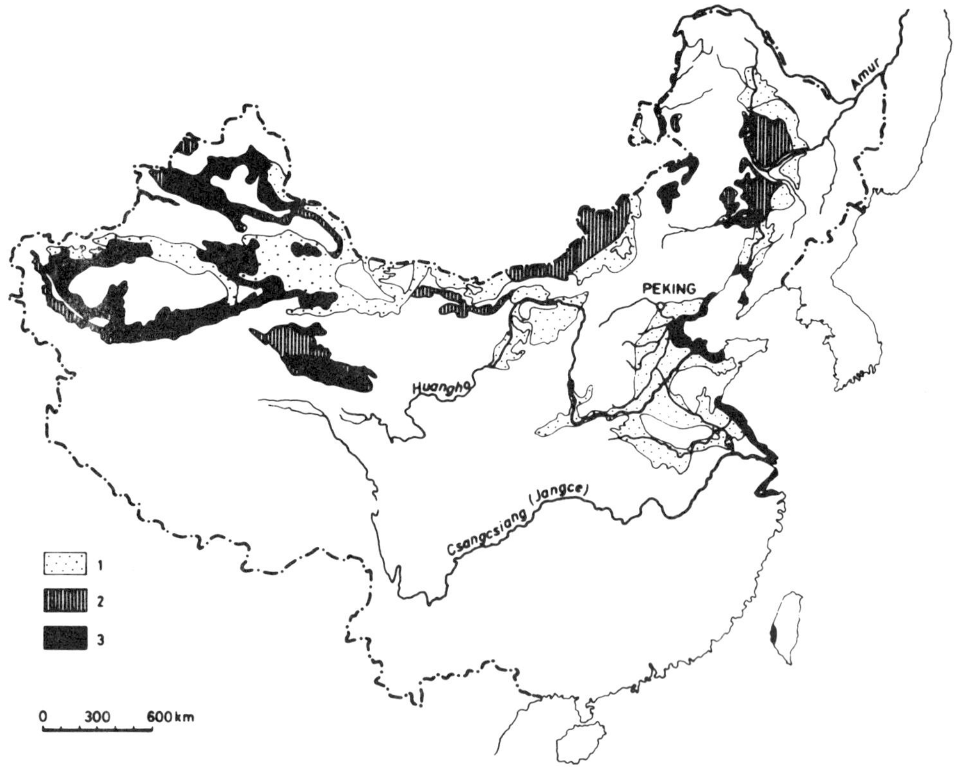

FIGURE 15. Distribution of salt-affected soils in China. (1) Potentially salt-affected soils; (2) alkali soils; (3) saline soils. (After Kovda et al.[1])

1. The coastal humid and semihumid region with seawater submergence salinization — In this region salt accumulation is caused by seawater both in soils and groundwater, preceding even the soil-forming process. The degree of mineralization varies and the effect appears either in the form of a continuous ribbon, as in the seashore north of the Hangzhou Gulf or in the form of a discontinuous ribbon, as in the estuary south of the Hangzhou Gulf. The phenomenon occurs parallel with the coastline, and the nearer the seashore, the higher the salt content of the soil and groundwater. With the exception of the acid sulfate soils found sporadically south of the River Ming and a lesser number of soda salinized soils existing at the mouths of the Rivers Luan and Liao, the salt composition is predominated by chlorides. This region can be divided into three subregions:

- Taiwan Strait — South China Sea coastal area
- East China Sea — Yellow Sea coastal area
- Bohai Sea coastal area

2. Northeastern semihumid and semiarid steppe region with meadow salinization — In this region the alkali-saline soils are mostly soda solonetz and meadow solonized soils which are primarily derived from meadow bog soil, dark meadow soil, meadow chernozem, black soil, etc., and are mostly found in the form of spots among the original soils.

This region can also be divided into three subregions, as follows:

- Son Liao Plain subregion
- Three River Plain semihumid subregion with sporadical salinization
- Da, Xiao Hinggan Ling and ChangBei Mountains subregion

3. Huang-Huai-Hai Rivers alluvial plain, semihumid and semiarid region with cultivated, salinized meadow soil. — On the alluvial fan and the upper part of the plain, bicarbonate and carbonate zones are most frequent, whereas in the central and lower parts of the plain, chloride-sulfate or sulfate-chloride zones can be found. In the region of water-salt convergence within the plain and in the region bordering on the coastal plain, there are chloride accumulation zones.

In terms of genesis, the alkalized and salinized soils in this region have developed essentially on the fluvo-aquic soil which was formed from the alluvial parent materials and has since been subjected to dry farming. On the basis of the composition of their salts, the soils can be divided into:

1. Chloride salinized soils
2. Sulfate-chloride or chloride-sulfate salinized soils
3. Soda-chloride salinized soils
4. Tile-like crust solonetzes (belonging to the alkalized meadow soils) formed as a result of the desalinization of the soils mentioned above

This region can be divided into four subregions according to the bioclimate and the salinization regime, as follows:

- Lower Huan-Huai-Hai Rivers Alluvial Plain
- Fen-Wei River valley
- Shanxi-Hebei mountain basins
- Yi-Meng mountain area

4. Mongolian plateau, arid and semidesert. (Salinized steppe region) — In this region the alkali and saline soils are composed mainly of soda-saline soils, meadow solonetzes, and steppe solonetzes. Soda-saline soils are often found in the river and lake terrace areas of river-terrace depressions and the lowlying areas of contemporary river valleys; meadow solonetzes are generally scattered in the form of spots over the higher positions of the microrelief and lowlying terraces, while steppe solonetzes lie on the higher terraces or the higher flat lands, scattered in spots among the zonal soils.

This region may be divided into three subregions, as follows:

- Eastern arid steppe
- Western semi-desert
- Da Hinggan Ling subregion

5. Middle and upper Huanghe River semiarid and semidesert region with salinity — This region is known as the principal salt source of the Huang-Huai-Hai alluvial plain, with varied types of salinization and an extensive accumulation of sulfates. On the loess plateau of the Wei-Gin-Lo River valley, the loess soils and dark loess soils contain carbonates and sulfates in their substrata, which constitute a potential threat of salinization, while in the narrow valleys, mainly sulfate-saline soils can be found.

Various kinds of contemporary saline soils are scattered extensively on the Huanghe alluvial plain with notably horizontal differentiations of the soluble salts in the soil and groundwater. On the piedmont diluvial fan and high terraces, there are large quantities of carbonate and bicarbonate; bicarbonate and sulfate are frequent on the upper part of the plain, chloride and sulfate on the middle part, while sulfate and chloride dominate on the lower part. There is also a great deal of soda in the soils of depressions and their peripheries, and sulfate solonchak and chloride solonchak are predominant in the wasteland.

Starting from the west of Lake Ulanshuai in the Houtac area of the River Huanghe, along the line ranging from the Anbei-langshan to Xidatan on the Yingchuan plain, vast quantities of takyr-like solonized soils occur. In the frozen alpine region of the upper Huanghe river to the west of Lanchou, saline soils in which chloride-sulfate is predominant are sparsely scattered in the mountain valleys.

This region can be divided into four subregions, as follows:

- Weijing-Ro loess plateau
- Middle Yellow River Erdos loess plateau
- Middle Yellow River alluvial plain
- Upper Yellow River frigid semidesert subregion

6. Gansu-Xinjiang desert, salinized region — With the exception of the hydrographic net of the River Ertix, this is a river-sealed inland region, where all the water streams either disappear in the Gobi desert or converge into inland lakes.

The soils of this region not only contain a large amount of salt, but are also characterized by a significant accumulation of gypsum or magnesium carbonate. In addition to the wide distribution of crust solonchak, primarily associated with chloride-sulfate, there is also an extensive area of takyr-like solonized and magnesium solonized soils. Based mainly on the salinization-alkalization regime, this region can be divided into four subregions, as follows:

- Hexi Corrider-Alxa desert
- Jumggar Basin desert
- Qilian Mountain salt source subregion
- Tian-Shan-Allay mountains subregion

7. Chinghai-Xinjiang, extremely arid desert, saline region — The most important saline geochemical characteristics of this region are as follows: they have a long history of salinization; Tertiary-Old Quaternary salt-bearing strata are widely exposed to the open along the front range and in the submountainous diluvial zones; the soils have very high salt contents; various contemporary active solonchaks are covered by a thick, hard salt crust on the surface, with salt contents ranging from ten to more tens of percent; great diversity is observed in the processes of salinization; in addition to the process of salinization caused by groundwater, very extensive processes of residual salinization and diluvial salinization can be observed, as can the formation of lacustrine salt, mud, and salt crust; pronounced gypsum desertification and weak alkalization can be observed. In this region there is a variety of saline-type soils. With the exception of chloride, sulfate and carbonate-bicarbonate, the accumulation of nitrate or borate is predominant in the soils of some areas, and in certain regions, salts of boron, lithium, and potassium are accumulated in an available form and to an extent large enough for exploitation.

This region may be divided into three subregions, as follows:

- Tarim Basin — Shule River Valley
- Qaidam Basim gypsum pan
- Tianshan-Kunlunshan Altunshal salt source subregion

8. Xizang Plateau, frozen alpine desert, saline region — This is known as the highest-lying salinized region above sea level. Only minor amounts of water and salt flow out of the region by way of the Rivers Yarlung Zangbo, Shiquan, and Xiangquan, while the rest converge into numerous isolated salt lakes through countless inland drainages. Around these lakes the intensity of salinization is much lower than in the two major regions of saline soils in Chinhai-Xinjiang and Gansu-Xinjiang. This region is mostly characterized by meadow bog with indistinct saline geochemical differentiation but distinct regional differentiation. On the Quantang Plateau, there is abundant soda-chloride as well as boron salt accumulation in the soils. In the soils of the Shiquan, Xingquan, and Yarlung Zangbo River Valleys, sulfates are predominant. Hence this region can be divided into three subregions, as follows:

- Quangtang Plateau
- Valleys of the Northwest Tizang high mountains
- Yarlung Zangbo River valley

According to Yu Ren Pei and Xione Yi,[160] not only saline soils but alkali soils, too, are widely distributed in China, ranging from the Hulun Buir Plateau of the Inner Mongolia Autonomous Region in the north to the Huang Huai-Hai Plain which lies north of the Yangtze river, and from the coastal region in the east to the Junggar Basin in the Uygur Autonomous Region of Xinjiang in the west. They are even to be found on the Tibetan Plateau at an altitude of more than 4000 m above sea level. Regions where large areas of alkali soils are distributed in a fairly centralized manner are: the Song-liao Plain in the northeast, the Huang-Huai-Hai Plain, and Da, Xiao Heihe River Basin in Inner Mongolia, the Datong and Yanggao Basins in Shanxi, and the Yanqi and Junggar Basins in Xinjiang. Alkali soils are dispersed as spots in other soils, and often coexist with saline soils. This indicates that their formation is closely related with that of saline soils. Large stretches of alkali soils can be found in the semihumid, semiarid, arid, and even warm temperate zones of China, particularly north of the Yangtze river and in areas influenced by the monsoon climate of the Pacific Ocean in

Eastern Asia. They are common in the north-eastern provinces of the country where their formation is closely related to hydrological conditions. The groundwater of these regions usually contains sodium bicarbonate, or sometimes even sodium carbonate, the influence of which is decisive for the formation of alkali soils.

Song Daquan and Hu Siming report that the salt-affected soils in the tidal flat regions of China include saline, alkali, and acid-sulfate soils, with the saline type prevailing.

In regions adjoining Mongolia and the U.S.S.R., takyric types of salt-affected soils are widespread but they also occur in other areas of the country under arid conditions.

5. India

In India the extension of salt-affected soils is considerable and their occurrence is related to different regional soils in arid, semiarid, semihumid and even humid territories of the country. As J. S. Kanwar states,[161] they occur in wide belts or as patches of varying expanse, often associated with hydrological or topographical features. At times they are extremely variable, and salinity conditions frequently change even over relatively short distances. The presence of salts on the surface is also a seasonal phenomenon. The salts become more conspicuous during the dry season, particularly during the winter in North India. When the deterioration of the soil has become extreme and alkalinization has taken place, the salts may or may not become evident on the surface.

Figure 16 illustrates the distribution of salt-affected soils in India. On the map saline, saline-alkali, acid sulfate and potentially salt-affected soils are included. In India, as in many other Asian countries, the saline group prevails. Different subgroups of saline soils are also demonstrated on the map.

Distribution of saline-alkali soils — The principal areas of saline-alkali soils in India are the arid alluvial tracts of Punjab, Haryana, and Rajasthan in the north and west; the Ganges Valley west of a longitude of about 80°E in the states of Delhi and Uttar Pradesh, (representing the alluvial soils of semiarid regions); the arid and semiarid uplands of the Deccan Plateau, especially between the Rivers Tapti, Godavari, and Bhima in Maharashtra, Mysore and Andhra Pradesh; coastal saline soils in the deltas of the Rivers Ganges, Godavari, Krishna Cauvery, and Mahanadi in the states of West Bengal, Orissa, Andhra Pradesh and Madras (Tamil Nadu); coastal salt flats along the Rann of Cutch; and acid saline soils in the backwaters of Kerala State.

It is estimated that in Uttar Pradesh about four million acres lie waste because of saline and alkaline conditions. This is the largest area in any single state in India. These lands are mostly located in the region south-west of the Ganges. The worst affected districts are Mainpuri, Aligarh, Etah, Farrukhabad, Etawah, Kanpur, Unnao, Farehpur, Allahabad, Rae-Barely, Lucknow, Partapgarh, Sultanpur, and Hardoi. There is also a stretch of lowlying land extending to the semihumid parts of the state, comprising the districts of Azamgarh, Western Ballia, Northern Ghazipur, and the Bhadhoi taluk of Varanasi district, where salinity is due to a rise in the water table.

The Rajasthan, Pali, Bhilwara, Bharatpur, Ajmer, Alwar, Jaipur, Jodhpur, Jolor, Tonk, Nagor, Sirchi, and Chittorgarh districts are reported to contain saline and alkali soils. Salinity is more prevalent than alkalinity, but due to irrigation the alkali problem is becoming more serious. In Gujarat, the districts of Banas Kantha, Mehsasna, Ahmedabad, Kaira, Surat, Broach, Amroti, and some coastal areas, have salinity and alkalinity problems.

In Maharashtra State, the Khandesh, Stara, Sholapur, Thana, Kolaba, and Ratnagiri districts are reported to have salinity problems. Except in the first three districts, the problem is mainly due to the incursion of seawater.

Mysore State has a salinity problem in Dharwar, Bijapur, Raichur, Bellary, and Mandya.

In Andhra Pradesh, the Krishna Delta area, the Nizamabad, and Nalgonda districts have salinity problems.

Tanjavur district in Madras (Tamil Nadu), Balasore in Orissa and Sunder Bans in West Bengal also suffer from salinity. In the Delhi territory both salinity and alkalinity are found.

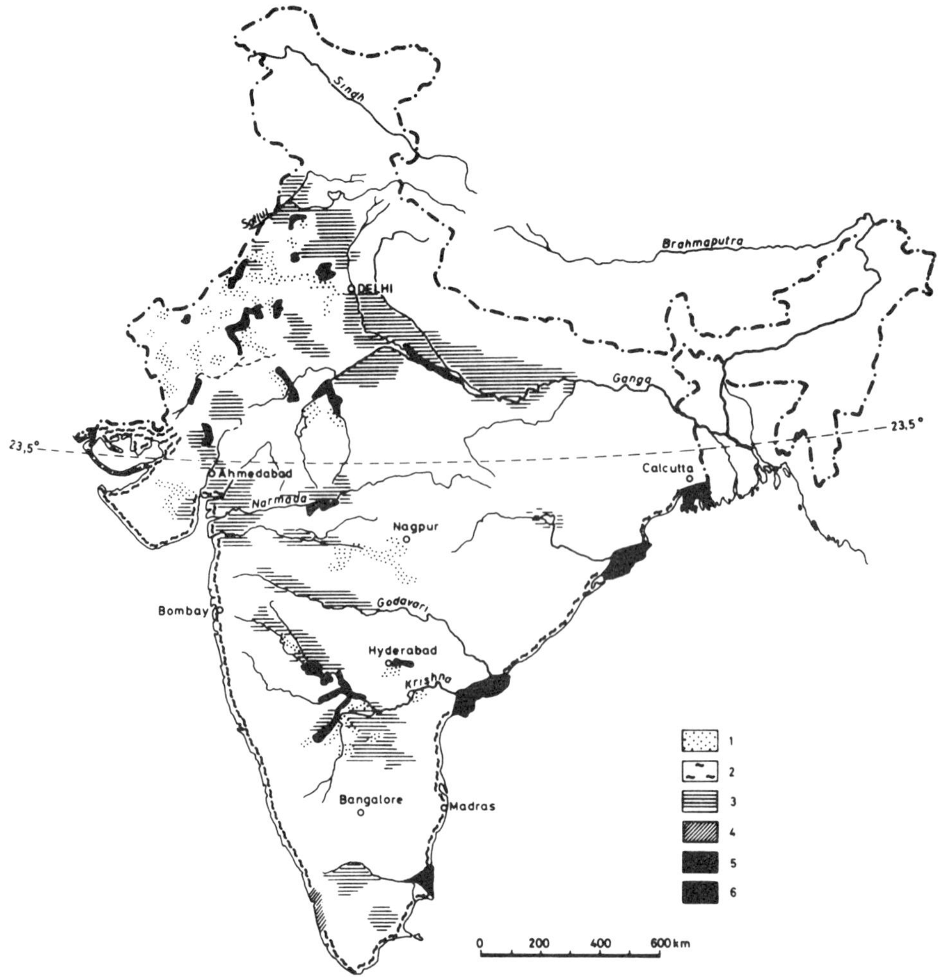

FIGURE 16. Distribution of salt-affected soils in India. (1) Potentially saline soils; (2) coastal saline soils; (3) saline alkali soils; (4) acid sulfate soils; (5) deltaic saline soils; (6) saline soils. (After Bhumbla et al.[1])

Potential salinity: — In many soils in arid areas, salts occur at depth in the profile. As a result of the introduction of irrigation, with inadequate leaching and low soil permeability, perched water tables may develop and bring salt to the surface. Thus, as a result of the extension of irrigation, more and more areas are becoming salt-affected.

Based on different sources Kanwar characterizes the main salt-affected districts in India as shown in Table 7. As the table clearly shows, salt-affected soils are extensive in most states of India at very different altitudes and under various climatic conditions.

Yadav characterizes the main salt-affected areas in India as follows:[30]

Salt-affected soils in the Indo-Gangetic Plains — This is the biggest salt-affected area of the country where salt-affected soils cover nearly 3 million ha (not including potential salinization). The soils of this region are mostly found in the states of Uttar Pradesh, Haryana, and Punjab and are generally sodic, except for a few pockets which are only saline. The soluble salts mainly comprise carbonates and bicarbonates of sodium, which are capable of alkaline hydrolysis. Owing to the almost flat topography of the area and to impeded drainage, the problem has become more pronounced and about 40% of all the salt-affected soils in the country are found in this region. The occurrence of a zone of calcium carbonate nodules at a depth of about 1 m is a common feature; in places it forms an indurated layer.

Table 7
MAIN SALT-AFFECTED DISTRICTS IN INDIA

Name of state	District or area of major saline and alkali problem	Elevation (in feet)	Relation to river system	Climate: Rainfall (inches)	Climate: Mean annual temp. (°F)
Punjab	Gurdaspur Alluvium	200—500	Between the Ravi and Beas Rivers	25—40	
	Amritsar District	200—500	Between the Ravi and Beas but intercepted by a large number of main canals	21—30	
	Ferozepur District	<500, 0—200, 200—500	All between the Sutlez and the natural drain or canal parallel to it in the district	15—24	
	Kapurthala	200—500	Between the Beas and the Beins	22—30	
	Ludhiana	200—500	Between the Sutlez and local rivulets and canals	22—29	76
	Jullundur	200—500	Between the White or East Bein and Sutlez	24—35	
	Patiala	200—500	All between Ghagger and Saraswati and other local rivulets	26—34	
Haryana	Hissar	200—500	Between the Ghagger and other local rivers	18—23	77
	Rohtak	200—500	Between the Jamna and other local rivulets and canals	23—28	
	Gurgaon	200—500	Between the Jamna and local canals	25—28	
	Karnal	200—500	Between Jamna and other local rivulets	27—35	
Uttar Pradesh	Muzaffarnagar	200—500	Between Jamna and Ganges and Hindon rivers	28—40	
	Meerut	200—500	Between Ganges river and the Jamna river	28—34	75.8
	Bulandshahr	0—500	Between Ganges River and the Jamna River	28—32	
	Aligarh	0—200	Between Ganges River and the Jamna River	28	
	Part of Mathura	0—200	Between Ganges River and the Jamna River	<28	
	Etah	0—200	Between Ganges River and the Jamna River	28—32	
	Mainpuri	0—200	Between Ganges River and the Jamna River	28—32	77.9
	Farrukhabad	0—200	Between Ganges River and the Jamna River	32—36	
	Etawah	0—200	Between Ganges River and the Jamna River	30	
	Kanpur	0—200	Between Ganges River and the Jamna River	30—32	77.8
	Fatehpur	0—200	Between Ganges River and the Jamna River	32—36	
	Allahabad	0—200	Between Ganges River and the Jamna River	40	78.2

Table 7 (continued)
MAIN SALT-AFFECTED DISTRICTS IN INDIA

Name of state	District or area of major saline and alkali problem	Elevation (in feet)	Relation to river system	Climate: Rainfall (inches)	Climate: Mean annual temp. (°F)
Uttar Pradesh	Hardoi	0—200	Between the Ganges and Gomati Rivers	36	
	Unnao	0—200	Between the Ganges and Gomati Rivers	32—36	
	Lucknow	0—200	Between the Ganges and Gomati Rivers	36—40	77.9
	Rai Bareilly	0—200	Between the Ganges and Gomati Rivers	36—40	
	Sultanpur	0—200	Between the Ganges and Gomati Rivers	40—42	
	Partapgarh	0—200	Between the Ganges and Gomati Rivers	36—40	
	Jaunpur	0—200	Between the Ganges and Gomati Rivers	40	
	Azamgarh	0—200	Between Gomati and Ghagra rivers	41—46	
Bihar	Banaras	0—200	Between Gomati and Ganges	<40	78.2
	Saran	0—200	Between Ghagra and Gandak and Ganges	41—44	
	Muzaffarpur	0—200	Between Buhri Gandak and the Ganges	44—46	
	Darbhanga	0—200	Between Burhi Gandak and the Ganges and Kamla Rivers	46—50	77.1
Rajasthan	Pali	200—500	Several tributaries of Luni River	16—22	
	Bhilwara	200—500	Between tributaries of Banas River	21—26	
	Bharatpur	200—500	Local rivers of Jamna and Gambhir	27	
	Ajmer	200—500	Kheri River and other tributaries of Banas River	24	76.7
	Alwar	200—500	Sabi and other local rivers		
	Jaipur	200—500	Salt lake and local rivers and tributaries of Banas River	24	77.3
	Jodhpur	200—500	Tributary of Luni River	14.2	79.2
	Jalor	0—200	Several tributaries of Luni Rivers		
	Tonk	200—500	Banas River and its tributaries		
	Nagor	200—500	Salt lake and tributaries of Luni River		
	Sirohi	200—500	Tributaries of Luni		
	Chittorgarh	200—500	Banas River and its tributaries		

Gujarat	Banas Kantha	0—200	Banas and Luni Rivers	<25	
	Mehsana	0—200	Between Banas Sabarmati and their tributaries	27.5	
	Ahmedabad	0—200	Sabarmati River and small local rivers	30	81.4
	Kaira	0—200	Between Sabarmati and Mahir Rivers	30	
	Surat	0—200	Tapti River	50	
	Broach Coastal Area	0—200	Sea and Tapti River sea coast	40	
	Amreli Coastal Area	0—200	Sea and Tapti River sea coast		
Maharashtra	Khandesh	0—500	Tapti and Narmada	30	
	Surat	0—500	Sea coast	40—60	81.0
	Thana	0—500	Sea coast	60—90	
	Kolaba	0—500	Sea coast	100	
	Ratnagiri	0—500	Sea coast	100	
Mysore	Sholapur Western	200—500	Bhima River and tributaries	25	80.6
	Ghat		Bhima River and tributaries		
	Dharwar	200—500	Tungabhadra and Malaprabha Valley	25—50	
	Bijapur	>200—500	Bhima and Dhona Rivers	20	79.4
	Raichur	200—500	Tungabhadra and its tributaries	25	82.2
	Mandya	500—100	Cauvery River and reservoir		
Andhra Pradesh	Krishna Delta area	0—200	Sea coast	42	
	Nizamabad area	200—500	Godavari and its tributaries	28—33	
	Mahbubnagar	200—500	Krishna River and tributaries	29	
	Nalgonda	200—500	Musi River and other tributaries of Krishna River	30—34	
Tamil Nadu	Tanjore	0—200	Cauvery River and sea	50—60	
Orissa	Balasore	0—200	Sea coast	57—60	
West Bengal	Sundarbans Marshes	0—200	Sea coast	57—67	79.4
Delhi	Delhi	200—500	Jamna River and its canals	26	76.7

Table after Kanwar.[161]

Salt-affected soils in the arid regions of Rajasthan and Gujarat — The main soluble anions are chlorides and sulfates, though bicarbonates are also met with in some cases. Due to low rainfall and the scarcity of irrigation water, the reclamation of these soils is by no means easy. The poor quality of the underground water imposes another constraint in this regard.

Salt-affected soils in the arid and semiarid regions of the Black Cotton Soil Group (Vertisols) — These soils are mainly found in the states of Madhja Pradesh, Gujarat, Maharashtra, Karnataka, Andhra Pradesh, and Tamil Nadu, and are characterized by black clays dominated by montmorillonite clay minerals with nodular lime in the subsoil. They swell on wetting and crack on drying and show a very poor infiltration rate when wetted, thereby rendering the leaching process rather difficult unless the physical conditions of the soil are first improved through the addition of a suitable amendment. The quality of the underground water is questionable in some cases and limits its use for irrigation and leaching.

Coastal Saline Soils — Coastal saline soils are found along a long coastline of about 35,000 miles. The problem occurs in varying degrees in the states of West Bengal, Orissa, Andhra Pradesh, Tamil Nadu, Kerala, Maharashtra, and Gujarat. The most important tracts of coastal saline land are the Sunder Bans of West Bengal, the deltas of the Rivers Krishna, Godavan, the Kari soils of Kerala, the Khar lands of Maharashtra and Gujarat, and the salt flats of the Rann of Cutch. The saline soils of the Sunder Bans in West Bengal have resulted from inundation by brackish seawaters. In Kerala, the Kari soils are peaty in nature, containing 10 to 40% organic matter. These are low-lying and remain submerged during the monsoons.

While chlorides and sulfates are the dominant anions, Na is the dominant cation, though Mg is also present in appreciable amounts in some cases. The rainfall is moderate to high in the eastern and southeastern areas and low in Gujarat; therefore, the reclamation and management of the two areas necessitate the employment of different techniques.

The Kari soils in Kerala, which are heavy clay in texture, black in color and highly acidic in reaction, contain a high amount of organic matter, nitrogen, and soluble salts of chlorides and sulfates. There is no horizon differentiation in the profile. The soils are extremely plastic and sticky and present difficulties for the cultivation of most crops except paddy.

The groups of salt-affected soils listed also include the types of jointly occurring salinity and alkalinity. Nevertheless, typical alkali soils are also widespread in India under varying climatic conditions extending from Rajasthan to West Bengal and from the north to the south.

Alkali (sodic) soils — Alkali soils are predominantly found in the states of Punjab, Haryana, and Uttar Pradesh in the Indo-Gangetic plains, though they also occur in some other parts of the country. The important characteristics of these soils, which are responsible for very low yields or even for the complete failure of crops, are as follows:

- Excessive amounts of exchangeable and soluble sodium
- A high soil pH often exceeding 10
- A deficiency of available calcium
- An extremely low water intake rate and hydraulic conductivity, leading to poor moisture transmission characteristics
- The presence of calcium carbonate nodules called Kankar, often occurring at a depth of about 1 m
- The imbalance of plant nutrients
- Poor physical conditions leading to the hindrance of root penetration

Besides saline and alkali soils, gypsiferous soils are also common, particularly in the arid areas of India. They are widespread in the northeast of the country in the deserts of Rajasthan, but they are found in other regions as well.

Table 8
SALINITY AND/OR ALKALINITY OF SURVEYED AREAS IN PAKISTAN

			Chemical analysis		
Sites	Year of survey	Area surveyed	Low salts, low pH in profile	Low salts, low pH at surface	High salts, high pH at surface
Thal Doab	1941	2.655	0.502	0.913	1.250
Lyallpur	1941	1.000	—	0.204	0.785
Rakh Miani and Pakhowal	1942	0.002	0.001	0.000	0.001
Ghagh and Darkana	1943	0.150	—	0.104	0.045
Mailsi	1943	0.124	—	0.001	0.124
Jalalpur	1943	0.167	—	0.002	0.165
Rangpur	1944	0.065	—	0.007	0.058
Taliri and Garesh	1944	0.145	0.051	0.064	0.016
Taunsa	1946	1.860	—	0.372	1.400
Total		6.168	0.554	1.667	3.844

Note: Values represent millions of acres.

6. Pakistan

Salt-affected soils are very extensive in Pakistan owing to the hydrological and climatic conditions of the country. When comparing the area of salt-affected soils to the total territory of the country, Pakistan is one of the leading states both in Asia and in the world as regards the proportion of saline and alkali soils.

In deserts and semideserts, as well as in river valleys, saline, alkali, and saline-alkali soils occur sometimes in large areas and sometimes in small patches. In both western parts of the country and in the eastern part bordering India, soil salinity is fairly common. In the Plains of the River Indus both salinization and alkalization are extensive on the alluvial deposits. Salinization also develops on the coast. The conditions found in Pakistan are particularly favorable for salt accumulation. According to Nazir Ahmad the sources of soil salts can be enumerated as follows:[162]

1. Salts from the original seawater, minor remnants of which may still be in existence, particularly at deeper levels
2. Salts brought down with the soil as a result of the disintegration of rocks and deposited along with the alluvium
3. Salts brought with irrigation water and spread on the land during agricultural operations
4. Salts coming to the surface from underground formations due to rising groundwater or evapotranspiration
5. Salt accumulation as a result of the interchange of radicals and base exchange

Based on data reported by Nazir Ahmad[162] the salinity and alkalinity of the main affected areas in Pakistan are given in Table 8. The table shows that in practically every area, half or more of the total surface is affected by soil salinity or alkalinity.

Not only are saline soils widespread, but the degree of salinity is also fairly high in nearly all the salt-affected areas. The salinity conditions in the northern Punjab region of Pakistan, which is one of the typical salt-affected territories of the country, are reflected in Table 9. The table clearly shows that nearly half of the total area has a salinity of more than 0.2%.

In the Sind area, five salinity classes are distinguished, as listed in Table 10. In some places the degree of salinity is as high as 1.5%, which is considered ultra-saline.

Table 9
SOIL SALINITY IN NORTHERN PUNJAB REGION OF PAKISTAN

Name of area	Total area (acres)	S—1 (salts <0.2%)	S—2 (salts 0.21—0.5%)	S—3 (salts 0.5—1%)	S—4 (salts >1%)
Rechna Doab					
Chichoki Mallian Shadman	160,000	16,000	107,200	36,800	—
Shah Kot	247,200	178,500	49,700	19,000	—
Chuhar Kana	10,800	7,400	2,500	900	—
Pindi Bhattian	8,300	7,200	600	500	—
Khangah Dogran	121,800	74,600	21,800	25,400	—
Jaran wala	90,900	63,800	8,200	18,900	—
Hafizabad Thatta	131,300	81,200	30,200	18,700	1,200
Harse Sheikh	115,400	77,000	9,500	8,900	20,000
Sangl-a Hill	150,700	115,100	8,700	26,900	—
Beranwal-a	125,900	95,800	9,600	9,300	11,200
Zafarwal	204,200	137,000	24,900	42,300	—
Subtotal	1,366,500	853,600	272,900	207,600	32,400
Chaj Doab					
Lalian	198,000	167,200	16,900	10,900	3,000
Maggowal	45,400	25,700	13,600	3,700	2,400
Rest irrigated area of chaj less item 12—13	1,660,000	830,000	442,000	222,000	166,000
Subtotal, Lower Thal	1,903,400	1,022,900	472,500	236,600	171,400
Muzaffargarh	89,500	65,600	5,000	12,600	6,300
Lower Thal except Muzzaffargarh	1,200,000	600,000	300,000	180,000	120,000
Subtotal	1,289,500	665,600	305,000	192,600	126,300
Grand total	4,559,400	2,542,100	1,050,400	636,800	330,100
Percentage of total	—	55.6%	23.1%	13.9%	7.4%

Table 10
MAIN FEATURES OF THE FIVE CLASSES OF SALINE SOILS IN THE SIND AREA OF PAKISTAN

Salinity class	1 Saline	2 Slightly saline	3 Moderately saline	4 Highly saline	5 Ultra saline
Salt (%)	0.07	0.2	0.37	0.73	1.43
Elec. cond. (mmho/cm)	0—4	4—8	8—15	15	40
Salts (tons/acre)	6	18	33	65	111
Effect on crop	No effect	Noticeable effect	Serious effect	Unsuitable for crops	Difficult to reclaim

In Table 11 the distribution of salt-affected soils belonging to these five classes is demonstrated for the Sind area. The table shows that the area is saline in varying degrees.

Although saline soils make up the dominating group of salt-affected soils in Pakistan, the problem of soil alkalinity is also fairly considerable and alkali soils in association with saline soils are widely extended particularly in river plains and terraces. Magnesium soils occur in areas with alkali soils, while gypsiferous soils are common in the northeastern and eastern desert areas of the country. Acid sulfate soils, which also can be found in the tidal coastlines of the Arabian Sea, have minor importance in comparison with the saline and alkali soils of the country.

Table 11
DISTRIBUTION OF SALT-AFFECTED SOILS IN THE SIND AREA OF PAKISTAN, ACCORDING TO THE SALINITY CLASSES

		1		2		3		4		5	
	Total area	**Area**	(%)	**Area**	(%)	**Area**	(%)	**Area**	(%)	**Area**	(%)
Left bank nonperennial	1,715,800	65,200	3.8	236,780	13.8	224,770	13.1	549,056	32	639,993	37.3
Gaja	115,300	25,943	22.5	28,249	24.5	22,483	19.5	25,366	22	13,260	11.5
Tando Bago	37,200	2,304	7.0	5,282	14.2	5,840	15.7	14,359	38.6	9,114	24.5
Right Bank	603,800	19,322	3.2	82,117	13.6	82,117	13.6	224,613	37.2	195,631	32.4
Total	2,472,100	112,769	—	352,428	—	335,210	—	813,394	—	857,998	—
Larkana Shikarpur	578,000	168,200	29.1	165,300	28.6	96,100	16.6	84,400	14.6	63,000	10.9
Khairpur	650,000	192,000	32.0	165,000	27.5	84,000	14.0	170,000	26.5	15,000	4.3
Total	3,700,000	473,000	12.7	683,000	18.4	515,000	14.0	1,068,000	28.8	936,000	20.1

Note: Values represent acres.

Table 12
DISTRIBUTION OF SOLONCHAK AND SALT-MARSH SOILS IN VARIOUS PARTS OF IRAN[85]

Province	Total area (1000 ha)	Solonchak		Salt-marsh		Total	
		1000 ha	(%)	1000 ha	(%)	1000 ha	(%)
Gilan	3,800	—	—	—	—	—	—
Mazanderan	14,000	400	2.86	1,600	11.40	2,000	14.26
Azerbaijan	10,500	360	3.48	120	1.14	480	4.62
Kurdistan	3,122	—	—	—	—	—	—
Kermanshah	6,212	—	—	—	—	—	—
Khuzistan	13,466	1,000	7.54	1,260	9.47	2,260	17.01
Fars	17,420	1,640	9.40	120	0.67	1,760	10.07
Kerman	23,280	1,740	7.56	400	1.72	2,140	9.28
Khurasan	30,900	800	2.59	1,600	5.18	2,400	7.77
Esfahan	17,600	760	4.32	920	5.23	1,680	9.55
Baluchistan	18,500	520	2.81	1,240	6.70	1,760	8.51
Tehran	6,200	80	1.29	920	14.86	1,000	16.15
Total	165,000	7,320	4.45	8,180	4.97	15,500	9.42

7. *Iran*

In Asia the greatest extension of salt-affected soils after the U.S.S.R., China, India, and Pakistan can be found in Iran. The arid conditions of this country contribute to the formation of saline soils, which are very frequent in several areas. Alkali soils are also frequent and gypsiferous soils particularly so, furthermore, acid sulfate soils also occur. Both solonchak and solonetz soils are distributed in the arid, semiarid and dry sub-humid regions of Iran.[85] They are either poorly drained or have developed under poor drainage conditions. Solonchak soils contain large quantities of soluble salts, are usually light-colored and poor in organic matter, and have a slightly crusted, friable granular structure. The salts in solonchak soils are mainly dominated by chlorides, but sulfates are also present in considerble quantities, being found as sodium sulfate in saline soils and as calcium sulfate in gypsiferous soils.

Solonetz soils are the products of the partial leaching and alkalization of solonchak soils, often due to irrigation without proper drainage and management. Solonetz soils frequently develop in spots, scattered throughout solonchak or other soil types. They have a surface layer of light-colored leached material over a darker subsoil.

Solonchak soils are common in the Khuzistan region, in the southwest in Azerbaidzhan, in the northwest, and in the Doro Odzin area in Fars. Their texture may vary from silt loam to clay loam or clay, but in the subsoil clay — ranging from fine-textured to plastic — also may be found quite often. There is usually a typical salt crust on the surface and a granular layer just below. The total salt content may exceed 3% of the soil material. As a rule, all these soils are calcareous, with pH values varying from 8 to 8.5. Some solonchak soils are called "Shurzar", and are located mainly in the south of Khuzistan, on the shores of Lake Urmia.

The saline soils are closely associated with salt marshes and with their soil cover. The total extension of these kinds of salt-affected soils exceeds 15 million ha and occupies nearly 10% of the total surface area of the country. Table 12 gives the distribution of solonchak and salt marsh soils in various parts of Iran.[85]

The alkali or solonetz soils are sometimes also saline but they have columnar structure and their salt content is lower than that of solonchak soils. The clay content of the surface layer of alkali soils is high, up to 60%, while the organic matter contents of both saline and alkali soils are low.

Like solonchak soils, the solonetz soils are calcareous, but the calcium carbonate content of their upper layers is often lower than that of solonchak soils.

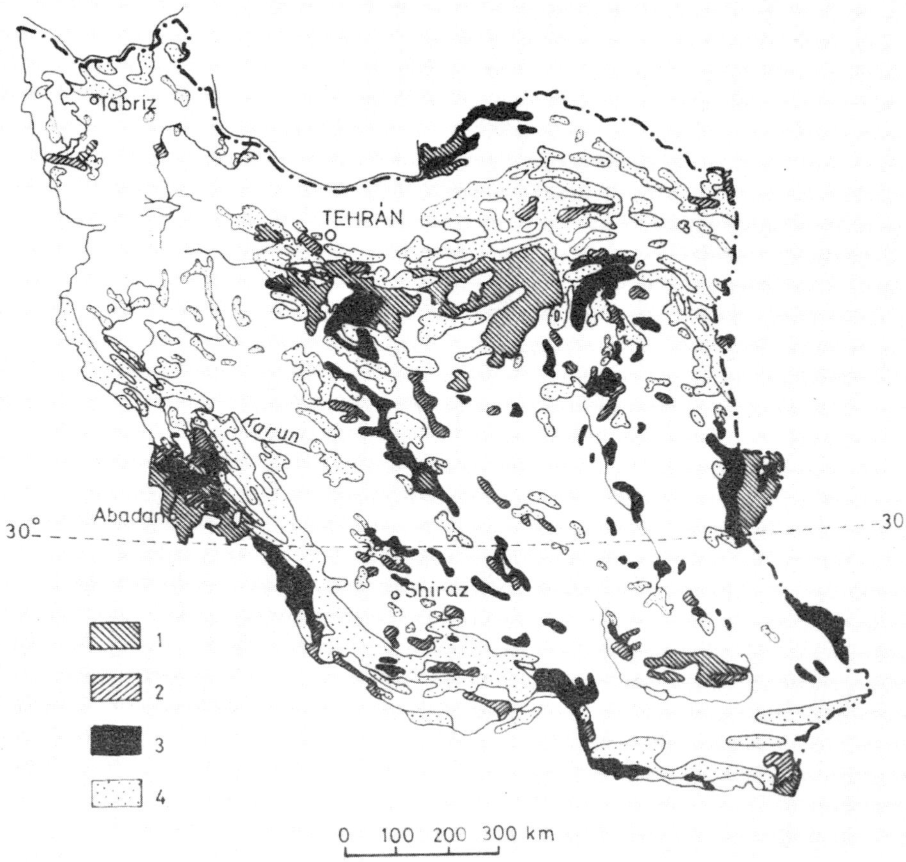

FIGURE 17. Distribution of salt-affected soils in Iran. (1) Salt-marsh soils; (2) saline alluvial soils; (3) solonchak and solonetz soils; (4) potentially salt-affected soils (After Dewan and Famouri.[1])

In Iran, salt-affected soils are particularly widespread in the alluvial zones of the rivers, where different groups of soils occur. Figure 17 illustrates the distribution of salt-affected soils in Iran.

8. Iraq

Iraq has one of the highest ratios of salt affected soils compared to the total soil cover. According to several authors, more than half of the total territory of this country is seriously affected by salts. The valleys and interflows of the River Tigris and Euphrates in the territory of Iraq — the famous Mesopotamian plain — is strongly salinized, partly due to natural processes and partly in consequence of several thousand years of irrigation.

Figure 18 shows the extension of salt-affected soils in Iraq.

As in some neighboring countries (Iran, Pakistan), various groups of salt-affected soils can be found in Iraq. The dominant group is saline soils, to which the major part of the salt-affected soils belong, but saline-alkali, gypsiferous, and alkali soils are also frequent. The principal soil formation process in central and southern Iraq is salinization. The main chemical type is sodium chloride, but sodium sulfate is also often met with. Besides these two salts, calcium chloride also occurs, as do sodium sulfate and magnesium sulfate. Particularly important is gypsum, which is present in most soils in Iraq and is responsible for vast territories of gypsiferous soils. In some places its concentration is so high that it prevents water uptake by the plants due to high osmotic pressure.

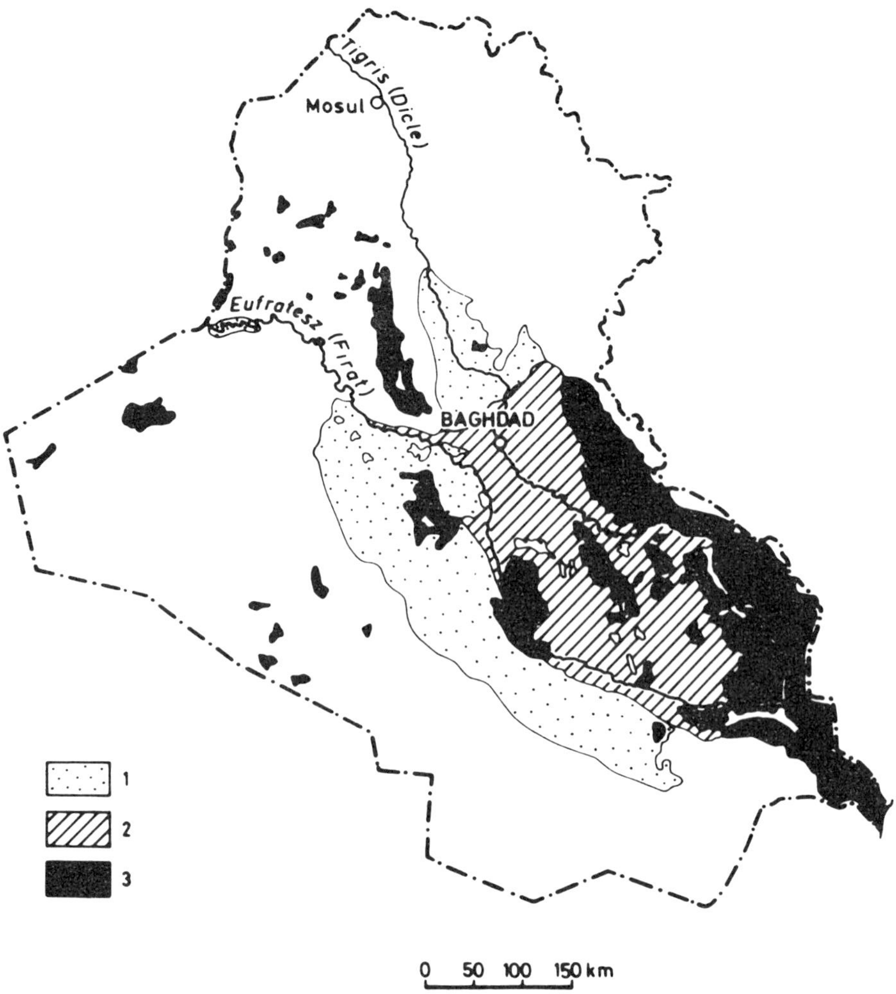

FIGURE 18. Distribution of salt-affected soils in Iraq. (1) Potentially salt-affected soils; (2) Salt-affected soils constituting less than 50% of the area; (3) Salt-affected soils constituting more than 50% of the area. (After Buringh.[1])

In the arid regions of Iraq, sodium chloride represents about half of the total salinity and in areas inundated by seawater it may account for nearly 90% of the total salt contents. In the latter territories maritime saline soils dominate. The accumulation of salts is partly due to the arid climate and to the salinity of the ground as well as of the groundwaters, but irrigation has also contributed quite considerably to the salinization of the soils. In consequence of the dry climate, geochemical salt accumulation and the proximity of the sea, airborne salts from the seawater also contribute to soil salinity, particularly in the southernmost part of the country.

The following local names are in use for salt-affected soils:

1. Shura soils, which are saline with a wide salt crust or efflorescence
2. Sabakh soils, which have a dark brown color and a high content of deliquescent salts

The latter are similar to the solonchak soils referred to in the Russian classification as "puffy" solonchaks where, in consequence of the presence of calcium and magnesium chlorides, the upper layer and the salt efflorescence on the surface are deliquescent. This is illustrated by Table 13, in which the chemical composition of three Sabakh soils is given.

Table 13
CHEMICAL COMPOSITION OF SOLUBLE SALTS OF THREE SABAKH SOILS[163]

Salt	Soil 1	Soil 2	Soil 3
$CaCl_2$	115	187	197
$MgCl_2$	91	53	83
NaCl	12	—	14
$MgSO_4$	—	16	—
Na_2SO_4	24	1	22
$NaHCO_3$	0.4	0.4	0.4
$NaNO_3$	4	17	16

Values are milliequivalents per liter.

Salty groundwater plays a dominant role in the formation of such soils.[86]

Saline-alkali soils also are located in many parts of the saline areas. Alkali soils with low salt contents are relatively scarce, but do occur in some areas. They do not develop during the leaching of saline soils since, as a rule, most Iraqi saline soils are more or less gypsiferous. However, in some depressions in the lower Mesopotamian plain, where the gypsum and salt contents are low, alkali soils may develop.[86]

Practically all groups of salt-affected soils can be found in Iraq, including magnesium soils and acid sulfate soils.

9. Afghanistan

In Afghanistan, as in neighboring countries with similar climatic and geochemical conditions, salt-affected soils are extensive in many areas, particularly in terraces and closed depressions.

The dominating type is that of saline soils, where sodium chloride and sodium sulfate constitute the major part of the water-soluble salts. As gypsum, too, frequently accumulates in soil horizons, gypsiferous soils are also widespread in several regions of the country.

10. The Arab Peninsula (Saudi Arabia, Yemen and South Yemen), the countries on the East Coast of the Mediterranean (Syria, Israel and Jordan), and the Gulf States

The types, morphological features, and chemical and physical properties of the salt-affected soils are very similar in the countries, but the extent of the problem is quite different. In some of these states a considerable proportion of the soil cover is affected by salinity, alkalinity, or both (e.g., in Syria and Israel), while in other countries the occurrence of these soils is not very frequent (e.g., in Saudi Arabia, Yemen, or Jordan). In a few of them the extension of salt-affected soils is negligible (e.g., in South Yemen and in some of the Gulf States).

The dominating group of salt-affected soils in the countries listed above is saline soils, where the main part of the salts consists of sodium chloride and sodium sulfate. Saline soils are frequent in the arid regions of Syria, the valley of the Euphrates, and the Negev Desert in Israel, and they constitute a serious hindrance to the utilization of the land. The total salt contents of the solonchak soils in this region may exceed several percent on the surface of the soil, and salty groundwaters in association with dry climatic conditions are responsible for their development.

In Syria,[91] and even more so in Israel,[89] solonchak soils are so widespread that several national and international projects have been carried out for their reclamation. The main methods of reclamation are irrigation and drainage, applied to remove the excess of salts and to improve soil fertility.

Alkali soils are also extensive. Alkalinity mainly occurs jointly with salinity, particularly in the solonetz soils of the region.

Gypsiferous soils are very frequent, either in association with saline soils or in individual massifs. The gypsum content may be as high as 15 to 20% in the upper soil layers; consequently its presence dominates several soil properties and makes the soil very vulnerable to the infuence of water during irrigation and drainage. Besides calcium sulfate, sometimes magnesium sulfate is also present in the soil horizons and its concentration may surpass the threshold of toxicity in several places.

Calcium carbonate is not a water-soluble salt and consequently has not been considered as a direct agent of soil salinity and alkalinity. However, in some regions, for instance in the countries mentioned above, it is widespread and its concentration in the soils is very high. Due to this fact, highly calcareous soils have peculiar physical and chemical properties (poor structure, alkali reaction, rapid decomposition and disposition of organic materials, etc.), which are disadvantageous for the agricultural utilization of the land. Highly calcareous soils with 20 to 30% or more calcium carbonate in the soil material often appear in association with saline soils, magnesium soils or gypsiferous soils.

Acid sulfate soils seldom occur in the region in question; they can only be found in some tidal marshes on the seashores.

Figure 13 (demonstrated earlier in this chapter) is a schematic map of the extent of soil salinity in the countries of the Near and Middle East. The map shows the extension of several types of salt-affected soils on areas ranging from the west coast of the Mediterranean, and the south coast of the Caspian Sea, to the River Indus in the east. Among the salt-affected soils, special emphasis is given to the saline variations.

11. Southeast Asia (Peninsula of Burma, Thailand, and Indo-China)

In the previous parts of this chapter, the Asian countries north-west of this Peninsula were characterized with respect to salt-affected soils. When speaking of the salt-affected soils of Indo-China, Burma, Thailand and the neighboring territories (including some parts of Indonesia and Japan) a substantial difference should be pointed out in their formation and occurrence in comparison with other Asian regions. While in the countries described previously saline soils form the main group of salt-affected soils and other groups, if they occur at all, have only secondary importance, in southeast Asia, where saline-alkali and to some extent gypsiferous and magnesium soils are also frequent, the group of acid sulfate soils covers vast territories in several states of the Peninsula and constitutes a major problem not only for the physico-geographical features of the area but also for land use and the national economy.[29] Large numbers of publications are available in the literature on acid sulfate soils, a great part of which refer to the Peninsula comprising Burma, Thailand, Malaysia, and Indo-China. Acid sulfate soils are particularly extensive in Thailand, Vietnam, and Burma, but they can also be found both in other states of the Peninsula, and also in Malaysia, Indonesia, Sri Lanka, and even on some seashores in Japan.

In southeast Asia, a clear geographical distribution of the different groups of salt-affected soils can be found. Saline soils, which are first in extension followed by acid sulfate soils, can be found under comparatively dry conditions or in river terraces or directly adjoining the seashore. In some regions, (e.g., in North Thailand or in South Vietnam) they occupy quite considerable territories, and sodium chloride prevails among the salts causing salinity.

Alkali soils also occur in southwest Asia, as do magnesium soils, but their extent and their importance is less significant than that of saline and acid sulfate soils. The same can be said for gypsiferous soils.

Acid sulfate soils account for several percent of the soil cover in Thailand; and this ratio is much higher when the territory of acid sulfate soils is compared to that of the arable land. A similar ratio can be found in Vietnam and a somewhat lower one in Malaysia. Acid sulfate soils dominate the soil cover particularly in the tidal marshes and related areas, and when swamps and waterlogged territories are desiccated so that they can be used for agricultural purposes their extension sharply increases. More than half of all the rice paddies in these countries is or has at some time been influenced by processes related to the formation of salt-affected soils.

12. The Korean Peninsula

The extension of salt-affected soils in the Korean Peninsula is very similar to that in southeast Asia. As in the latter region, several groups of salt-affected soils occur in the Korean Peninsula, but alkali soils and magnesium soils are somewhat more frequent than in Indo-China, Thailand, and Burma, and the extent of saline soils is similar in the two areas. This general picture is true of both parts of the Korean Peninsula (north and south), though the territories of salt-affected soils are somewhat larger in South Korea than in North Korea. The majority of salt-affected soils can be found in the comparatively dry areas in river alluviums, at least as far as saline, alkali, and magnesium soils are concerned.[95]

Acid sulfate soils are widespread in the estuary of the big river Tae-dong-gang, as well as in the tidal marshes that follow the coastline of nearly the whole Peninsula except where there are rocky seashores. As in Thailand and other countries in southeast Asia, the problem of acid sulfate soils is not only theoretical, but also constitutes a severe practical problem, threatening the production of rice in paddy fields on the coast. The degree of acidity of the acid sulfate soils in Korea is somewhat lower than in Thailand, and the rate of acidification in the tidal marshes is also slower. This is why the pH values of the acid sulfate soils in the tidal marshes of Korea are much higher than in those of Thailand. This difference can be explained by the lithology and mineralogy of the parent material in the tidal marshes: in Korea they quite often contain calcium carbonate or related materials which mitigate the acid reaction and the development of acidification processes. Consequently the formation of acid sulfate soils, which usually follows the desiccation of tidal marshes, is also often retarded. At the same time, in rice paddies established on territories formerly covered by tidal marshes, soil degradation and secondary solod forming processes may occur on a large scale as the result of irrigation. This area of salt-affected soils covering former tidal marshes is known as the "Kanzuchi area".

Needless to say, in Korea, as in Thailand and some Indo-Chinese countries, the accumulation of sodium chloride is quite frequent near the coast in soils directly affected by seawater, and in such places sodium chloride solonchak soils develop.

C. Africa

All possible groups of salt-affected soils can be found in Africa, where they cover considerable territories in more than thirty countries. Interestingly enough, the distribution of the different groups of salt-affected soils can be clearly distinguished on this continent according to geographical and climatic zones. For this reason, and due to the limited information available on the soil conditions in some African countries, it is expedient to discuss these soils as they occur in typical regions instead of describing them in connection with every country where they have some importance.[65,66]

In the regions north of the Sahara and in the Sahara itself the dominating type of salt-affected soils is the saline group, which is closely associated with the arid and semiarid conditions prevailing in several countries such as Morocco, Algeria, Tunisia, Libya, Egypt, and Sudan, as well as Chad, Niger, Mali, etc.[65,67] The saline soils in these territories, which are indicated by El-Gabally on the Soil Map of the Near and Middle East[74] (Figure 13), are very similar to those described in connection with arid Asian countries. These soils are only shown schematically because they occupy more extensive areas than indicated on the map. A great proportion of the above-mentioned countries is covered by saline soils to such a

Table 14
SOLUBLE SALTS (%) IN SALINE SOILS FROM "PLAINES DU CHELIF" ALGERIA[164]

Depth (cm)	Total salt	Cl^-	SO_4^{2-}	CO_3^{2-}	Ca^{2+}	Mg^{2+}	Na^+	K^+
0—2	7.37	3.45	0.8	0.02	1.04	0.27	1.33	0.02
2—10	3.02	1.19	0.6	0.02	0.45	0.10	0.47	0.02
10—24	0.8	0.3	0.2	0.03	0.08	0.02	0.13	0.01
24—41	0.9	0.3	0.2	0.08	0.09	0.03	0.19	0.01
41—57	0.9	0.3	0.3	0.03	0.10	0.04	0.25	0.01
58—76	0.8	0.2	0.3	0.03	0.08	0.03	0.13	0.01
76—91	0.5	0.1	0.3	0.03	0.04	0.02	0.14	0.01
91—107	0.4	0.1	0.4	0.04	0.03	0.02	0.10	0.01
107—120	0.4	0.1	0.4	0.04	0.03	0.02	0.06	0.01
120—134	0.4	0.1	0.4	0.04	0.03	0.01	0.10	0.01
134—179	0.4	0.1	0.3	0.03	0.04	0.02	0.08	0.01

Table 15
CHEMICAL COMPOSITION OF SALINE AND GYPSIFEROUS SOILS FROM THE SAHARA[67]

	Depth (cm)	Humus (%)	$CaCO_3$ (%)	Gypsum (%)	pH
Gypsiferous soil	0—33	0.12	0.60	13.28	8.1
	33—57	0.34	1.58	14.56	8.1
	57—80	—	1.00	17.16	7.9
	80—122	0.40	1.78	12.32	8.9
	122—200	0.63	1.18	19.18	7.7
Saline soil	0—10	0.15	4.6	2.35	7.4
	10—49	0.50	3.2	2.52	7.5
	49—70	—	—	5.05	7.5
	70—100	—	—	0.26	7.5
	100—140	—	1.21	0.35	7.5

degree that a major part of the land is barren desert and only limited spots and oases can be used for agricultural purposes, and here, too, irrigation is obligatory.

It is generally accepted that most of the Sahara and the semi-desert north of the Sahara are covered by saline soils which contain, as a rule, high amounts of sodium chloride. In Table 14 chemical analyses are given for soils of this type in the Plaines du Chélif (Algeria).[164] The table shows that the majority of the soluble salts in saline soils consists of sodium chloride and to some extent magnesium and calcium chlorides, followed by the sulfates of the same cations. It can also be seen that the highest salt contents can be found on the soil surface and that the quantity of water-soluble salts diminishes as the depth increases. This proves that the soils are nonhydromorphic.

The next most frequent group of salt-affected soils in the above-mentioned areas in gypsiferous soils, which are very frequent in the northern part of Africa, particularly in Morocco, Algeria, and Tunisia. As in the Middle East, these constitute a serious problem for the utilization and irrigation of the land. In Table 15 the chemical composition of a solonchak soil and a gypsiferous soil from the Sahara is demonstrated.[67] The Table shows that pedological similarities exist between solonchak and gypsiferous soils in the Sahara, although the gypsum content is evidently much higher in the latter.

In the deserts of North Africa a very common phenomenon is the formation of "lunettes", temporary waterlogged depressions which become gypsiferous efflorescences on the soil surface after desiccation. The chemical composition both of the salty water and the salt efflorescences is characterized by high amounts of calcium sulfate. In almost all the arid countries of North Africa such formations can be found to different extents.[65,70-72]

While saline soils are extensive in the arid parts of the continent, alkali soils can be found under arid, semiarid, semihumid and humid conditions, mostly in areas where saline and gypsiferous soils occur rarely, if at all. One exception to this is when alkali soils with a high calcium carbonate content, also called saline-alkali soils or alkali soils with a structural B horizon, often appear jointly with saline soils, e.g., in the delta region of the Nile in Egypt or in several places in Tunisia. They are also located in the valleys of large rivers, especially in the Nile Valley in Sudan and on the banks of the River Niger. Soil formations of this type are also frequent in salt-affected depressions such as Wadi el-Natrun in Egypt or on the shores of Lake Chad.[70,71]

Alkali soils are frequent south of the Sahara. Up to quite recently little information has been available on these soils under tropical conditions, but lately more and more publications have mentioned the presence of this group of salt-affected soils in many countries even near the equator, e.g., in Kenya, Tanzania, Chad, Nigeria, Ghana, Zambia, Zimbabwe, Mozambique, Burundi, etc. In spite of the insufficiently detailed information it would seem that the majority of the alkali soils are similar to the solonetz type described in this book. The main characteristics of these soils are the lack of a high salt concentration and the existence of high ESP values in the upper layers and the alkali reaction in the profile. The majority of such soils develop under the influence of slightly mineralized groundwater by means of hydromorphic processes. They occur at different elevations ranging from near sea level to the piedmontal areas of the Kilimanjaro, the Atlas Mountains (where the influence of the desert has not reached the Mediterranean strip of North Africa) and other places. Alkali soils are also extensive in river valleys with a humid climate.

Magnesium soils, which can be found in many places in Africa, are closely related to the lithological and mineralogical conditions of the given areas. This group of salt-affected soils is frequently associated with the occurrence of alkali soils, e.g., in Sudan. Some authors have studied these two groups of salt-affected soils together. The regular features of the regional extension of saline and alkali soils described above can only be interpreted as a general rule, because saline soils also can be found under more humid conditions while alkali soils can sometimes be found under dry conditions, particularly if we consider that several authors include the saline-alkali soils in the saline group. This is partly true of the map of the salt-affected soils in Africa given in Figure 19. This map distinguishes three groups:[165,166] potentially salt-affected soils, alkali soils, and saline soils. The last group includes some of the acid sulfate soils which are widely extended along the seashores of many countries in Africa.

Acid sulfate soils are very common in the tidal marshes of the west coast of Africa, particularly in Senegal, and the Ivory Coast, but generally in the Gulf of Guinea, in Nigeria, in the deltas of the Niger, the Congo and other big rivers, on the southern shores of the continent, in Mozambique and on the western coast of the island of Madagascar. They can also be found in many other places, along practically all the seashores of the continent. The morphology and properties of the acid sulfate soils in Africa are very similar to those described above for southeast Asia.

In areas of Africa which are near the sea (Mediterranean, Red Sea, Atlantic Ocean, Indian Ocean) or near salty lakes, seawater has a direct influence on the soil-forming processes, irrespective of the climatic zones, and promotes the development of coastal saline soils. In many of these areas the salt content of the groundwater is very high and contributes to this process.[167-170]

If salts migrate with artesian waters through tectonic cracks, salinity may develop as a result of evaporation under arid conditions, as in the Nile Valley in Egypt. The formation of saline soils in these regions may be either hydromorphic or nonhydromorphic.

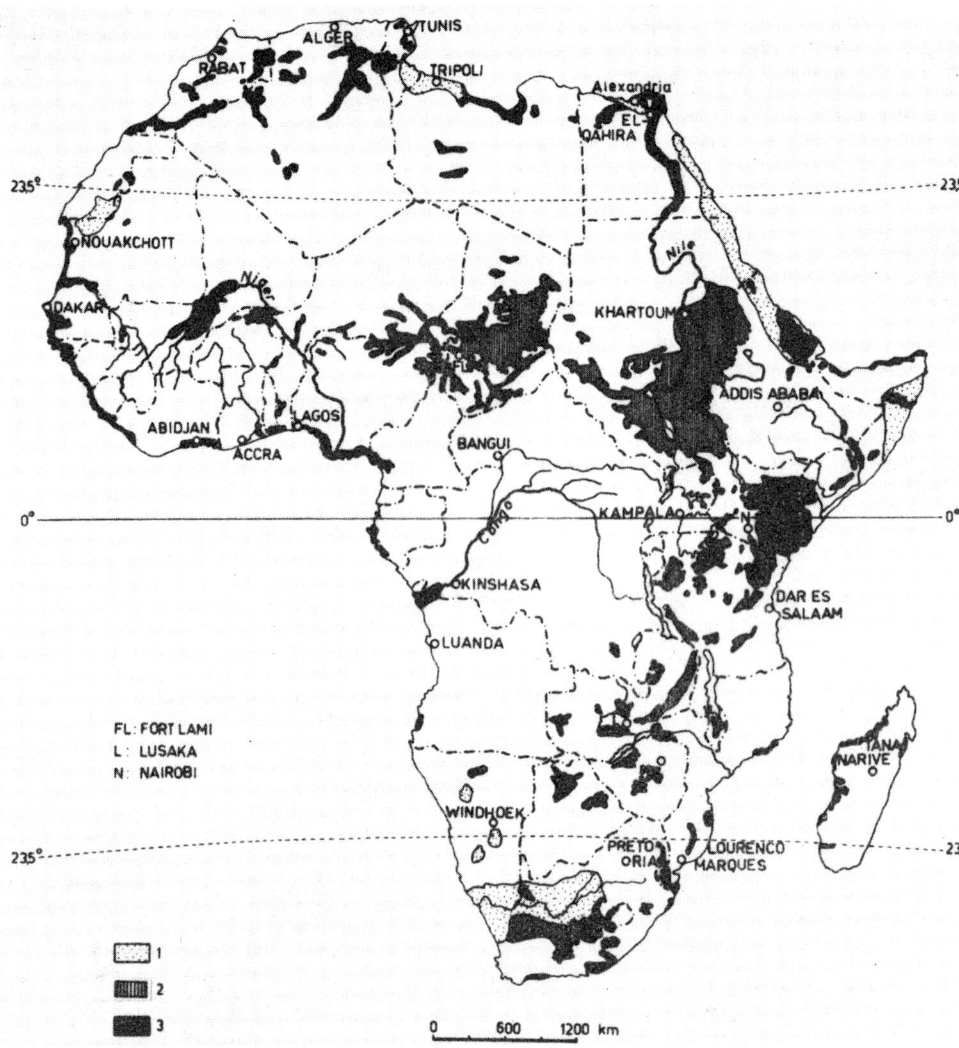

FIGURE 19. Distribution of salt-affected soils in Africa. (1) Potentially salt-affected soils; (2) alkali soils; (3) saline soils. (After Aubert.[1])

D. North America

Salt-affected soils are widespread in all three countries of North America but in each of them the distribution of the various groups of salt-affected soils is different. In Figure 20 a schematic map is presented of the distribution of salt-affected soils in North America (this map is based on data taken from the FAO/UNESCO Soil Map of the World).

1. Canada

Saline soils occur rarely in Canada with the exception of a few places where mineralized groundwater has an intensive influence on the soil-forming processes. Even in these cases salinity is usually associated with alkalinity due to the presence of a considerable concentration of sodium carbonate.[58,59]

The occurrence of gypsiferous soils is also rare in Canada, which can be explained by the climatic conditions of the country. As there are no arid or semiarid areas, the climatic and geochemical conditions are less favorable for the formation of these two groups of salt-affected soils than in the southern countries of the continent.

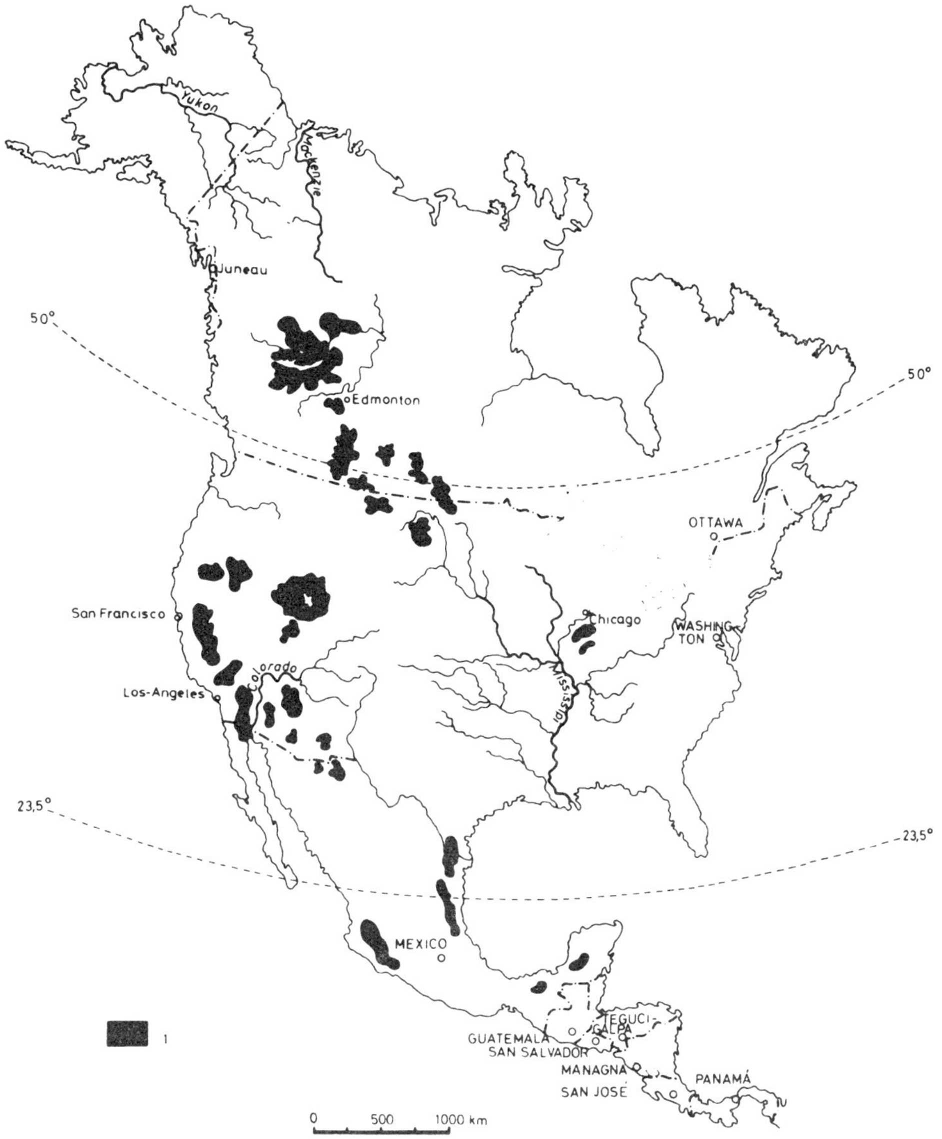

FIGURE 20. Distribution of salt-affected soils in North America. (1) Areas of salt-affected soils. (Compiled from the data of the FAO/UNESCO Soil Map of the World.[1])

Canada is the only country in the North American continent where the dominant group of salt-affected soils is alkali. The environmental conditions of soil formation, the morphology, and properties of the alkali soils developed and the possibilities for utilizing these soils are similar in many respects to those in Europe and to some extent in Australia.

In Canada the decisive role of mineralized groundwater in the formation of solonetz soils is even more pronounced than in Europe or in Australia. Due to the geological history and recent geochemical conditions of this region the landscape of the northern great plains of North America ranges from rolling to undulating. The bedrock is shallow and consists of marine and nonmarine shales and sandstones.[171] This area includes not only the great plains of Canada, found in Alberta, Saskatchewan, Manitoba, and part of Winnipeg, but also some northern states of the U.S., such as Montana and North and South Dakota. In this region

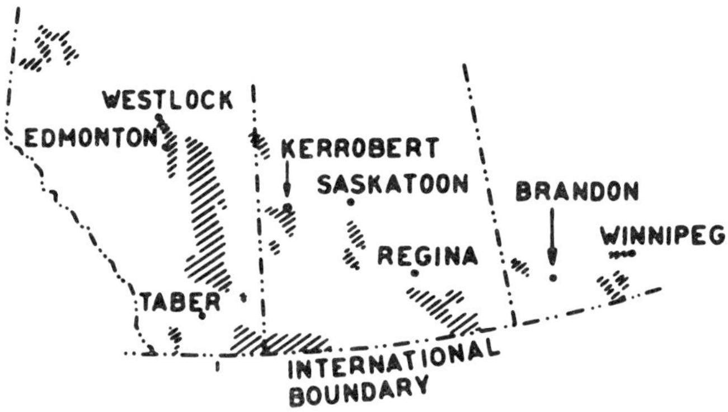

FIGURE 21. Solonetzic areas in Western Canada. (After Cairns.[58])

"saline seepage", which occurs under both dryland and irrigated conditions is extensive. Van der Pluim defines this formation as follows: "Recently developed saline soils in non-irrigated areas that are wet some of or all of the time often with white salt crusts and where crop or grass production is reduced or eliminated." It should be noted that the term "saline" in the above situation refers not only to saline soils but also to some other groups of salt-affected soils.[171]

Canadian authors estimate the territory of such soil formations to be about 8 million ha in dryland areas in the great northern plains of Canada, where the process is highly intensive, while another 3 million ha are affected to a lesser extent. In some areas of this region, as much as 16% of the arable land is affected by dryland saline seepage. The extent of dryland seepage is on the increase due to the progressive influence of saline groundwater.

The processes briefly characterized above are chiefly responsible for the formation of the different saline and alkali soils in Canada, which, according to Cairns,[58] occur spotwise or in greater massifs. As can be seen in Figure 21, the majority of the salt-affected soils in Canada are made up of various solonetz soils. As in Europe and Australia, a great variety of different solonetz and solod soils exists in Canada, the main properties of which are similar to that of their European counterparts but which also have their own peculiarities due to local environmental conditions. (Soap holes with alkalinity and waterlogging on the surface).

The processes leading to the formation of Canadian solonetz soils are jointly influenced by salinization and desalinization as described later in this book. The combination of these soil development processes has given rise to the horizon and profile characteristics that are used by Canadian soil scientists to classify solonetz and solonetzic soils at the order, great group, and sub-order levels. At the order level, solonetzic soils are also defined as imperfectly drained mineral soils having solonetzic B horizons and saline C horizons. Under undisturbed conditions they usually have an A horizon of variable thickness and character, but this may be very thin or absent. Structurally, a solonetzic B is characterized by a columnar or prismatic macro-structure that can usually be broken down into a blocky meso-structure.

In Canada the solonetzic soils are defined in two groups: solonetz and solod, based on the different extents of solonization and solodization.

In Canada solonetz soils occur as the dominating components of map units comprising about 72,500 km², which is nearly 1% of the total area of the country. These soils are most frequently associated with brown, dark brown, and black chernozomic soils but may also be associated with dark gray chernozomic and gray luvisolic soils. The main areas of solonetzic soils are in the inland plains in Alberta and Saskatchewan. They occur to a lesser extent in the Peace River area in British Columbia and Manitoba. The majority of the solonetz soils in Canada are found under sub-humid conditions, followed by those occurring under sub-arid boreal and semiarid boreal conditions. Local circumstances influence the morphological, physical, and chemical properties of the various types of solonetzic soils.

As in other regions with a wide extent of solonetz soils, there are local names for the formations typical of the area. The term "brown solonetz" for instance, is used both in Australia and Canada. The different groups of Canadian solonetzic soils are listed below:

Solonetz soils
- Brown solonetz
- Dark brown solonetz
- Black solonetz
- Alkali solonetz
- Gleyed brown solonetz
- Gleyed dark brown solonetz
- Gleyed black solonetz

Solodized solonetz soils
- Brown solodized solonetz
- Dark brown solodized solonetz
- Black solodized solonetz
- Dark gray solodized solonetz
- Gray solodized solonetz
- Gleyed brown solodized solonetz
- Gleyed dark brown solodized solonetz
- Gleyed black solodized solonetz
- Gleyed dark gray solodized solonetz
- Gleyed gray solodized solonetz

Solod soils
- Brown solod
- Dark brown solod
- Black solod
- Dark gray solod
- Gray solod
- Gleyed brown solod
- Gleyed dark brown solod
- Gleyed black solod
- Gleyed dark gray solod
- Gleyed gray solod

Magnesium soils can be found in Canada in association with alkali soils. Sometimes magnesium ions are present in the groundwater whereas in other cases they appear in an exchangeable form on soil colloids. Acid sulfate soils are very rare, occurring in a few tidal marshes on the Canadian Coast.

2. United States

The salt-affected soils of the U.S. are diverse, and each of the different groups has a distinct geographical distribution. In the northern states, where the climate is moderate, salt-affected soils are very similar to those in the great plains of Canada. Solonetzic types dominate among the salt-affected soils both north and south of the border.

Apart from the states already mentioned when describing the rare occurrence of saline soils and the high occurrence of alkali soils among the salt-affected soils of Canada, these features are also characteristic of the states of Idaho, Illinois, Ohio, and others where (as in the solonetzic soils of Canada), the low salinity and advanced alkalinity result in typical solonetz profiles with characteristic ABC horizons. The main difference between the solonetz soils in Canada and those in the northern states of the U.S. is the more distinct character of the lowering groundwater table in the latter. Consequently, types of solonetz soils turning into steppe formation are more frequent in these areas of the U.S. than in Canada. Solod-forming processes are also distinct, particularly in areas near the Canadian border.

Alkali soils are located not only in the above-mentioned states of the U.S. but also in the more arid areas of the country. However, in the arid western states, where the majority of the salt-affected soils of the country can be found, the dominating group of salt-affected soils is not alkali but saline.

Saline soils are very widespread in the western parts of the U.S., particularly in arid states such as Utah, Arizona, Texas, New Mexico, Nevada, California, etc. In this region the arid climate and the associated geological conditions have resulted in the formation of vast territories of saline soils. In some western states a great part of the territory is salt-affected to such an extent that the land cannot be utilized for agri-, horti-, or sylviculture without irrigation and drainage. The environmental conditions and soil cover in these areas are very similar to those in the deserts and semideserts of Eurasia, Africa, and Australia.[2,5,10,172]

The dominating group of salt-affected soils in such regions is the saline type, whereas the dominant salt is mainly sodium chloride. In some extreme cases even sodium nitrate may accumulate, but sodium sulfate and gypsum accumulation are next in importance to sodium chloride. The physical and chemical properties of saline soils in the western states of the U.S. are almost identical to those described for soils in Asia and Africa, and the processes by which they can be utilized and reclaimed are also similar.[173]

It is noteworthy that the processes of salinization and alkalization often develop jointly in these western states,[174] in consequence of which the soils are both saline and alkali, making their amelioration particularly difficult.

In Table 16 the chemical properties of two profiles of saline-alkali soils are demonstrated.[175] As the data in Table 16 show, salinity and alkalinity appear jointly in both profiles. In profile No. 1, the electrical conductivity of the upper layer shows a high salt accumulation consisting mainly of sodium chloride, while in the deeper layers the salt content is lower and the pH value increases due to the occurrence of free sodium carbonate.

In profile No. 2, salinity and alkalinity also occur jointly, but the salt accumulation in the top layer is much lower than in profile No. 1 in spite of the similarity of the pH in the two top layers. Sodium carbonate also appears in the profile, resulting in a high alkalinity.[176]

According to the mentioned properties of the two profiles, the ESP values are very high, particularly in profile No. 1.

The different classification systems used in the U.S. place the salt-affected soils at various taxonomical levels in different units. While some old classification systems, e.g., that of Hilgard, distinguished "white and black alkali soils",[10] in the soil taxonomy of the U.S. salt-affected soils are classified into various large groups at lower taxonomical levels.[23]

The magnesium soils in the U.S., like those in Canada, are extensive, partly as the result of a high magnesium content in the groundwater, and partly as the result of the high exchangeable magnesium content of the soil colloids.[176,177]

In the arid western states of the U.S., the accumulation of gypsum is quite frequent, leading to the widespread occurrence of gypsiferous soils, e.g., in Texas, New Mexico, Utah, etc.[177] They can often be found in association with saline soils.[178,179]

Acid sulfate soils can be located in several places in the tidal marshes of U.S. coasts.

3. Mexico

Due to the environmental conditions in Mexico, the majority of the salt-affected soils in this country are represented by saline soils (which are widespread on the seashores and in river estuaries, as well as in dry areas). The main salt causing soil salinity is sodium chloride,

Table 16
CHEMICAL PROPERTIES OF SALINE-ALKALI SOILS[175]

Profile no.	Depth (cm)	pH	E.C. (mmho/cm)	Ca^{2+} (meq/100g)			Mg^{2+} (meq/100g)			Na^+ (meq/100g)		CEC	ESP (%)	Anions (meq/100g)			
				Soluble	Exchang.	as $CaCO_3$	Soluble	Exchang.	as $MgCO_3$	Soluble	Exchang.			CO_3^{2-}	HCO_3^-	SO_4^{2-}	Cl^-
1.	0—1.5	6.1	46.8	0.7	0.7	3.5	1.9	0.8	3.1	20.5	3.5	9.8	36	—	0.1	21.7	2.4
	1.5—5	8.4	8.7	0.1	2.4	3.5	0.1	2.6	8.2	9.2	18.8	29.0	65	0.1	0.4	5.7	2.9
	5—18	9.1	11.8	0.1	1.6	10.2	0.1	1.4	15.9	16.0	30.0	37.2	81	0.2	0.4	13.1	2.5
	18—41	9.5	6.9	—	0.7	5.9	—	0.9	20.9	11.2	29.6	40.0	74	1.1	0.6	7.7	3.3
	41—56	9.5	5.4	—	0.3	4.6	0.1	1.6	23.3	10.6	26.6	34.4	77	1.2	0.6	6.5	3.5
	56—76	9.5	4.7	—	0.3	5.7	0.1	2.5	29.1	10.1	24.3	32.4	75	1.0	0.5	5.8	3.4
	76—92	9.5	4.5	—	0.2	6.6	0.1	3.8	26.2	9.6	22.8	30.0	76	1.2	0.6	4.8	3.5
	92—114	9.5	4.5	—	0.2	6.6	0.1	3.7	36.0	8.7	19.3	29.4	66	1.1	0.6	4.4	3.4
	125—	9.3	3.6	—	1.4	90.7	0.1	4.5	44.8	6.6	18.8	25.4	74	0.6	0.6	3.0	2.8
2.	0—2.5	5.8	0.5	—	6.2	1.6	0.1	20.8	—	0.1	0.7	32.0	2	—	0.1	0.1	—
	2.5—10	6.0	0.3	—	7.3	1.8	0.1	25.6	—	0.1	1.0	37.2	3	—	0.1	0.1	—
	10—30	6.5	0.3	—	6.2	1.2	0.1	28.8	—	0.1	1.7	37.8	4	—	0.1	0.1	—
	30—48	7.9	1.2	—	4.8	1.0	0.1	27.8	0.9	0.7	4.5	38.0	12	—	0.3	0.3	0.3
	48—69	8.2	3.2	0.1	4.4	4.6	0.4	24.6	3.8	2.2	7.2	37.2	19	—	0.3	1.4	1.1
	69—97	8.2	7.2	0.2	2.7	2.9	1.2	23.1	6.1	5.3	8.7	35.6	24	—	0.2	3.7	2.6
	97—104	8.2	7.7	0.2	2.9	10.1	1.4	21.5	1.0	5.3	8.8	33.6	26	—	0.2	4.2	2.7
	104—130	8.3	7.7	0.2	4.2	103.1	1.3	18.2	30.0	5.2	8.3	30.4	27	—	0.1	3.8	2.7
	140—	8.4	5.9	0.1	4.8	100.6	0.7	17.5	23.8	5.4	8.3	30.0	28	—	0.1	2.5	2.5

which sometimes effloresces on the soil surface, while in other cases reaches a maximum in the deeper layers of the soil profile.

Some gypsiferous soils and a very low extent of alkali soils can be found in Mexico. However, soil alkalinity mostly occurs in combination with soil salinity.

Acid sulfate soils also occur in some tidal marshes of the country.

E. Central America

In Central America the majority of salt-affected soils are found not on the mainland but on the islands. Salt accumulation in this region is not a consequence of the climate (which is far from arid) but rather of hydrogeological and geological conditions.[180] The type of salt accumulation is more alkali than saline; consequently, most of the existing salt-affected soils belong to the alkali group. Acid sulfate soils also occur in the tidal marshes of Central America. The country with the greatest area of salt-affected soils in Cuba, where 20% of the land is affected by salinization and alkalization, partly as a consequence of irrigation.[181]

Particularly in the eastern part of the island (in the Province of Guantanamo, in Oriente), salt-affected soils are very widespread and contain high amounts of both sodium chloride and sodium carbonate. Based on this fact, the majority of these soils can be characterized as saline-alkali. The typical causes of salinization are the salinity of the groundwater and the poor drainage of the territory. Both temporary waterlogging (partly due to irrigation) and geochemical processes also result in the accumulation of salts capable of alkaline hydrolysis in the soil layers.[182]

Besides Cuba, similar formations of salt-affected soils can be found in Haiti and on smaller islands such as Puerto Rico. Among the saline soils in these areas, marine saline soils also occur in some coastal districts, while acid sulfate soils are to be found on some tidal marshes with mangrove vegetation.

F. South America

Salt-affected soils occur in most countries of South America to a greater or lesser extent. Basically two types of environment can be distinguished in this respect:[63,64]

(1) Under arid or extremely arid conditions (e.g., along the western Pacific coastline of Peru and particularly that of Chile) saline soils are very common. The salt accumulation in this region is the result of the dry climate and the failure of water-soluble products to be leached from the soil. On the western coast of Chile (in the Antofagasta Desert) even sodium nitrate accumulates in the soil, showing how extremely arid the conditions are. The majority of the saline soils are of the sodium chloride type, with the frequent occurrence of sodium sulfate accumulation, and sometimes that of magnesium sulfate and, of course, gypsum.

(2) Saline soils are also extensive on the western coast of Peru. According to Zavaleta,[64] along the numerous alluvial valleys found in the arid coastal area of this country, as much as half of the arable soil may be affected by salinity. It should be noted that the widespread use of irrigation contributes very substantially to the further development of salinity in Peru. Table 17 shows the extension of salt-affected soils in different regions of Peru.[64] The table clearly demonstrates that in some areas nearly all the soils are affected by these adverse processes.

Both in Chile and in Peru saline soils and alkali soils are widespread. This group can be found east of the Andes, where the climate is moderate. Consequently, the alkalization develops not as the result of the arid climate, but rather under the influence of saline groundwaters.

In Chile, alkali soils are extensive west of Santiago and have both alkali and saline properties. Tables 18 and 19 show that the dry residue of the aqueous extract of a saline-alkali soil from Lampa (Chile) contains a considerable amount of water-soluble salts while at the same time its pH value is rather high. The tables also show that a considerable percentage of exchangeable sodium, and also of magnesium, can be found in this soil.

Table 17
SOME SALT-AFFECTED SOILS OF PERU[64]

Region	Location	Hectares surveyed	Irrigation	Normal	Saline	Saline alkali	Non-saline alkali	Solod	Total	Poorly drained (%)
				Percentage of various soils						
Northern	Low Piura Valley	23,000	Old	25.0	21.2	37.5	15.8	—	75.0	67
Central	Santa Rose Irrigation	6,500	15 years old	79.5	2.0	6.0	12.5	—	20.5	18
Southern	Average of 14 Southern Alluvial Valleys	85,500	Old	16.9	4.0	59.9	18.4	—	83.1	22.7
Southern	Majes, Siguas La Joya, Irrigation Project	250,000	Non-irrig.	0.0	89.4	0.0	0.0	10.6	100.0	0.0

In Peru, the alkali soils rarely exhibit pH values in excess of 8.8. At the same time they always contain mainly sodium salts and have high ESP values (16 to 60%). Due to the joint occurrence of salinity and alkalinity in these soils they seldom have a typical solonetz profile, nor are they identical with the "black alkali" soils referred to in the former American classification.

In most cases chloride and sulfate are dominant among the anions and sodium and dominant among the cations in the saturation extract. The alkali soils of this region, as in many other places on the American continent, contain a comparatively high amount of boron.

Along the entire western coast of South America, extensive areas of level to gently sloping or occasionally rolling land are found, which are composed of unconsolidated marine sediments, such as stratified clays, silts, sands, and gravel, and associated with desert lithosols. Some of the sedimentary layers are salty, gypsiferous and/or calcareous. In some tidal marshes on the coast, acid sulfate soils also can be located as well as marine solonchaks affected directly by the sea-water.

Saline soils also can be met with in other countries in South America. They are particularly extensive in arid regions of Brazil, in the north-eastern part of the country.

In Argentina the most common group of salt-affected soils is in the Pampas, and to some extent in Patagonia and other areas the most common group is alkali soils.[62] These soils occupy huge areas in several regions of the country, existing also in areas stretching over several millions of hectares even in wet regions (i.e., in the lowlands of Buenos Aires province, in the south of Entre Rios, in the middle of Santa Fe, in the west of Chaco, and Formosa, etc.).

These soils (totally or partially unproductive), spread out more and more due mostly to the over-grazing which is currently practised in nearly all the country. The soils are of varying textures, generally with little or no structure and having in all cases an impermeable layer in some horizon. This gives them the general characteristics of lands which easily flood, especially during periods of heavy rain. In wet regions these soils can be found in the low-lying poorly drained areas. The presence of salts such as chlorides or sulfates is generally low and very often these salts are found only in traces. However, the pH is very high. In the superficial layers of soils during the period of drought, pH values as high as 11.6 can be found. Between 20 to 50 cm there is generally a hard layer very rich in calcium carbonate.

The climate of this area is subhumid; consequently, the accumulation of water-soluble products is not the result of the arid climate but rather the consequence of the salt content of the groundwater.

Table 18
CHEMICAL ANALYSIS OF THE 1:5 AQUEOUS EXTRACT OF A SALINE-ALKALI SOIL FROM LAMPA (CHILE)

Sampling depth (cm)	pH	Residue (dry)	Residue (ignition)	Water-soluble humus (%)	Alkalinity (%/meq) Na_2CO_3	Alkalinity (%/meq) Alkali metal	Alkalinity (%/meq) Alkali earth metal	Alkalinity (%/meq) Total HCO_3	Cl^- (%/meq)	SO_4^{2-} (%/meq)	Ca^{2+} (%/meq)	Mg^{2+} (%/meq)	Na^+ (%/meq)	K^+ (%/meq)
0—20	8.3	0.19	0.12	0.016	0.001	0.099	0.024	0.123	0.004	0.018	0.004	0.002	0.038	0.002
					(0.024)	(1.625)	(0.396)	(2.021)	(0.113)	(0.377)	(0.200)	(0.158)	(1.652)	(0.051)
40—50	8.8	1.78	1.68	0.023	0.033	0.131	0.037	0.168	0.211	0.747	0.005	0.009	0.530	0.006
					(0.619)	(2.146)	(0.603)	(2.749)	(5.957)	(15.547)	(0.255)	(0.740)	(23.047)	(0.161)

Note: Parentheses indicate meq.

Table 19
EXCHANGEABLE CATIONS IN THE SALINE-ALKALI SOIL FROM LAMPA (CHILE)

	Ca^{2+}	Mg^{2+}	Na^+	K^+	S	Ca^{2+}	Mg^{2+}	Na^+	K^+
Surface	12.400	5.345	6.955	1.061	25.761	48.13	20.74	26.99	4.11
At 50 cm depth	6.800	6.167	37.166	0.665	50.798	13.38	12.14	73.16	1.31

Note: The first five values are milliequivalents per liter; the next four are percent of S (sum of exchangeable cations).

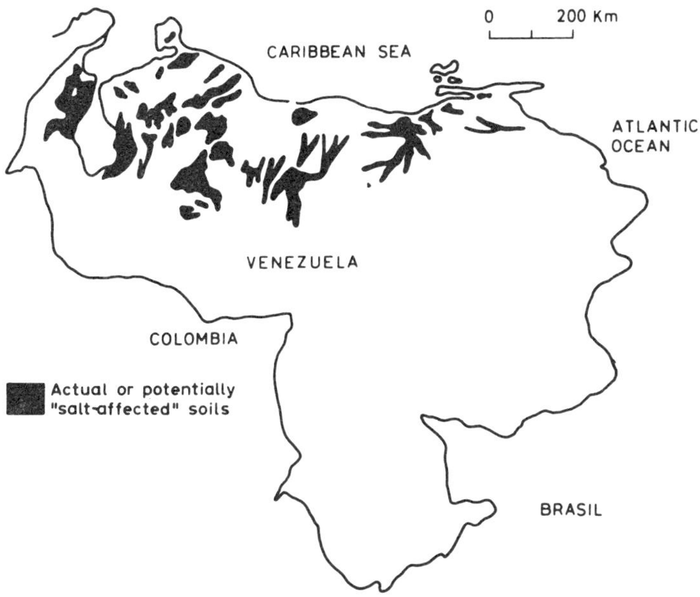

FIGURE 22. Actual or potentially salt-affected soils in irrigated areas of Northern Venezuela. (After Pla.[63])

The largest area of alkali soils in South America can be found in Argentina. The morphology and the physical, chemical, and biological properties of most of the alkali soils in Argentina are very similar to those in Europe, North America, or Australia. The majority of the alkali soils belong to the solonetz and solodized solonetz types and some of them are more or less saline. In the Pampas, the precipitation in some places reaches 1000 mm/year. Consequently, the leaching of the soil profile would be possible except in places where poor drainage impedes the removal of salts. In Argentina, gypsiferous soils are not widespread but magnesium soils can be found in several cases in association with alkali soils.

In Uruguay and Paraguay, salt-affected soils occur to a lesser extent than in Argentina but the properties of the salt-affected soils in these countries are closely related to those in Argentina. The same is true of Bolivia.

In countries near the Caribbean, salt-affected soils are fairly common, particularly in Colombia and Venezuela. They can be found under dry farming conditions but are particularly associated with the extensive use of irrigation.

In Venezuela, where several million hectares of salt-affected soils occur, the main properties of those soils are generally both saline and alkali. Apart from the occurrence of water-soluble salts, sodicity or high ESP values are also common in the soil profiles. The area of salt-affected soils in this country is increasing mainly due to the ever wider use of irrigation. In Figure 22 the actual or potentially salt-affected soils are demonstrated in irrigated areas of North Venezuela.[63] It can be seen from the figure that salt-affected soils and potentially salt-affected soils are closely associated with practically all the irrigation systems in this area.

In Venezuela,[63] as in Colombia and Ecuador, salinity and alkalinity appear in many places not on the top of the soil profile but at a certain depth.[64] It is evident that when irrigation is applied the salinity can easily reach the upper layer of the soil profile and cause the salinization and alkalization of this layer. This phenomenon may also occur under rainfed agricultural conditions in years or seasons with a high quantity of precipitation.

In this region, as in Argentina, the formation of sodium carbonate and bicarbonate (which are the main components of alkali soils) often develops partly as the result of biological processes, for instance during the soda-forming anaerobic conditions.

The occurrence of magnesium and gypsiferous soils is similar to that described in relation to Argentina.

At the estuaries of the Amazonas, and north of this region on the Caribbean coast, both saline-alkali and saline soils occur. In the tidal marshes of this region acid sulfate soils are also frequent. In these same areas and in many other places along the coast, marine saline soils also can be located.

In Figure 23 a general map is presented of the distribution of salt-affected soils in South America. The map was based mainly on the information available from the UNESCO/FAO Soil Map of the World.

G. Australia

Australia has the highest ratio of salt-affected soils in relation to the total surface of the continent. About one third of this continent is covered by different types of salt-affected soils, as demonstrated in Figure 24. In the map, the grouping system used by the Subcommission on Salt-Affected Soils of the ISSS is applied, though the Australian soil classification employs a somewhat different grouping system.[100,101]

The large extension of salt-affected soils in Australia can be explained partly by the fact that this is the driest continent of the world. Due to the climatic conditions, 87% of the average annual rainfall, which totals 420 mm, is lost into the atmosphere due to evaporation and transpiration. Consequently, the geochemical accumulation of salts in the soils is very pronounced.[102]

A great part of the Australian continent consists of deserts and semideserts, where saline soils are very widespread. Such areas are the central part of the continent and the northwestern and western territories.[184] Table 20 shows the areas of different salt-affected soils in Australia according to the categories applied by Northcote and Skene.[100] The table shows that the majority of Australian salt-affected soils consist of different types of alkali and sodic soils.

1. Saline Soils

In the western part of the continent saline depressions of earlier geological origin occur, where the soils have a high content of salts, mostly sodium chloride. In this area, and to the east of these territories, salt lakes are also common and are associated with poor drainage conditions. In southwest Queensland salt pans can be found adjacent to major rivers.[185]

The origin of solonchak soils is partly residual and partly the result of recent salt accumulation processes. In many cases the salty sediments are the results of deep weathering during ancient geological epochs. As a consequence of the very arid conditions, the salts remain in the groundwater and pores for a very long time.

In Australia numerous literary sources describe not only the *in situ* accumulation of soils but also the processes caused by atmospheric agents. A considerable amount of salts, mainly sodium chloride, accumulates on the saline areas of Australia due to annual deposition from the rainwater. Particularly along the coastline this process contributes to the formation of saline soils.

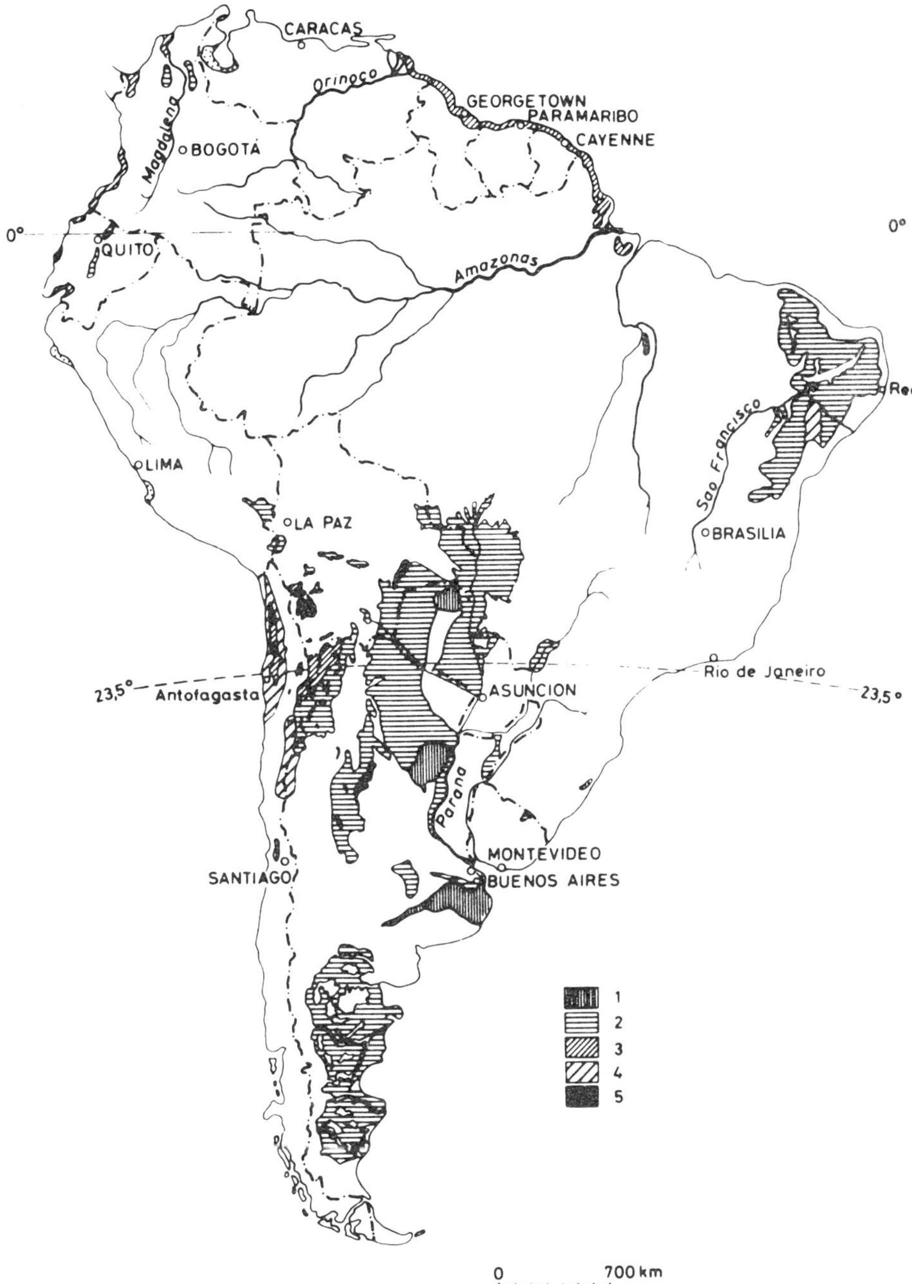

FIGURE 23. Distribution of salt-affected soils in South America. (1) Alkali soils constituting more than 50% of the area; (2) alkali soils constituting less than 50% of the area; (3) saline soils constituting more than 50% of the area; (4) saline soils constituting less than 50% of the area; (5) salt flats. (After Zavaleta, Dudal, and Szabolcs.[1])

Another factor in the airborne accumulation of salts is the salt content of strong winds. According to various authors, in the desert areas of Australia wind-blown salts contribute considerably to the accumulation of sodium chloride, and partly to that of other salts, in the surface layers of the soils. Saline soils also develop under the influence of mineralized groundwater in irrigated areas.

Northcote and Skene,[100] in accordance with Loveday,[183] characterize the salinity of Australian soils as follows:

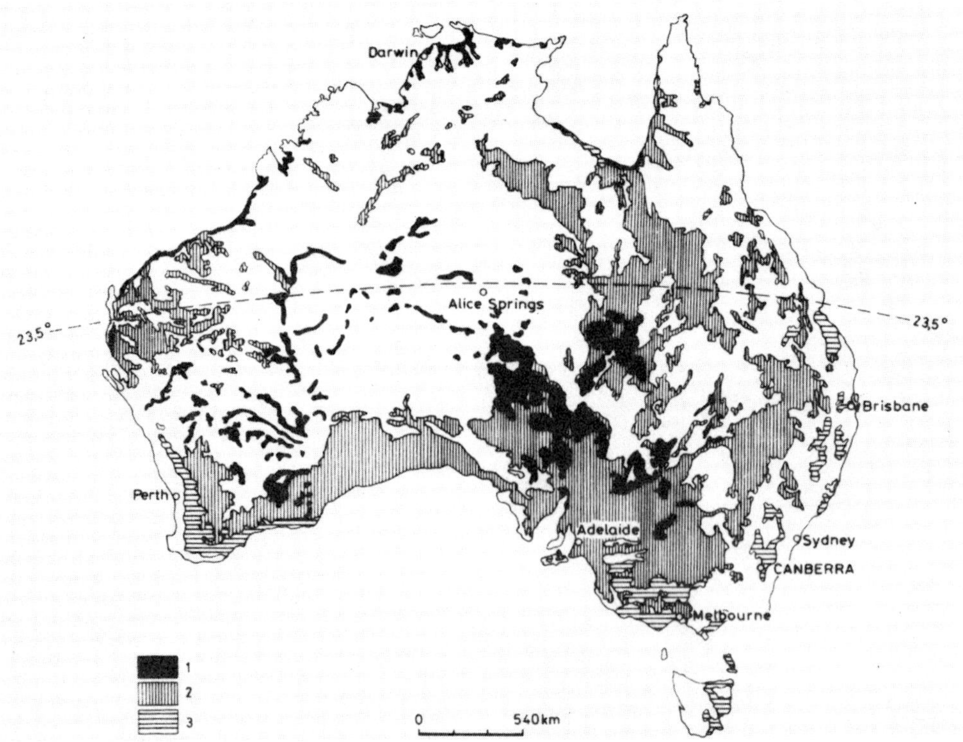

FIGURE 24. Distribution of salt-affected soils in Australia. (1) Saline soils; (2) alkaline sodic soils; (3) nonalkaline sodic soils (After Wright and Schuster.[1])

Table 20
AREA OF SALT-AFFECTED SOILS IN AUSTRALIA[100]

	Area	
Salt-affected soil category	**(km^2)**	**(% total)**
Saline soils	386,300	5.3
Alkaline strongly sodic to sodic clay soils with uniform texture profiles	666,400	9.2
Alkaline strongly sodic to sodic coarse- and medium-textured soils with uniform and gradational texture profiles	600,700	8.3
Alkaline strongly sodic to sodic duplex soils	454,400	6.3
Nonalkaline sodic and strongly sodic neutral duplex soils	134,700	1.9
Nonalkaline sodic acid duplex soils	140,700	1.9

"Limits have been chosen for sodium chloride in the surface soil which are considered to have a bearing on the performance of many, although not all, agricultural crops. Some allowance for the effect of soil moisture on the concentration of salts in the soil solution has been made by setting a higher limit for fine-textured than for coarse- and medium-textured soils. The values adopted approximate to an electrical conductivity of the saturation extract of 4 mmho/cm at 25°C, the value specified for saline soils by the U.S. Salinity Laboratory Staff.

Soluble salts in the subsoil in the absence of surface salinity are regarded as potentially,

Table 21
CLASSES OF SALT-AFFECTED SOILS IN TERMS OF SALINE, SODIC, AND ALKALINE PROPERTIES[100]

Class	Salinity category (Cl)	Sodicity category (Na)	Alkalinity category (pH)	Description	Occurrence
				Nonsodic soils	
0	0	0	0	"Normal" soils	Widespread
1	0	0	1	Nonsaline, alkaline	Widespread
2	0	0	2	Nonsaline, strongly alkaline	
3	1	0	0	Saline	
4	1	0	1	Saline, alkaline	
5	1	0	2	Saline, strongly alkaline	No soils recorded
6	2	0	0	Saline subsoil	
7	2	0	1	Saline subsoil, alkaline	
8	2	0	2	Saline subsoil, strongly alkaline	
				Sodic soils	
9	0	1	0	Nonsaline	Widespread
10	0	1	1	Nonsaline, alkaline	Very widespread
11	0	1	2	Nonsaline, strongly alkaline	Locally important
12	1	1	0	Saline	Rare
13	1	1	1	Saline, alkaline	Rare
14	1	1	2	Saline, strongly alkaline	No soils recorded
15	2	1	0	Saline subsoil	Few
16	2	1	1	Saline subsoil, alkaline	Rare
17	2	1	2	Saline subsoil, strongly alkaline	No soils recorded
				Strongly sodic soils	
18	0	2	0	Nonsaline	Widespread
19	0	2	1	Nonsaline, alkaline	Very widespread
20	0	2	2	Nonsaline, strongly alkaline	Widespread
21	1	2	0	Saline	Few
22	1	2	1	Saline, alkaline	Few
23	1	2	2	Saline, strongly alkaline	Rare
24	2	2	0	Saline subsoil	Few
25	2	2	1	Saline subsoil, alkaline	Few
26	2	2	2	Saline subsoil, strongly alkaline	Rare

if not actually, significant to agricultural crops. However, a higher limit is imposed than in the surface soil because not all the plant root zone is salt-affected and only part of the salt present is likely to move into the surface layer with moisture fluctuations.

Category O — non-saline: No chloride salinity in either the surface soils or subsoil as defined for categories 1 and 2.

Category 1 — surface salinity: Soils containing in their A horizons, or in the surface 20 cm if either the A and B horizons are undifferentiated or the A horizon is less than 10 cm thick, more than 0.1% sodium chloride in loams and coarser soils and more than 0.2% in clay loams and clays.

Category 2 — subsoil salinity: Soils lacking surface salinity but containing more than 0.3% sodium chloride in the B horizon, or below 20 cm if the A and B horizons are undifferentiated."

The same authors compiled a more detailed classification for the different salt-affected soils on the Australian continent than that given in Table 20. This is demonstrated in Table 21.

Saline soils occur along the coastline of the continent where low coastal plains are favorable for the accumulation of salts directly from the seawater. The particle size composition of saline soils in Australia varies but coarse-textured soils prevail.

Apart from sodium chloride, sodium sulfate, and sometimes even magnesium salts also are present in Australian saline soils.

The salinity processes in Australia often occur jointly with other soil-forming processes and have a closer genetic relationship with brown soils, red soils, and other soil types than elsewhere. As in other arid areas, in the deserts and semideserts of Australia, gypsiferous soils often can be found in association with saline soils. They are extensive in the central part of the continent, often forming "lunettes" in temporarily waterlogged arid areas similar to such formations in North Africa. In Australia saline soils are also closely associated with alkali soils.

2. Alkali soils

Except for Europe, Australia is the only continent where the extension of alkali soils exceeds that of saline soils. In Australia various kinds of alkali soils occur and different names and classification systems exist for their characterization. As in many other places, the terms alkali soils and sodic soils are used alternatively — however, depending on the author, a slight genetic difference exists between the terms.

While saline soils can be found mainly in desert and semidesert areas, the occurrence of alkali soils is more closely associated with semiarid and subhumid conditions. The alkali soils in Australia are very similar to those described when European salt-affected soils were discussed earlier in this chapter. At the same time, the Australian alkali soils show even greater variety than their European counterparts and, in consequence of the climatic, geological, and other environmental conditions of this continent, are quite different in some cases from equivalent soils in Europe. The solod-forming process is particularly widespread in Australia and consequently, as in some places in Europe, the majority of solonetz soils are affected by this process. According to Northcote and Skene:[100]

> "Whilst pH expresses the alkalinity of the soil, pH itself is affected by too many factors to have more than limited value as a specific criterion for the classification of alkali soils. Values in excess of about pH 8.5 (1:5 suspension) reflect significant amounts of exchangeable sodium on the clay, and of bicarbonate, and at the higher pH carbonate, in the soil solution. Above pH 9.5, soils are invariably strongly sodic and carbonate is present in water extracts. High pH, therefore, is useful to indicate sodicity when exchangeable cation data are lacking. However, some sodic soils have pH values much below 8.5 and cannot be recognized as such from these pH values. When this is the case the pH is (1) depressed due to the presence of excess soluble salts and (2) influenced materially by the presence of exchangeable hydrogen. In (2), soluble salts may or may not be present but lime (calcium/magnesium carbonate) is absent.
>
> The pH limits for the three categories of alkalinity are based largely on the above considerations, except that pH 8.0 has been adopted instead of pH 8.5 to separate soils significantly alkaline from those that are not significantly alkaline. The lower pH allows greater flexibility in the grouping of alkaline soils, although it may bring some normal soils such as nonsodic calcareous soils into an alkaline classification.
>
> **Category 0 — acid or insignificantly alkaline:** pH <8.0.
> **Category 1 — alkaline:** pH 8.0—9.5.
> **Category 2 — strongly alkaline:** pH >9.5.
>
> **Sodicity** — The ESP reflects genetic influences as well as having agricultural importance. Values less than 6 are associated with normal soils, but values of 6 or more are taken to indicate sodium salts as an important genetic factor. The view has been taken that soils with an ESP of 6 or more are sodic and those with an ESP of 15 or greater are strongly sodic. The latter figure, originally proposed by the U.S. Salinity Laboratory Staff is widely accepted as a value above which soils exhibit adverse properties due to the dominance of exchangeable sodium."

The Australian soil classification, like many other classification systems, uses the terms *solonetz* and *solod* for the characterization of the respective alkali soil formations. Practically, most morphological, physical, and physico-chemical properties of Australian solonetz soils are identical to those on other continents.

Solonetz soils occur mainly under moderately arid or humid conditions and are often of hydromorphic origin. Such formations are widely extended in the southeastern part of the continent, particularly in the states of Victoria, New South Wales, and South Australia. They are also found to some extent in Western Australia.

Solonetz and weakly solodized solonetz soils are more frequent than typical solod soils, a situation similar to that described for Europe. However, solod soils occur in Australia more frequently than in any other continent and can be characterized by their well-developed, typical soil profiles and low fertility.

Solod soils develop, again similarly to those in Europe, on the weathering products of basaltic rocks with a high content of sesquioxide in the lower A and B horizons. Solod soils are widespread in the moister subhumid regions of southwestern Australia and throughout eastern Australia, mainly as small occurrences associated with solodized solonetz or even podzolic soils. Solid soils are often associated with brown soils and even with red soils, a unique phenomenon occurring only in Australia. The great diversity of alkali soils in Australia results in different approaches to their study and utilization.

Some of the alkaline sodic clays are calcareous and these soils do not exhibit irreversible dispersion of the clay following wetting. Such soils crack and break into aggregates on drying and where the aggregates are fine, the soils are called "self-mulching". Soils of this character are utilized for cereals where the climate permits, as in the Wimmera district of Victoria, the Darling Downs of Queensland, and northern New South Wales. Sodicity presents no problems to the cultivation of these calcareous soils provided that they are worked at a suitable moisture content.

Alkaline sodic gray and brown clays are irrigated in northern Victoria and in the Riverina district of New South Wales. The calcareous self-mulching varieties support both perennial and annual pastures without difficulty and are utilized for dairying and raising fat lambs. The soils not naturally self-mulching exhibit degrees of low permeability depending on their clay content, degree of sodicity, and whether gypsum is present in the subsoil. Little has been achieved towards improving these soils by ameliorative methods apart from the use of gypsum in the irrigation water to achieve surface flocculation.

In the mid-Murray areas of Victoria and South Australia, these alkaline sodic soils are irrigated for fruit growing. Secondary salinization (considered later) is a common problem in most of these horticultural settlements and there has been considerable loss of production of dried fruits and citrus over the last 50 years. These soils occur mainly under an annual rainfall of about 550 mm in Western Australia, Victoria, Tasmania, New South Wales, and southern Queensland, with much smaller areas in northern Australia. Small areas are under cultivation, but mainly the soils are given to the grazing of annual clover pastures by sheep for wool and fat-lamb production.

The soils are subject to erosion in hilly landscapes. Tunnel erosion leading to gullying in northeastern Victoria is an example of erosion due to sodic properties in duplex soils. Such erosion occurs not only on the neutral sodic soils but also on alkaline and acid sodic soils. Dispersion and washing away of the sodic B horizon appear at points of water entry such as rabbit burrows, followed by collapse of the overlying A horizon.

Such phenomena are also common in Europe and Canada, where the surface of alkali soils is very often affected by the process of erosion, which results in a drastic removal of the upper layer and the loss of organic matter and plant nutrients from the soils.

Magnesium soils are found in Australia partly in association with saline soils in dry areas, where magnesium sulfate and, to some extent, magnesium chloride can be found in the upper soil layers. In other cases the magnesium ions are associated with alkali soils, particularly with solonetz and solod soils. In river valleys, lowland terraces, and other places where the groundwater is linked with the soil profile, magnesium ions may accumulate in the soils due to hydromorphic processes.

Magnesium salts play a diverse role in several soil types in Australia as a result of their mineralogical composition and of various soil-forming processes (other than those leading to the formation of salt-affected soils).

Acid sulfate soils, found in the tidal marshes of all the continents, are also extensive along the coast of the Australian continent and Tasmania. As a rule they develop on the narrow strip of tidal marshes forming the typical profile of this soil type. However, their extent is much inferior to that described in southwest India or in some African coastal districts. Acid soil formation can be found on the coast of New South Wales in estuarine sediments rich in sulfide and exposed at the surface in bog swamps. The comparatively moderate extension of acid sulfate soils in the estuaries of rivers in South Australia compared to those of Southeast Asia can be explained partly by the effect of river sediments of alluvial origin overlying the sulfide-rich sediments. The majority of the acid sulfate soils on this continent are situated in northern Australia, where large areas of mangroves occur and provide all the conditions necessary for the development of this soil type, namely an environment rich in organic matter and sulfide, waterlogging, etc.

Chapter 2

LANDSCAPE GEOCHEMISTRY OF SALT-AFFECTED SOILS

I. GEOCHEMICAL CHARACTERISTICS OF ELEMENTS, WITH PARTICULAR REGARD TO THEIR IMPORTANCE IN THE FORMATION OF SALT-AFFECTED SOILS

As was mentioned in the first chapter, it has long been taken for granted that salt-affected soils occur under arid and semiarid conditions, and the interrelation of soil salinity and environmental conditions is often only considered with respect to the climate.[186,187] Although aridity does indeed promote the process of soil salinization, and the possibility of salt accumulation in soils and waters should always be taken into consideration under arid and semiarid conditions, the accumulation of electrolytes in a soil is nevertheless a complex process which cannot be interpreted as a simple consequence of climatic effects.[1,187]

The type of climate does not only influence the general tendency of geochemical salt accumulation. Parallel with increasing aridity and decreasing leaching possibilities, salts — some of which are formed *in situ* as weathering products of rocks and some of which are transported to the given area by water, wind, etc. — accumulate in the underground layers of soils and in the ground and surface waters. Besides the aridity index, the seasonal dynamics of salts in soils and underground waters are also responsible to a great extent for the diverse properties and global distribution of salt-affected soils under different climatic conditions.[188]

The different patterns for the seasonal dynamics of salts in soils are schematically illustrated according to the type of climate in Figure 25. The figure clearly indicates that the annual duration and seasonal rhythm of salt accumulation is different in the various regions, and that under certain conditions the accumulation or leaching of salts is promoted by one season, while under other circumstances it is promoted by another season. Such regularities sometimes have a dominating influence not only on the types of salt-affected soils which develop, but also on the way in which they can be utilized.[189]

Modern landscape geochemistry, which has been developed from the classical geochemical aspects of pedology and the achievements of Clarke,[190] Fersman,[191] Vernadsky,[192] Polynov,[193] and others, represents a holistic approach to the mass and energy flows which lead to the formation of salt-affected soils. According to this approach (besides the influence of the climate) the geochemical and biogeochemical processes determining the mass flow that affects soil formation should also be studied with particular care.[194]

Comparatively few elements play a decisive role in the development of salt-affected soils. Their geochemistry must be described and the accumulation or leaching in the landscape must be studied, so that the preconditions for the occurrence of the salinity and/or alkalinity affecting the soils, waters and vegetation in a given territory can be clearly established.

Polynov considered vegetation, soils, underlying rocks, water, geomorphology, and geology to be interrelated and carried out complex investigations to discover the qualitative and quantitative relationships between the factors making up the geochemistry of the landscape.[193] He considered the landscape as a dynamic system, constantly changing to result in natural processes having different and often opposing directions. Relict and newly arising factors take part in these changes and determine the direction of soil forming processes as well as the shape of the landscape as a whole. Polynov's interpretation of a landscape includes the interrelationships and interactions of the lithosphere, the hydrosphere, and the daylight surface of these geospheres in contact with the atmosphere. He also emphasizes the effects of living materials (the biosphere) on the development and appearance of the landscape. All these factors, the matrix of their interrelations and influences, determine the migration of elements in a given place.

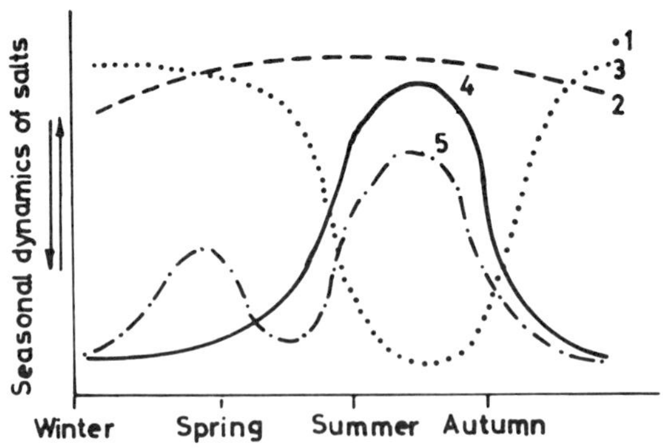

FIGURE 25. Seasonal dynamics of salts in the soils depending on the type of climate: (1) continental tropical, (2) continental subtropical, (3) monsoon tropical and subtropical, (4) mediterranean and continental subboreal with a mild winter, (5) continental boreal and subboreal with a frosty winter. (After Kovda.[3])

Table 22
AVERAGE CHEMICAL COMPOSITION OF THE EARTH'S CRUST

Components	Percentage	Components	Percentage
SiO_2	59.07	Na_2O	3.71
Al_2O_3	15.22	K_2O	3.11
Fe_2O_3	3.10	TiO_2	1.03
FeO	3.71	P_2O_5	0.30
CaO	5.10	MnO	0.11
MgO	3.45	H_2O	1.30

After Clarke (cited in Fortescue[190]).

In order to interpret the main geochemical laws governing the formation and distribution of salt-affected soils, it is necessary to characterize those elements of the earth crust which play a primary or sometimes secondary role in these processes.[195,196]

The chemical composition of the earth crust is given in Table 22. Although the oxides of silicon and aluminum are well known to constitute the greatest part of the globe, the elements directly responsible for the formation of salt-affected soils are also present at a medium rate.[197]

Besides the chemical characteristics of the earth crust (including the atmosphere and hydrosphere) it is necessary to study the chemical composition of rocks and soils. Table 23 shows that the elements which result in soil salinization and alkalization occupy a transitional place between the dominating and rare elements in rocks and soils.[198] A comparison of their chemical compositions leads to the interesting finding that practically all the cations which cause soil salinity and/or alkalinity can be found in higher concentrations in rocks than in soils.[199] The ratio in favor of rocks is nearly 3 for Ca, more than 3 for Mg and Na, and nearly 2 for K. This would seem to be a paradox, because — as will be explained later — these cations are the most mobile during the soil formation processes. On the other hand, Table 23 shows that tremendous amounts of substances which contribute to soil salinization and alkalization are present in rocks, but in immobile forms.[200-202]

Among the elements listed in Table 23 and their compounds, only a few play a decisive role in the salinization and alkalization of grounds, soils, and waters. These are as follows:

Table 23
CHEMICAL COMPOSITION OF ROCKS AND SOILS

Element	Rocks (%)	Soils (%)	Element	Rocks (%)	Soils (%)
O	47.2	49.0	C	0.1	2.0
Si	27.6	33.0	S	0.09	0.085
Al	8.8	7.13	Mn	0.09	0.085
Fe	5.1	3.8	P	0.08	0.08
Ca	3.6	1.37	N	0.01	0.1
Na	2.64	0.63	Cu	0.01	0.002
K	2.6	1.3	Zn	0.005	0.005
Mg	2.1	0.6	Co	0.003	0.008
Ti	0.6	0.46	N	0.0003	0.001
H	0.15	?	Mo	0.0003	0.0003

After Vinogradov (cited in Beus[198]).

Table 24
SEQUENCE OF ION EXTRACTION DURING WEATHERING

Sequence of extraction	Ions	Coefficient of energy	Sequence of extraction	Ions	Coefficient of energy	Sequence of extraction	Ions	Coefficient of energy
I.	Cl^-, Br^-	0.23	II.	Na^+	0.45	III.	SiO_3^{2-}	2.75
	NO_3^-	0.18		K^+	0.36			
	SO_4^{2-}	0.66		Ca^{2+}	1.75	IV.	Fe^{3+}	5.15
	CO_3^{2-}	0.77		Mg^{2+}	2.10		Al^{3+}	4.25

Sequence according to Fersman.[191]

Cations	Anions
Ca^{2+}	Cl^-
Mg^{2+}	SO_4^{2-}
Na^+	HCO_3^-
K^+	CO_3^{2-}
Al^{+++}	SiO_3^{2-}
Fe^{+++}	S^{2-}
H^+	

In spite of their practical importance, this list does not contain certain toxic elements which are present in different types of saline and alkali soils. This is because some well-known toxic elements, such as boron, do not take part in the formation of various types of salt-affected soils. Others, such as Cl and others, which take part directly in the formation of these soils, have of course been listed.[203,204]

The ions and compounds, the migration and accumulation of which lead to the formation of salt-affected soils, behave diversely during the weathering processes taking place in the earth's crust.[205,206]

In Table 24, compiled after Fersman,[191] the places occupied by the dominant elements in the sequence of extraction are demonstrated. As the number of the sequence rises, the mobility of the elements during the weathering processes decreases. According to Fersman,

Table 25
THE RELATIVE MOBILITY AND AVERAGE DISTRIBUTION OF SOME ELEMENTS AND IONS IN ROCKS AND WATER

Component or ion	Average composition of igneous rocks (%)	Average composition of the mineral residue of river waters (%)	Relative mobility of elements and compounds (%)
SiO_2	59.09	12.80	0.20
Al_2O_3	15.35	0.90	0.02
Fe_2O_3	7.29	0.40	0.04
Ca^{2+}	3.60	14.70	3.00
Mg^{2+}	2.11	4.90	1.30
Na^+	2.97	9.50	2.40
K^+	2.57	4.40	1.25
Cl^-	0.05	6.75	100.00
SO_4^{2-}	0.15	11.60	57.00
CO_3^{2-}	—	36.50	—

According to Polynov.[193]

the energy coefficient can be calculated on the basis of known lattice energies in inorganic salts.[191] These values are called experimental energy coefficients and are considered to be the most reliable figures available. The energy coefficients given by Fersman are closely related to the sequence of ion extraction from minerals, to the rate of ion migration, and to the ability of the ions to accumulate in sediments and soils.

In recent times the weathering of rocks has been the primary source of soluble salts entering natural waters, sediments, and soils.[207] The geochemistry of salts in any given place is determined by the mobility of the compounds formed and by the sequence of precipitation of the weathering products. The mobility of the rock-forming elements depends on the following factors:

a. The stability of the crystalline network
b. The radius of ions formed during weathering
c. The charge of the ions formed during weathering

It is evident that the possibility of translocation for weathering products depends mainly on their mobility.

From Table 24 it follows that the elements and compounds which play a dominant role in salinization and alkalization are mainly found in sequences I and II; in other words, they are capable of intensive migration. In spite of this, very diverse values can be measured for the mobility of these compounds, and their presence in rocks and waters is similarly diverse. In Table 25 the relative mobility and ratio of nine elements and ions are set out.

From Table 25 it is clear that there are great differences between the quantitative occurrences of these compounds and ions in rocks and waters as well as between their relative mobilities. The global situation with respect to this problem is even more complex if the average chemical composition of river water on different continents, and of the oceans and seas are determined and studied.[196,205]

In Table 26 the chemical compositions of ocean- and seawaters are given, and it may be seen that there are considerable differences in the values. The Aral Sea contains only one third of the average salt content of the oceans, while the Baltic Sea is even more diluted.

Table 26
THE CHEMICAL COMPOSITION OF OCEANS AND SEAWATERS

		Percentage of total ion concentration							
Name	Salt (%)	Na^+	Mg^{2+}	Ca^{2+}	K^+	CO_3^{2-}	SO_4^{2-}	Cl^-	Br^-
Oceans (average)	3.3—3.7	30.6	3.7	1.2	1.1	0.2	7.7	55.3	0.2
Aral Sea	1.07	22.1	5.4	4.5	0.1	0.1	31.3	35.6	—
White Sea	2.6—3.0	30.7	3.75	1.2	0.9	0.1	7.9	55.2	0.1
Caspian Sea	1.27	24.5	5.8	2.6	0.6	0.9	23.8	41.8	0.05
Gulf of Mexico	3.55	30.8	3.6	1.2	1.1	0.3	7.5	55.2	0.2
Baltic Sea	0.72	30.5	3.5	1.7	0.1	0.1	8.0	55.0	0.1
Mediterranean Sea	3.73	30.6	3.65	1.2	1.1	0.2	7.9	55.1	0.2
Black Sea	1.8—2.3	30.5	3.70	1.4	1.2	0.5	7.5	55.1	0.2

Keeping in mind the very different environmental conditions of these seas (the Aral Sea is surrounded by a desert and the Baltic Sea has a humid climate), the importance of geochemical factors other than the climate must be recognized.

Table 26 shows rather high SO_4^{2-} contents in the Caspian and Aral Seas while in the other seas the SO_4^{2-} concentration is very close to the average level for this compound in the oceans. It should be noted that both seas with high SO_4^{2-} contents are situated in areas of intensive sulfate accumulation in rocks, waters, and soils.

Remarkable differences can also be found in the chemical composition of river waters on the different continents. The data in Table 27 show that river waters in South America and Australia are twice as diluted as the global average, and nearly three times as diluted as the waters in Europe. Even greater differences can be observed in the concentration of certain cations and anions in the river waters. However, it is also clear from this table that the river waters constitute good quality irrigation water on all continents, at least as far as their average values are concerned. From the above-mentioned geochemical characteristics, it can be concluded that the elements causing salinity on the continents are more concentrated in the seas than in the river waters.

On the other hand, sharp differences exist in the chemical compositions of the waters of different rivers. In Table 28 the salinity levels of the Volga and Amu Daria rivers are compared. While the latter flows through desert and semidesert regions, the Volga collects its water from mainly nonarid territories. Table 28 clearly shows that tremendous amounts of soluble salts can be found in the water of some rivers traversing deserts and semideserts. The water quality problems must be studied individually in such cases in order to control the chemical composition of irrigation waters.

Table 28 also indicates that in many regions the quantity of airborne salts should also be taken into consideration. However, when comparing this amount with that of the salts collected and transported by the waters, it is found that, with the exception of some coastal districts, the quantity of airborne salts is negligible.[209]

Among the possible sources of salts, Kovda also mentions the biological processes (particularly in arid regions) where the ash of halophytes may contribute to the salinity of soils and waters.[210] However, it is difficult to say whether it is a cause or a consequence when speaking in terms of landscape geochemistry, because halophytes grow as a result of the intensive salinity of their environment.[211]

During the weathering of rocks and the translocation of the products of this process, the quantity of water-soluble substances is still insignificant. The average chemical composition of some igneous rocks is given in Table 29. From the table it can be seen that the different mobile compounds causing salinity may be found in various igneous rocks in different proportions. Magnesium, for instance, is present mainly in ultrabasic rocks, while sodium is more frequent in granitic rocks.

Table 27
AVERAGE CHEMICAL COMPOSITION OF RIVER WATERS ON DIFFERENT CONTINENTS

Continent	Dry residue (ppm)	Ca^{2+} (meq/ℓ)	Mg^{2+} (meq/ℓ)	Na^{+} (meq/ℓ)	K^{+} (meq/ℓ)	HCO_3^- (meq/ℓ)	SO_4^{2-} (meq/ℓ)	Cl^- (meq/ℓ)	Na (%)	Residual carbonate (%)	SAR
North America	142	1.05	0.42	0.39	0.04	1.11	0.42	0.02	20.5	—	0.46
South America	69	0.36	0.13	0.17	0.05	0.58	0.10	0.14	24.1	0.095	0.35
Europe	182	1.55	0.47	0.24	0.04	1.56	0.50	0.20	10.4	—	0.24
Asia	142	0.92	0.47	0.24	—	1.30	0.18	0.25	14.8	—	0.29
Africa	121	0.63	0.32	0.48	—	0.69	0.28	0.39	34.3	—	0.70
Australia	59	0.20	0.23	0.13	0.06	0.52	0.05	0.28	21.0	0.098	0.28
Global average	119	0.78	0.34	0.27	0.05	0.96	0.26	0.21	20.8	—	0.37

Table 28
SALTS COLLECTED ANNUALLY IN THE CATCHMENT AREAS OF RIVERS

River	Catchment area (km^2)	Airborne salts ($tons/km^2$)	Discharge to sea (tons/year)
Volga	1,401,949	2	8,000,000
Amu Daria	308,804	10	226,000,000

According to Kovda.[3]

Table 29
AVERAGE ELEMENTAL COMPOSITION OF SOME IGNEOUS ROCKS[a]

Element	Ultrabasic rocks	Basaltic rocks	High Ca granitic rocks	Low Ca granitic rocks
Si	205	230	314	347
Al	20	78	82	72
Fe	94	86	30	14
Ca	25	76	25	5
Mg	204	46	9	2
Na	2	20	28	26
K	0.04	8	25	42
Ti	0.30	14	3.40	1.20
Mn	1.60	1.50	0.54	0.40
P	0.22	1.10	0.92	0.60

[a] Concentration in ppm.

In Table 30, igneous and sedimentary rocks are compared with respect to their contents of different elements. The table shows that the chemical elements playing a leading role in salinization occur in an even lower percentage in sedimentary rocks than in igneous rocks. Furthermore, both types of rocks completely lack the most common anion taking part in the process of salinization: chloride.

When studying clay minerals which are considered as comparatively mobile substances within the soils, it is found that their chemical composition also show a very diverse picture. The sodium cations which are dominant and common in salt-affected soils are present in very low percentages, as can be seen in Table 31. On comparing the data in Table 31 with the average chemical composition of the earth's crust (shown in Table 22) it is found that the percentages of such elements as sodium and calcium, which are important for salinity, are much lower in most clay minerals than in the earth's crust.

It follows from the above that the elements and compounds causing salinity must be mobile in order to accumulate in soils and waters. The average soil composition is always associated with certain amounts of sodium compounds in a nonmobile form. The percentage of nonmobile sodium is usually the same in salt-affected and nonsalt-affected soils. The mobile (exchangeable and water-soluble) sodium compounds, however, are present in different ratios in the saline and nonsaline soils.

In Table 32 the quantities of various forms of sodium compounds are shown in four types of soil. The first three, a solonetz and a solonchak-solonetz from Hungary, and a salorthid from Syria, are salt-affected. The fourth, a calciorthid from Syria, is not salt-affected.

Table 30
AVERAGE CHEMICAL COMPOSITION OF IGNEOUS AND SEDIMENTARY ROCKS

	Igneous rocks			Sedimentary rocks		
Component	Granite (%)	Grandiorite (%)	Basalt (%)	Sandstone (%)	Clay stone (%)	Limestone (%)
SiO_2	70.77	65.69	51.55	79.63	61.16	5.24
TiO_2	0.39	0.57	1.48	0.15	0.68	0.06
Al_2O_3	14.59	16.11	14.95	4.85	16.21	0.82
Fe_2O_3	1.58	1.76	2.55	1.09	4.23	0.55
FeO	1.79	2.68	9.10	0.31	2.58	0.55
Fe	2.50	3.31	8.86	0.99	4.96	0.43
MnO	0.12	0.07	0.20	Traces	Traces	Traces
MgO	0.89	1.93	6.63	1.18	2.57	7.96
CaO	2.01	4.47	10.00	5.59	3.27	42.97
Na_2O	3.52	3.74	2.35	0.46	1.37	0.05
K_2O	4.15	2.78	0.89	1.33	3.41	0.33
P_2O_5	0.19	0.20	0.30	0.08	0.18	0.04
CO_2				5.11	2.77	41.93

Table 31
COMPOSITION OF SOME CLAY MINERALS

Component	Allophane (%)	Kaolinite (%)	Halloysite (%)	Mont-morillonite (%)	Vermiculite (%)	Illite (%)	Chloride (%)
SiO_2	33.96	45.44	44.08	51.14	35.92	49.26	26.68
Al_2O_3	31.12	38.52	39.20	19.76	10.68	28.97	25.20
Fe_2O_3	traces	0.8	0.1	0.83	10.94	2.27	—
FeO	—	—	—	—	0.82	0.57	8.7
MgO	—	0.08	0.05	3.22	22.0	1.32	26.96
CaO	2.26	0.08	—	1.62	0.44	0.67	0.28
Na_2O	—	0.66	—	0.04	—	0.13	—
K_2O	—	0.14	0.2	0.11	—	7.47	—
H_2O	33.12	14.2	16.18	22.80	19.84	9.25	11.7

Consequently, the exchangeable and water-soluble substances occur in fairly high quantities in the first three, while their level is low in the last soil profile. It is also clear that the nonmobile sodium compounds (total) are much the same in all four soils.

The distribution and fate of the chemical elements on the earth are very characteristic when studying terrestial matter, crops, animals, and the human body. In Table 33 the average chemical composition of terrestial matter (including the atmosphere, lithosphere, hydrosphere, and pedosphere), of a maize plant and of the human body are given. Table 33 shows that those chemical elements which dominate in rocks, air, and waters can be found in lower percentages both in plants and in the human body. The same is true of elements present at a medium rate with the exception of calcium. The so-called bioelements (such as carbon, phosphorus, sulfur, and others) are present in human organs in quantities sometimes several orders of magnitude greater than in terrestial matter. It is interesting to note that plant tissues in general occupy a transitional position between terrestial matter and human tissue, but evidently, as living organs, they are closer to the chemical composition of the human body.

All the primary cations which are directly responsible for the formation of salt-affected soils are present in much lower concentrations in plant tissues than in terrestial matter. The difference is about one order of magnitude. This rule is not valid for the anions. As related

Table 32
THE DISTRIBUTION OF SOIL SODIUM COMPOUNDS WITH DIFFERENT MOBILITIES

Soil type	Depth of sampling (cm)	Na⁺ (meq/100 g soil)		
		Total	Exchangeable	Water soluble
Solonetz	0—5	69.6	8.5	8.6
(Hungary)	10—21	126.0	11.5	19.0
	38—48	71.9	12.6	13.5
	60—68	141.0	14.7	30.7
	80—90	121.5	7.7	2.3
Solonchak-solonetz	0—2	138.5	1.1	6.1
(Hungary)	2—11	150.0	6.3	6.8
	11—26	96.0	6.5	14.3
	32—47	122.0	4.3	10.4
	55—67	104.5	4.9	3.7
	78—89	116.0	3.5	1.9
Salorthid	0—4	203.0	4.5	19.5
(Syria)	20—30	232.0	1.0	20.1
	60—70	219.0	12.9	9.4
	145—155	326.0	1.5	2.2
Calciorthid	0—3	126.0	0.2	0.2
(Syria)	6—12	165.0	0.2	0.1
	25—30	184.0	0.7	1.1
	50—55	123.0	1.0	2.0
	80—85	168.0	1.5	3.5

to the human body the only exception is calcium, which occurs in low concentrations in plant organs, while in the human body its quantity is practically equal to that in terrestrial matter.[212]

II. THE ACCUMULATION OF SOLUBLE SALTS AND THE FORMATION OF SALT-AFFECTED SOILS IN DIFFERENT LANDSCAPES

Continental sediments have primary importance in the formation of salt-affected soils. These sediments, which are distributed diversely on geophysical elements and are transported mainly by water, occupy different places with different petrographical and chemical composition along the slope. Figure 26 demonstrates the general pattern for the distribution of continental sediments. The displacement and accumulation of different chemical compounds can be seen from the massive primary rock downwards to the groundwater table and to the catchment area.

When describing the weathering processes and the geochemistry of continental sediments, Polynov[193] distinguishes three characteristic types for the distribution of weathering products in saline depressions. Three types are not enough, of course, to account for all the possible patterns, but they are good examples of the different courses of salt accumulation under salt-affected conditions.

Polynov[193] used the landscapes of the Gobi desert to demonstrate the different types of distribution of weathering products.[193] In Section I of Figure 27, alluvial material with a $CaCO_3$ content covers the elevated points of the watershed, while sediments containing sulfates and chlorides occur in the depressions or in the saline lakes. The author explains that this is typical not only of the Mongolian regions of the Gobi, but also of several districts of Armenia, the Arab Peninsula, Chile, Western Australia, and the Kalahari desert in Africa.

Section II in Figure 27, siallitic orthoeluvia, which was discovered by the author in river terraces in Northern Mongolia, illustrates the accumulation of calcareous sediments in slopes

Table 33
THE AVERAGE CHEMISTRY OF TERRESTRIAL MATTER, CORN, AND HUMAN BODY

Chemical element	Average composition of terrestrial matter[a]	Average composition of a corn plant[b]	Average composition of a human body[c]
O	46.48	44.57	29.13
Si	27.60	1.17	0.00
Al	8.05	0.11	0.00
Fe	5.03	0.88	0.01
Ca	3.63	0.20	3.75
Na	2.72	0.10	0.40
K	2.56	0.92	0.90
Mg	2.07	0.18	0.10
H	0.145	6.26	8.40
Ti	0.696	0.00	0.00
Cl	0.095	0.14	0.40
P	0.152	0.18	2.50
C	0.149	43.70	45.00
Mn	0.116	0.00	0.00
S	0.100	0.17	0.60
F	0.030		0.14

[a] Values are given in percent, including atmosphere, lithosphere, and hydrosphere (the soil mass).
[b] Values are given in percent, dry weight.
[c] Values are given in percent dry weight.

Modified from Kovda.[213]

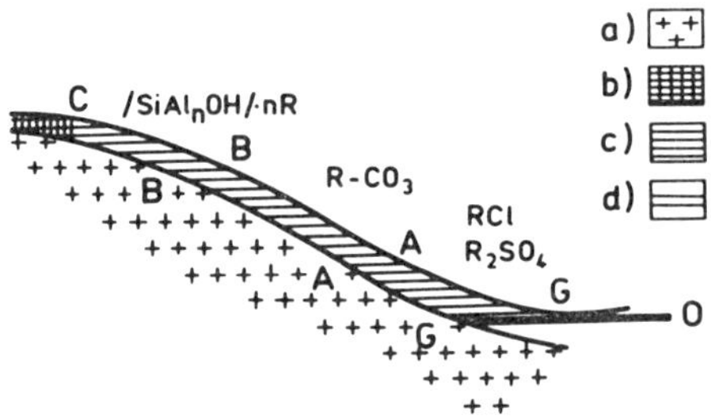

FIGURE 26. Distribution of continental sediments: (a) massive primary rock, (b) allitic eluviate, (c) eluvial overlying layer, (d) level of groundwater table. G-O = level of catchment area, AA-GG = region of sulfate-chloridic salinization, BB-AA = region of carbonate accumulation, C-BB = region of siallitic type of accumulation. (After Kovda.[3])

overlying the sulfates and chlorides which can be found in the depressions. This type is very characteristic in the environment of saline lakes, and in desert, and semidesert regions.

In Section III of the figure, the accumulation of the salts associated with lateritic rocks on the most elevated places of the landscape is demonstrated. According to Polynov, this type develops not only on some regions in the Gobi, but also in Africa in the valleys of the Rivers Congo, and Zambezi, and in the Kalahari desert.[193] Here, besides the accumulation of sulfates and chlorides in depressions covered by carbonate sediments, a siallitic ring has also formed on the slope.

The geomorphology of the landscape plays a governing role in the distribution of materials in all the types mentioned. In most cases water is the main carrier, both on the surface of the soil and underground, transporting the salts toward the places where they will accumulate. It is also mainly water which removes the salts after the weathering processes from other territories, leaching the soils and sediments in this manner.

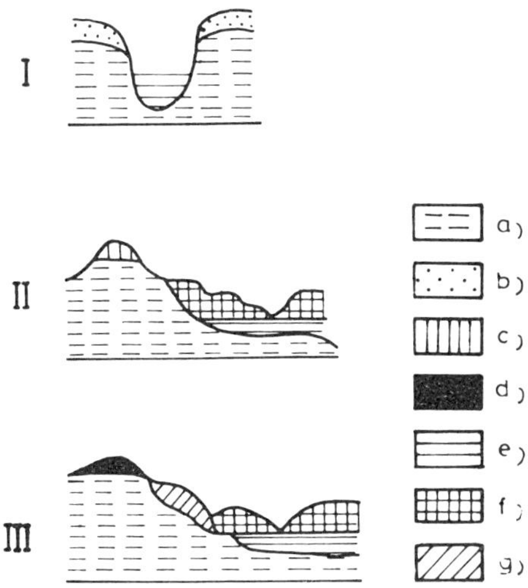

FIGURE 27. Distribution types of weathering products in saline depressions: (a) primary rocks, (b) calcareous eluvial rocks, (c) siallitic rocks, (d) allitic rocks, (e) chloride-sulfate sediments, (f) carbonates, (g) siallitic sediments. (After Polynov.[193])

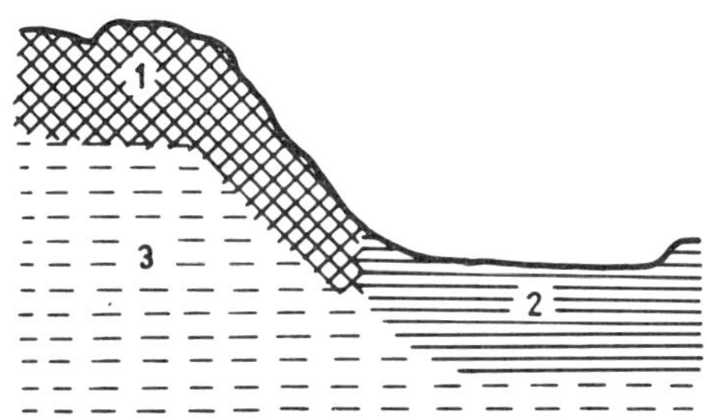

FIGURE 28. Sodic salinization of soils in the tropics: (1) allitized red crusts formed by weathering with kaolinite, gibbsite, and hydrargillite; (2) montmorillonitic argillaceous soils with features of sodic salinity and with accumulation of $CaCO_3$; (3) basalts, lavas, ashes. (After Kovda.[213])

Kovda describes some regular features of the mass flow related to sodic salinization in the tropics (Figure 28)[213] and even in moderate climatic conditions also mentioned by Bazilevich.[214] Figure 28 demonstrates that under tropical conditions soil alkalization is dominant, due to the direct effect of sodium carbonate. It should be noted that the base rock in the given figure is basalt, which is a very frequent type of base rock under alkali soils both under tropical, subtropical, and even moderate climatic conditions.[214]

Similar conditions are reflected in Figure 29. The enrichment of soda-saline soils with montmorillonite takes place due to the metamorphism of clay minerals in the alkaline medium, with silica present in the soil groundwaters. Minerals with a relatively low silica content are subject to silication under these conditions. Kaolinite or illite are thus transformed into vermiculite or montmorillonite. The possibility of such a transformation was experimentally proved by Jackson,[215] who observed this process particularly in calcareous humic gley soils. Kornbljum obtained reliable proof of the transformation of kaolinite-illite into illite-vermiculite-montmorillonite in the soils of the Volga-Akhtuba flood plain.[211] The transformation is achieved through losses of potassium by the illite layers, through hydration, silication, and, to a lesser extent, ferritization. Based on an analogy of this case, the transformation of kaolinite into montmorillonite when an additional layer of silica-oxygen tetrahedrons combines with the two layers of kaolinite can be regarded as a quite feasible path for the process of montmorillonitization.

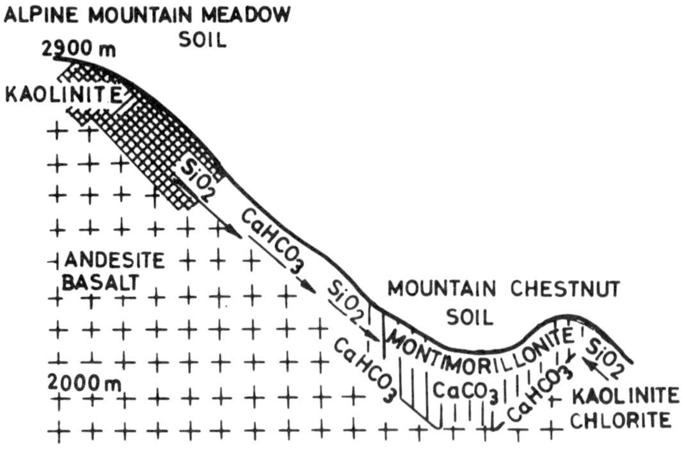

FIGURE 29. Scheme of soil clay fraction montmorillonitization in accumulative mountain landscape. (Modified from Kovda.[3])

Concepts developed by Jackson[215] on the genetic interrelations of clay minerals in watershed soils and in those of subdued depressions are of great interest. In all the climatic zones, according to the theories of both Jackson[215] and Polynov,[193] the most characteristic feature of weathering and soil formation in watersheds is desilication leading to the allitization of the eluvium. The liberated sesquioxides remain *in situ*, while silica is leached partly to the illuvial horizons and partly to depressions due to the groundwater flow. In depressions, where micas, kaolinite, or Al oxides are primarily present, the interaction of silica with these minerals leads to the formation of montmorillonite. A similar process is suggested by Jackson for the soil illuvial horizons. The montmorillonitization of soils on the peripheries of slopes and depressions was established by Edelman in Indonesian ash soils as early as the forties.[216]

Kovda[213] describes various landscapes where the geochemical flow leads to the formation of salt-affected soils in the U.S.S.R. and other countries. Based on his grouping and other studies, the following types of landscapes can be listed as the geochemical regions of salt accumulation:

A. Salt Domes

Particularly in arid and semiarid areas, salt domes may occur near the surface. One example of this phenomenon is the Caspian Lowland, where domes consisting mainly of sodium chloride occur and saline soils dominated by this salt have developed. A similar phenomenon can be found in the Central Asian republics of the U.S.S.R. and in Siberia in the basins of the Rivers Enisei and Lena. Such domes form, as a rule, near salt mines, as in the valley of the River Arak in the Caucasus, in the Pamir (Iran), and in other countries where NaCl is present in large quantities. These formations arose in various geological epochs. They may be of considerable size, sometimes measuring as much as 60 to 100 km^2.

B. Salt Accumulation of Maritime Origin

This group includes old, young, and recent maritime NaCl formations. Such saline soils occur in many countries, not only in the dry regions, but under practically all climatic conditions on sea shores, lagoons, etc. Several soil classification systems distinguish maritime saline soils whose chemistry and geochemistry are dominated by the influence of NaCl. As in the type described above, saline or solonchak soils develop in these areas.[217]

C. The Deluvial Type of Salt Accumulation

Salt-affected soils frequently occur on the slopes of mountains particularly if soluble salts

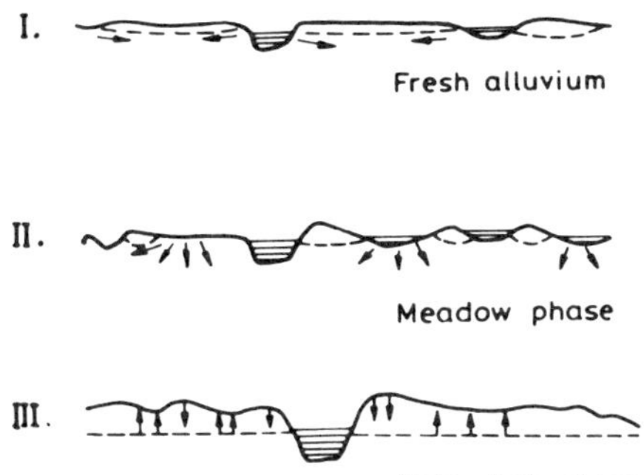

FIGURE 30. Development of deltas and salt accumulation. (- - -) Water table; arrows indicate the direction of salt movement. (After Kovda.[8])

form in considerable quantities during weathering and are then transported to lower-lying areas. Many authors have described this phenomenon, e.g., Kovda in Central Asia, Armenia, and Azerbaidzhan;[3] Filipovski and Ciric in Yugoslavia.[218] If deluvial salt-affected soils are to form, some barrier is necessary across the slope which prevents or hinders the removal of salts. This barrier may be formed by the specific shape of the slope itself or by impermeable subsoil layers. As a rule, the soils belonging to this group do not occur in big massives but rather spotwise. According to the chemical composition of the weathering products and waters, different types of salt-affected soils can develop. In Italy, for instance, in Volterra near Florence, a typical solonetz can be found in consequence of the phenomenon described above. The deluvial type of salt-affected soils may be found under practically any climatic conditions.

D. Salt Accumulation in River Deltas in Steppe and Desert Regions

There are vast territories covered by salt-affected soils in and near river deltas in deserts and semideserts as well as in some steppe regions.[219] It should be noted that in this case the term salt-affected soils does not include acid sulfate soils. Acid sulfate soils do occur in or near deltas in many regions, under the most diverse climatic conditions, for instance in humid tropics and in areas near the arctic circle, etc. The landscape geochemistry of the formation of acid sulfate soils is, however, completely different from other salt-affected soils discussed in this chapter.

Kovda describes three phases of delta development from the point of view of salt accumulation.[8] In Figure 30 this development is demonstrated.

During the formation and development of river deltas, both periodical and permanent changes take place in the base level of erosion. As a rule, this level gradually sinks, resulting in changes not only in the topography, but also in the distribution of waterlogged spots and dry places. The migration of water and salts is also altered due to this process.

During Phase I, young alluvium is formed, temporarily or permanently waterlogged spots develop, and the water table is close to the surface. Practically no salt accumulation takes place in this phase, because the salt content of the groundwater is nearly the same as that of the river water, which, under such conditions, is always fresh.

During Phase II, the topography of the delta changes. A certain differentiation and polarization of the waterlogged spots and the small "islands" of the developing delta ensues. The formation of small meanders and swamps begins. The formerly homogeneous alluvium changes into light and heavy sediments. This process does not yet result in salinization — though, as is indicated in Figure 30, the water mainly migrates downwards and meadow soils develop.

During Phase III, the topography changes again in and near the delta, as a consequence

Table 34
SALT CONTENT OF THE WATERS IN THE VOLGA DELTA

Origin of water	Dry residue	Cl^-	SO_4^{2-}	HCO_3^-	Ca^{2+}	Mg^{2+}	Na^+
Volga	310	20	80	80	60	10	20
Groundwater in delta	4910	320					

Note: Values given are ppm.

According to Kovda.[3]

of which the flow of water decreases. The stagnant water in the small lakes and swamps gradually makes contact with the groundwater, which also has rather poor outflow. Because of the lack of outflow, the salt concentration of the waters increases, the saline groundwater reaches the surface of the soil through capillary action and solonchak soils develop.

This phenomenon frequently occurs not only in river deltas in arid and semiarid regions but also in steppe areas. For instance, the River Volga can be characterized as nonsaline, according to the chemical composition of its water, before it reaches its delta near Astrakhan. In the delta, however, the groundwater is saline (Table 34).

E. Salt Accumulation in Dry Deltas in Dry Areas

In deserts and semideserts, dry deltas are a very common feature of the landscape. The formation of dry deltas is a consequence of former or recent river activity, and it occurs in the form of an alluvial fan. However, there are certain differences between the salt accumulation in maritime deltas and in dry ones. The most important of these differences is that while maritime deltas transport at least part of the dissolved salts into the sea or ocean, dry deltas broadcast the salts into the land surrounding them. In the alluvial fan, the granulometrically smallest particles occupy the most distant places. In the case of a maritime delta, all these particles are transported to the sea. From this it follows, that in maritime deltas close to the outlet there is a certain leaching zone, completely missing from dry deltas.

Rivers forming dry deltas in dry regions have a hydrological periodicity, but in ancient dry deltas the former rivers can sometimes hardly be traced, or only through a study of the alluvial fan, the stratification of the sediments, etc.

Where a dry delta is permanently or periodically supplied with water, its flow can be divided into three zones:

1. Formation zone
2. Transit zone
3. Evaporation and salt accumulation zone

A tremendous amount of salts can accumulate yearly in a dry delta. For instance, according to Kovda,[3,213] 264,000 tons of salt accumulate every year in the dry delta of the River Soch, while according to Vasilev,[3] this figure is 655,000 tons for the River Murgab.

In Figure 31, the similarities and differences observed in salt accumulations of maritime and dry delta origin are schematically represented. The figure shows that in dry deltas practically the whole salt content of the flow remains in the soil, while in other deltas certain parts of the dissolved materials flow into the sea or ocean. Salt accumulation is consequently much more intensive in dry deltas.

F. Salt Accumulation in Alluvial Plains and River Terraces

There are salt-affected soils on all continents in alluvial plains and river terraces. This

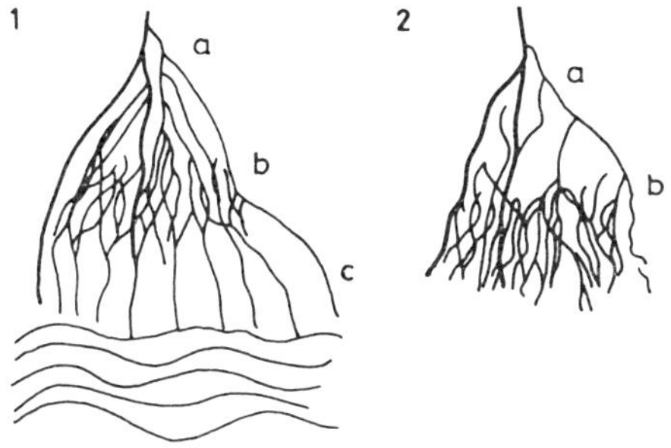

FIGURE 31. Salt accumulation in maritime (1) and dry (2) river deltas: (a) alluvium, (b) salt accumulation, (c) leaching. (Modified from Kovda.[3])

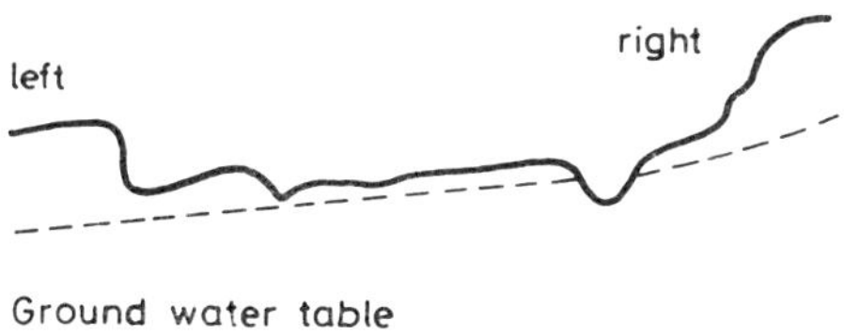

FIGURE 32. Scheme of left and right hand terraces of rivers in the northern hemisphere.

type of salt accumulation is probably the most common in the world. In the majority of literature the papers and books deal with this type of salinization and alkalization of soils. It is rational to discuss the problems of salt accumulation in alluvial plains and in river terraces together, because it is often difficult to divide them sharply. Genetically, the main regularities of salt accumulation are similar in alluvial plains and river terraces.

Salt-affected soils occur in alluvial plains under very diverse natural conditions. For instance, vast salt-affected alluvial plains can be found in Yakutia, above the arctic circle under permafrost conditions, and nearly the same type of salinization and alkalization appears near the equator in the humid tropics of Puerto Rico or Peru. Practically all the salt-affected soils in Hungary belong to this type, as well as the greatest part of other European salt-affected soils. Most of the alluvial formations of the big rivers of the earth, e.g., the Nile, the Volga, the Danube, the Lena, the Congo, the Mississippi, the Murray, the Rio Grande, the Ganges, the Tigris, the Euphrates, etc., are more or less salt-affected. This type of salt accumulation is important not only theoretically, but also practically. The methods for the utilization of salt-affected soils have mainly been elaborated and used under such conditions. It is also in the alluvial plains and river terraces where the secondary salinization and alkalization of irrigated soils are most frequent.[220,221]

In consequence of the great diversity of the salinization processes in alluvial plains and river terraces, the chemical composition of the accumulated salts may be quite different, as may the types of salt-affected soils in these areas. Solonchak soils dominated by sodium chloride or sulfate, or solonchak soils dominated by magnesium and calcium salts are common, but a large variety of solonetz or solod soils may also develop in this type of landscape. There are few general rules of salt accumulation related to the geomorphology or the climatic conditions of river valleys. The following are the most important:

(1) In the northern hemisphere salt accumulation and the formation of salt-affected soils in river terraces develop more frequently on the left banks of rivers, than on the right banks. The cause of this phenomenon is the shape of the terraces on the right and left banks, and their different salt and water regimes. The Ber law states that the shape of the terraces along the left and right sides of rivers forms differently in consequence of the rotation of the globe. This is valid if the river flows in the north-south direction. Figure 32 shows the schematic picture of river terraces in the northern hemisphere.

In the southern hemisphere the opposite phenomenon can be observed, as the terraces of the river Murray in Australia illustrate so well.

(2) In the terraces of the rivers and in their alluvial plains, salt accumulation is more frequent in the lower and middle courses of the rivers than along the upper course. This can only partly be explained by the different climatic conditions in the different parts of the river. For instance, the River Volga has its upper course in humid regions and its lower course in arid regions. But other rivers, like the Lena in Siberia, or the Ganges in India and Bangladesh, are good examples of salt accumulation mainly along the middle course.

In river terraces, as a rule, salt accumulation is slight or does not occur at all in the first terraces. (Exceptions are irrigated lands, where secondary salinization often appears even in the first terrace.) Saline or alkali soils can mainly be found in the second terraces of rivers. Topographically they can develop both in small depressions and on the more elevated elements of the micro- or mesorelief. The area of their occurrence depends on the local salt and water movements and on the results of the processes of periodical leaching and salt accumulation. Migrating solutions can sometimes concentrate in depressions and sometimes on the higher places encircling the depressions.

As in the various alluvial plains and river terraces, a very diverse chemistry of salts can be found and different types of salt-affected soils may develop. The local environmental conditions determine the formation and chemical composition of the soils. Depending on the river water and groundwater chemistry, the climatic, and lithological conditions, and the topography and shape of the terraces, solonchak or solonetz soils may develop.

Many types of salt-affected soils occur in the areas described above, however, the sodic types of soils should be specially mentioned. Most of these types may be found in alluvial plains and river terraces, as is the situation in Hungary. Mainly the same types dominate in piedmont regions, on the proluvial-alluvial plain of Yerevan, in the basin of the River Lena, in the valley of the Rio de la Plata in South America, in the valley of the river Murray in Australia, and in many other places.

G. Salt Lakes, Salt Marshes, and Swamps

This type of salt accumulation is related to the previous type, particularly when lakes and swamps are concerned. Temporarily or periodically waterlogged formations are frequent in the river valleys and alluvial plains. Salt lakes and swamps may transform into saline soils, but saline or alkali soils may also be temporarily waterlogged. The occurrence of saline soils is most common around salt lakes and swamps. Salt marshes are related to the salt accumulation types described in parts B, D, and E. Salt lakes, marshes, and swamps are located on all continents, particularly in arid or semiarid regions. There are some very large salt lakes, e.g., Lake Balkhash in the U.S.S.R. or the Great Salt Lake in Utah, U.S. There are small salt lakes in many deserts and alluvial plains, as well as in river valleys. Their formation depends mainly on the hydrological conditions; both climate and hydrology influence their development or their desiccation. While saline ground is under water, it can hardly be named soil. The environment of saline lakes and swamps is, as a rule, covered by salt-affected soils. Often, these lakes and swamps desiccate partly or wholly and their saline bottoms appear.[222]

The chemical composition of salts in the formations described above can be diverse, depending on the local environmental conditions. There are lakes where $NaCl$ or Na_2SO_4 dominates, whereas in others Na_2CO_3 prevails.

Partial or total desiccation often takes place in the alluvial plains. This is currently happening in the Baraba plain in Siberia and the process was partly characteristic of the Hungarian Plain until the greatest part of it, the valley of the River Tisza, was ameliorated and desiccated.

Some of the salt or saline lakes and swamps may have another origin. They can often be found in tectonic valleys, like Lake Fertö in Austria and Hungary, or may be of volcanic

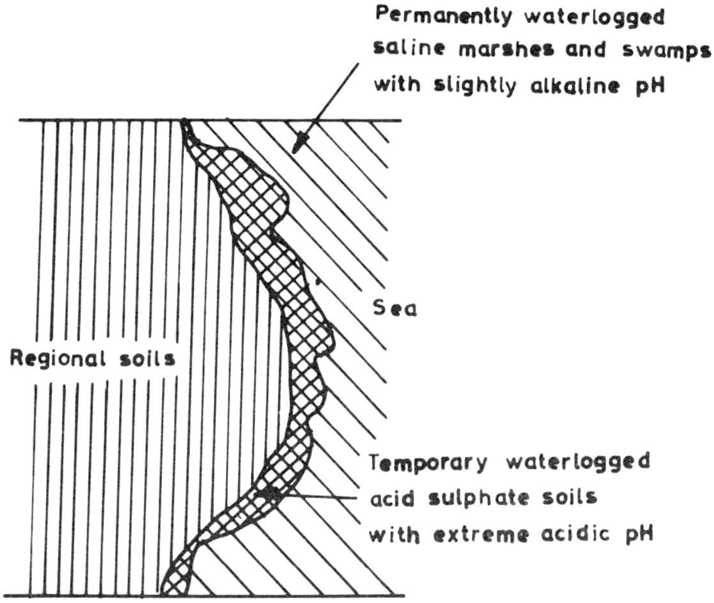

FIGURE 33. Scheme of distribution of acid sulfate soils on seashores.

origin. Many authors share the opinion that artesian waters can also contribute, directly or indirectly, to the formation of saline lakes.[16]

Although the types of landscapes discussed above do not present a full picture of all the possibilities of soil salinization and/or alkalization, the formation of saline and alkali soils is, however, related to one or more of these types in most cases.[223]

The study of environmental conditions and the quantitative description of salinization and alkalization in a given area lead to the elaboration of up-to-date, economic methods for the utilization and reclamation of salt-affected soils.

H. Accumulation of Sulfates in Acid Sulfate Soils

On the coastlines of all continents wherever sulfur-containing sediments accumulate in tidal marshes, swamps, or mangrove ecosystems (either from pyrite and related minerals or from organic material of both) the slightly alkaline parent material for acid sulfate soil formation is present. When affected by oxygen during temporary or permanent desiccation of the surface, the bivalent of sulfur oxidizes and iron and/or aluminum sulfates form. This process results in an extremely acidic pH. The acid sulfate soils, widely distributed on the coasts of the continents, may have a pH of 2 to 3. This rapid oxidation of sulfuric compounds and the sharp increase in the pH not only influence the soil-forming processes, soil mineralogy, chemistry, and physics, but also constitutes serious problems in relation to the utilization of the soils. Figure 33 shows the schematic distribution of acid sulfate soils on seashores and how their formation is related to tidal marshes and waterlogging.[224]

Acid sulfate soils are not related to a specific climatic or geographical zone or region, but may occur anywhere from the tidal marshes of the arctic circle to tropical seashores, on the coasts of all the continents. New information indicates that they exist even on some coastal regions in the Antarctic. In some coastal regions, however, acid sulfate soils are particularly widespread. In Europe, these soils may be found on the coasts of the Baltic and the North Sea, as well as on some parts of the Atlantic coast. In Africa (the coastline of Senegal), on the Gulf of Guinea, the east coastline of the continent, and coastal areas of Madagascar are especially affected.[29]

Acid sulfate soils are met with on the coast of the Arabian Gulf and are widespread in

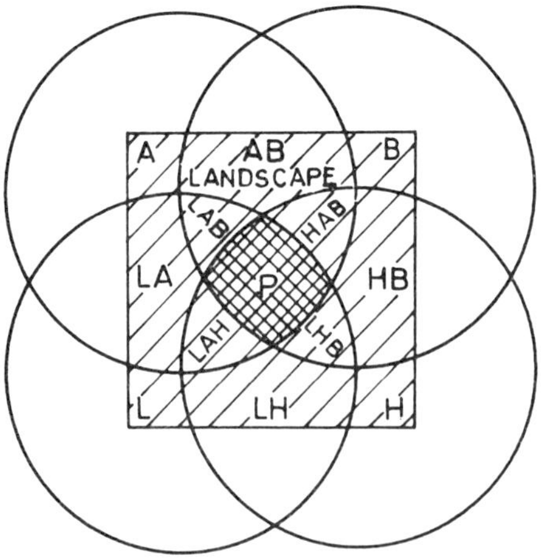

FIGURE 34. Schematic diagram showing possible interactions between the lithosphere (L), atmosphere (A), biosphere (B), and hydrosphere (H). ABLH = landscape; P = pedosphere. (After Mattson.[226])

tidal marshes on the coast of the Indian subcontinent and Sri Lanka. They are particularly common in the sea marshes and swamps of Indo-China. This area is not only the richest in acid sulfate soils, but is probably the most important with respect to their agricultural utilization.

The same is true of the seashore of southeast China.[225] Apart from these areas, acid sulfate soils can be found on the salt marshes of many islands in the Pacific. The Australian mainland and Tasmania also have acid sulfate soils on some coastal areas.

In both North and South America acid sulfate soils frequently develop in tidal marshes. They are particularly extensive in Central America where practically all the mangrove ecosystems, except for some strongly calcareous sediments, are covered with such formations.

* * *

The geochemical patterns of salt accumulation described above are the basic processes in the formation of salt-affected soils. The development of such processes is closely correlated to the different components of the ecosystems and this interrelation determines not only the pedological type of the salt-affected soils developed, but also the processes and rate of mass flow in the soil-forming processes.

The lithosphere, atmosphere, biosphere, and hydrosphere all take part in the formation of salt-affected soils due to their influence on salt accumulation and leaching, and to the interaction of ions with the soil material.

In Figure 34 a schematic diagram is given of this interaction in the landscape, with the focus on soil formation.[226] The figure shows that the pedological process is rather complex and is affected by all the major elements of the environment. This is fully valid for salt-affected soils in spite of the fact that electrolytes play a decisive role in their formation.

When the dominant geochemical processes are interpreted in a certain environment in order to disclose the regular features of soil formation, a thorough study should be carried out, including not only a pedological survey but also other important elements of the ecosystem as well.[227] In order to demonstrate a study of this type, a series of maps will be presented based on the investigations of N. Florea.[1]

The study in question was carried out on the northeastern Danube Plain where salt-affected soils are very widespread, and the environmental conditions present an excellent opportunity for the analysis of their influence on soil salinization and alkalization, and of possibilities of land utilization. The orginal maps were prepared on a scale of 1:500,000.

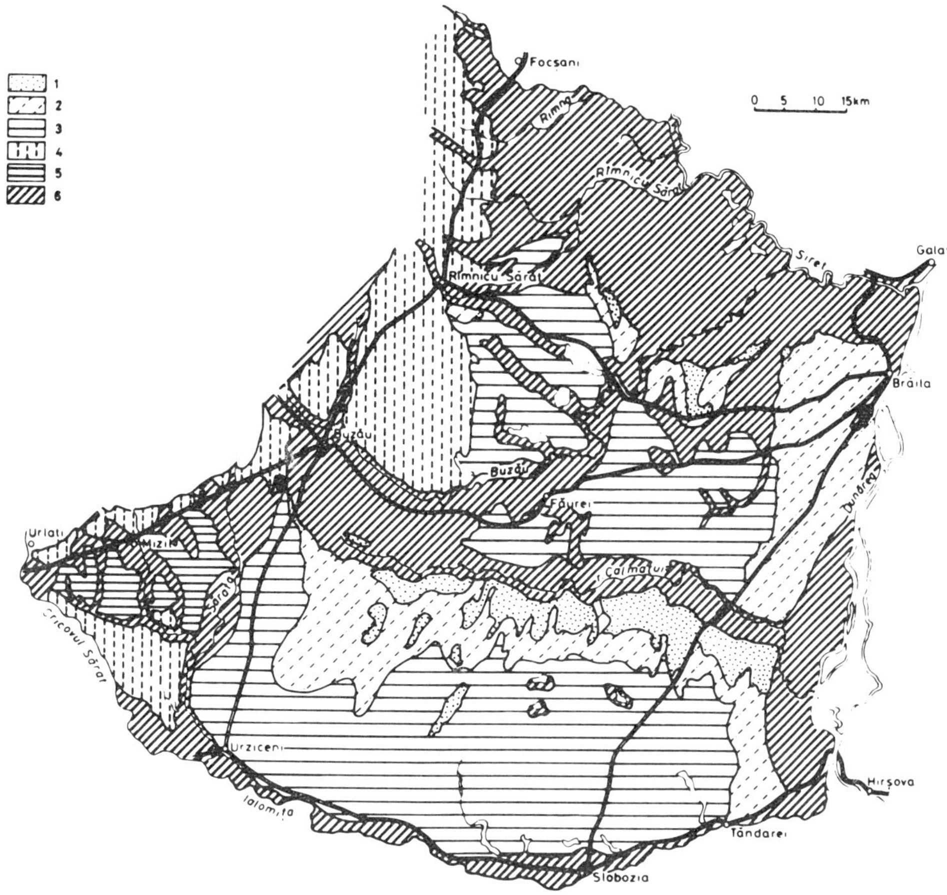

FIGURE 35. Deposits in the northeastern Romanian Danube Plain (1 to 5 are Upper Pleistocene). (1) Aeolian sandy and sandy-loam deposits; (2) sandy-loam and loam loess; (3) silty-loam loess; (4) silty-clay loam loess-like deposits; (5) clays (of different origin); (6) alluvial, alluvial-proluvial and proluvial; alluvial-lacustrine deposits (Holocene). (After Florea.[1])

Figure 35 shows the parent materials of soil formation in the northeastern Danube Plain and indicates the alluvial deposits, where the salt accumulation can be quite intensive due to the regular features of the geochemical processes, described earlier in this chapter.

The chemical composition of the surface waters of the area can be seen in Figure 36. Due to the geochemistry of the area the surface waters play an important role in salt accumulation, and their chemical type sometimes determines the direction of the soil-forming processes as well.

The influence of the groundwater is even more important whenever it surpasses the so-called "critical depth", which means it can play a direct role in the salt accumulation of the soil profile. This fact is characterized in Figure 37. The Figure indicates that in most places the groundwater table is near or above the critical depth, so that salinization and/or alkalization are actual or potential hazards. Apart from the depth of the groundwater, its mineralization is the other factor which influences salt accumulation processes in soils. The total salt content is illustrated in Figure 38, and the chemical composition of the groundwater is illustrated in Figure 39. Comparison of these three Figures (Figures 37-39) gives convincing proof that a close correlation exists between the depth and chemical composition of the groundwater, and of their total mineralization in most cases. The most hazardous ion, the sodium ion, is present in most groundwaters of the area, while calcium and magnesium ions

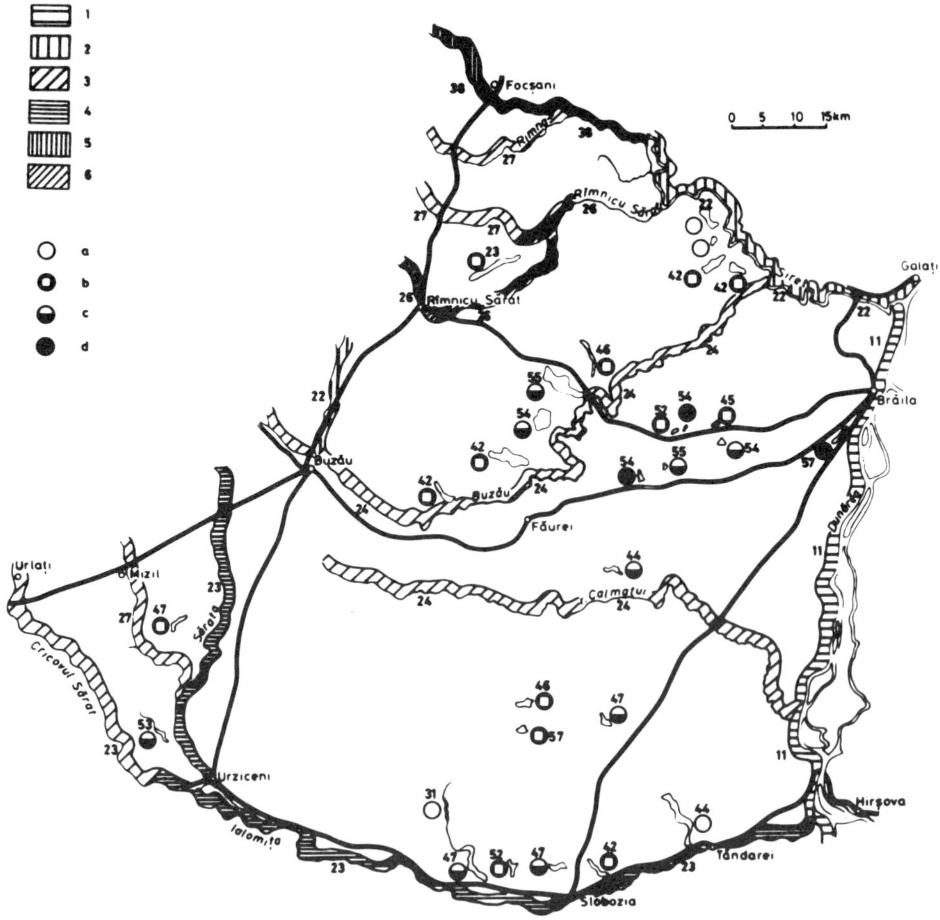

FIGURE 36. Mineralization and chemical composition of the river and lake waters of the northeastern Romanian Danube Plain.

Chemical composition: Cations	Chemical composition: Anions	Mineralization (g/ℓ)
River Waters		
1. Ca^{2+}, Mg^{2+}, Ca Mix	1. HCO_3^-, HCO_3 Mix	1. <0.4
2. Ca Mix, Na^+, Ca^{2+}	2. Cl^-, HCO_3^-, HCO_3 Mix	2. 0.4—0.9
3. Na Mix, Na^+, (Na^+, Ca^{2+})	3. HCO_3 Mix, HCO_3^-, Cl^-	3. 0.9—1.5
	4. HCO_3 Mix, Cl Mix	4. 1.5—2.0
	5. HCO_3 Mix, SO_4^{2-}, Cl^-	5. 2.0—3.0
	6. HCO_3 Mix, Cl^-	6. 3.0—4.5
	7. HCO_3 Mix, HCO_3^-, SO_4^{2-}	
	8. HCO_3 Mix, Cl^-, SO_4^{2-}	
Lake Waters		
1. Mg^{2+}, Na^+	1. HCO_3 Mix	a. 1.0—3.0
2. Na Mix	2. Cl Mix	b. 3.0—10.0
3. Na^+, Ca^{2+}	3. HCO_3^-, Cl^-	c. 10.0—35.0
4. Na^+, Mg^{2+}	4. SO_4^{2-}, Cl^-	d. >35.0
5. Na^+	5. Cl^-	
	6. SO_4 Mix	
	7. Cl^-, SO_4^{2-}	

Note: The first and second ciphers of the two-figure numbers on the map refer to the first and second columns in the key. (After Florea.[1])

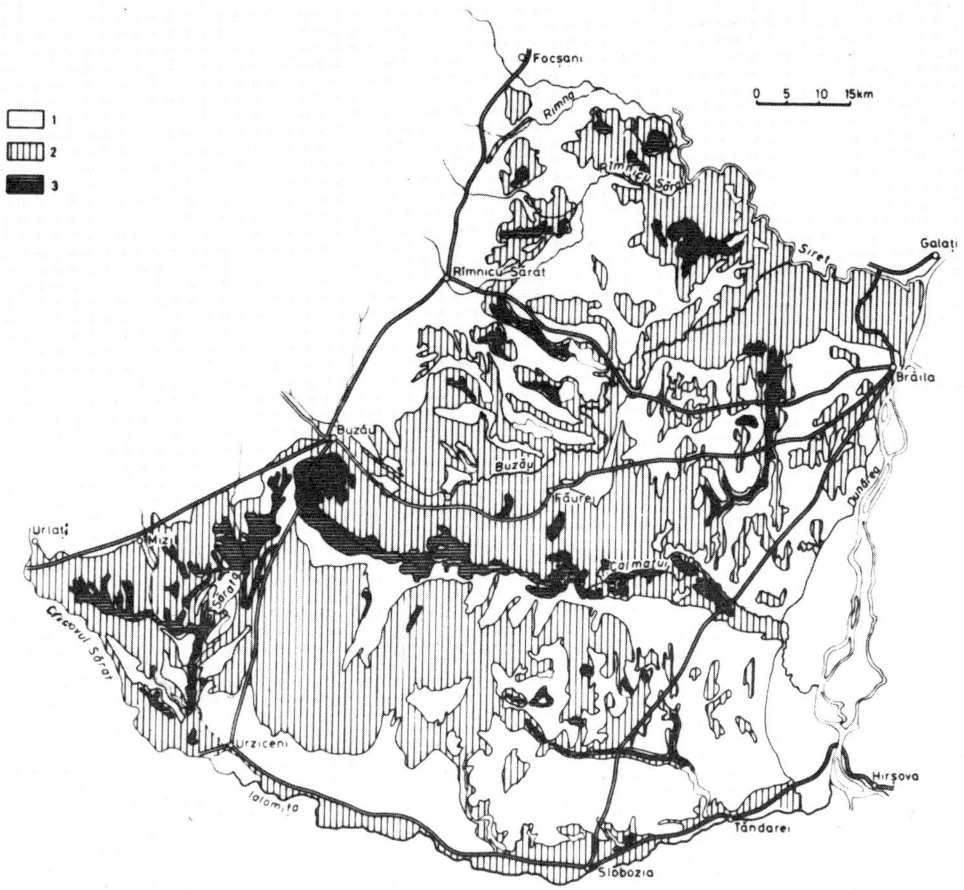

FIGURE 37. Groundwater levels in the northeastern Romanian Danube Plain: (1) below critical depth; (2) near critical depth; (3) above critical depth. (After Florea.[1])

are less widespread. Among the anions, bicarbonates, sulfates, and chlorides occur in different ratios in the groundwater, but in most cases, besides the neutral sodium salts, bicarbonates also can be found, showing the possibility that alkali soils will be formed. These are indeed found extensively in the area.

A further part of the study analyzed possibilities of regulating the hydrological and pedological conditions of the area, and the ways in which the land can be utilized. For this purpose, as shown in Figure 40, the territory was divided into zones in order to elaborate proper methods for pedo-hydrological improvements. In Figure 41, the texture of the upper soil horizons is demonstrated as the object of possible amelioration and utilization.

Based on previous maps in the series, Figure 42 shows the distribution of salt-affected soils in the area. It can be seen that both saline and alkali soils occur both alone and in combination. The map also indicates the potentially salt-affected soils, which are not salt-affected at present, but are under serious risk of salinization and/or alkalization due to the geo- and hydrochemical conditions of the area. A study of the series of maps reveals a close correlation between the different groups of salt-affected soils and the local conditions. For instance, comparing Figure 37 and Figure 42, it becomes clear that maximum salinization has developed where the groundwater table is the highest. Alkali soils develop mainly where the direct effect of groundwater with low alkalinity may be detected.

Finally, Figure 43 contains recommendations on the improvement of the hydro-saline regime of the area, including advice for the construction of drainage systems and groundwater control.

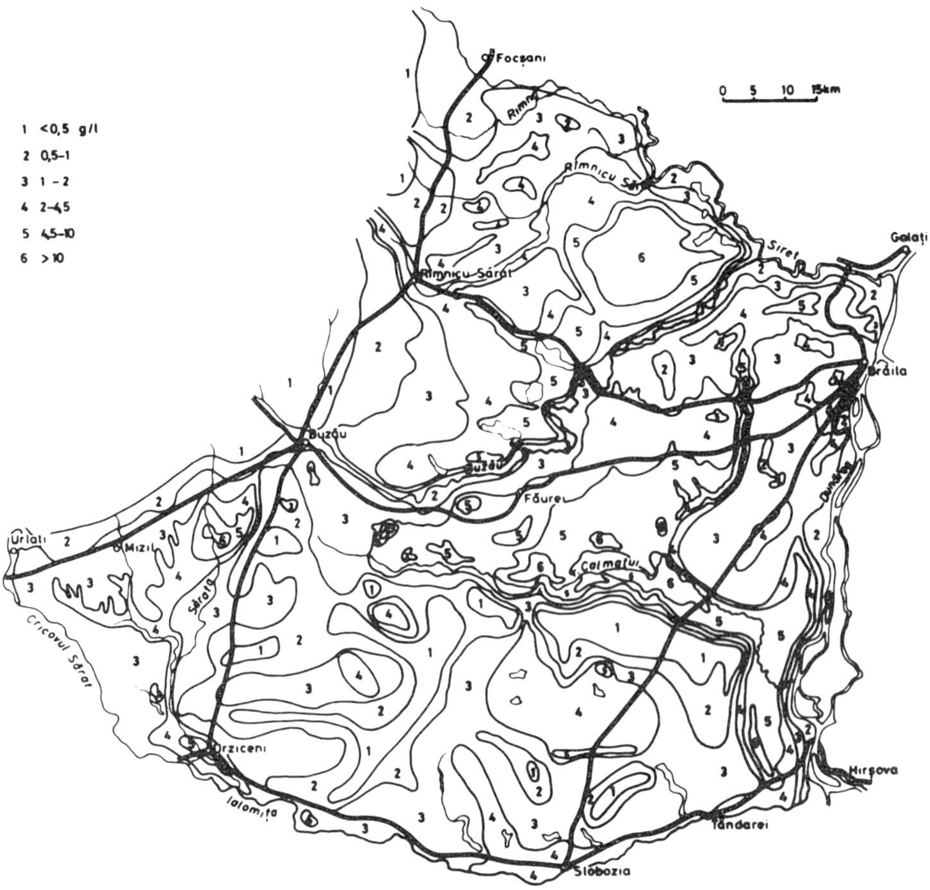

FIGURE 38. Groundwater mineralization (g/ℓ) in the northeastern Romanian Danube Plain. (1) <0.5; (2) 0.5 to 1.0; (3) 1.0 to 2.0; (4) 2.0 to 4.5; (5) 4.5 to 10.0; (6) >10.0. (After Florea.[1])

The model described above can be methodically applied in order to interpret the correlation of local geo-, hydro-, and pedochemical conditions with soil salinity problems, and to determine ways of combating them.

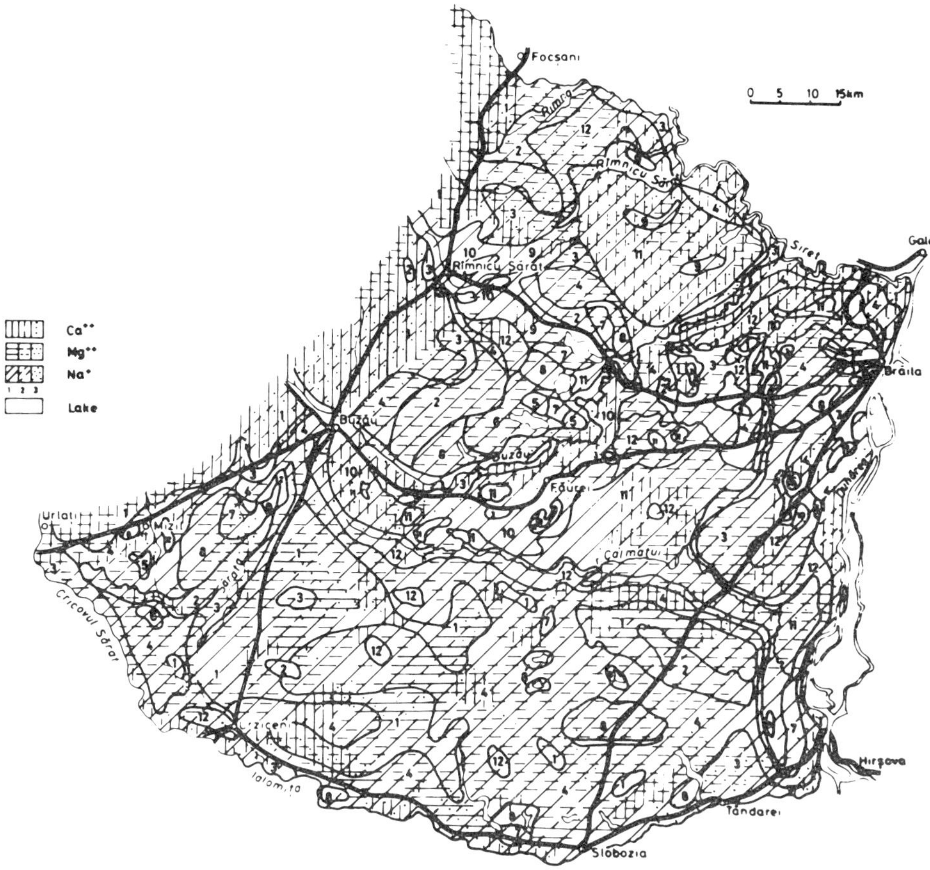

FIGURE 39. Chemical composition of the groundwaters in the northeastern Romanian Danube Plain. (After Florea.[1])

Anion composition		Cation composition	
1. HCO_3^-	7. Cl^-, SO_4^{2-}	Ca^{2+}	1. Dominant cation
2. SO_4^{2-}, HCO_3^-	8. Cl^-, HCO_3^-, SO_4^{2-}	Mg^{2+}	2. Second cation
3. Cl^-, HCO_3^-	9. Cl^-	Na^+	3. Third cation
4. Cl^-, SO_4^{2-}, HCO_3^-	10. HCO_3^-, Cl^-		
5. SO_4^{2-}	11. SO_4^{2-}, Cl^-		
6. HCO_3^-, SO_4^{2-}	12. SO_4^{2-}, HCO_3^-, Cl^-		

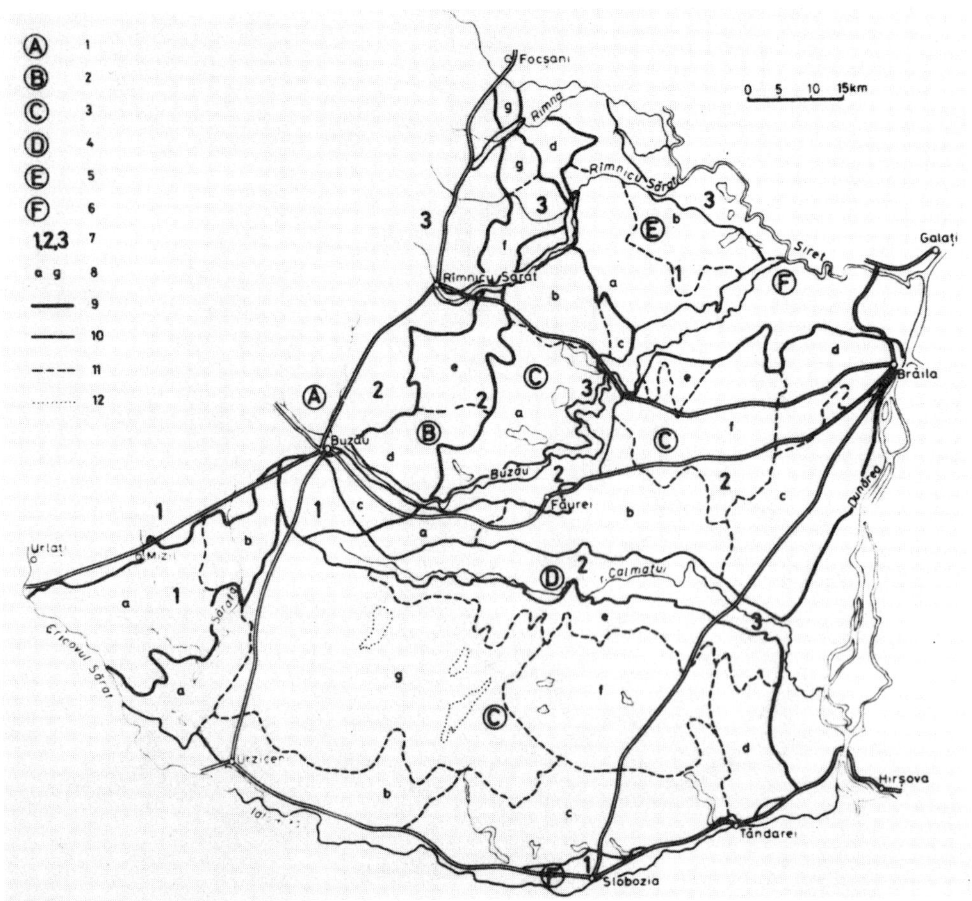

FIGURE 40. The zoning of the northeastern Romanian Danube Plain for pedohydrological improvement purposes: (1) subzone of the groundwater supply; (2) subzone of the groundwater rise; (3) subzone of the groundwater lowering; (4) Calmatui Valley; (5) Lower Sireth Plain; (6) flood-plains of the main rivers; (7) zoning units; (8) zoning subunits; (9) limit between main zoning units; (10) limit between zoning units; (11) limit between zoning subunits; (12) single areas with wind undulating relief. (After Florea.[1])

FIGURE 41. The texture of the upper soil horizon in the northeastern Romanian Danube Plain. (1) Sand; sand/loamy sand; loamy sand/light loam. (2) Loamy sand/medium loam; light loam; light loam/medium loam; light loam/clay. (3) Medium loam; medium loam/heavy loam; medium loam/clay. (4) Heavy loam; heavy loam/clay; clay. (5) Various textures. (After Florea.[1])

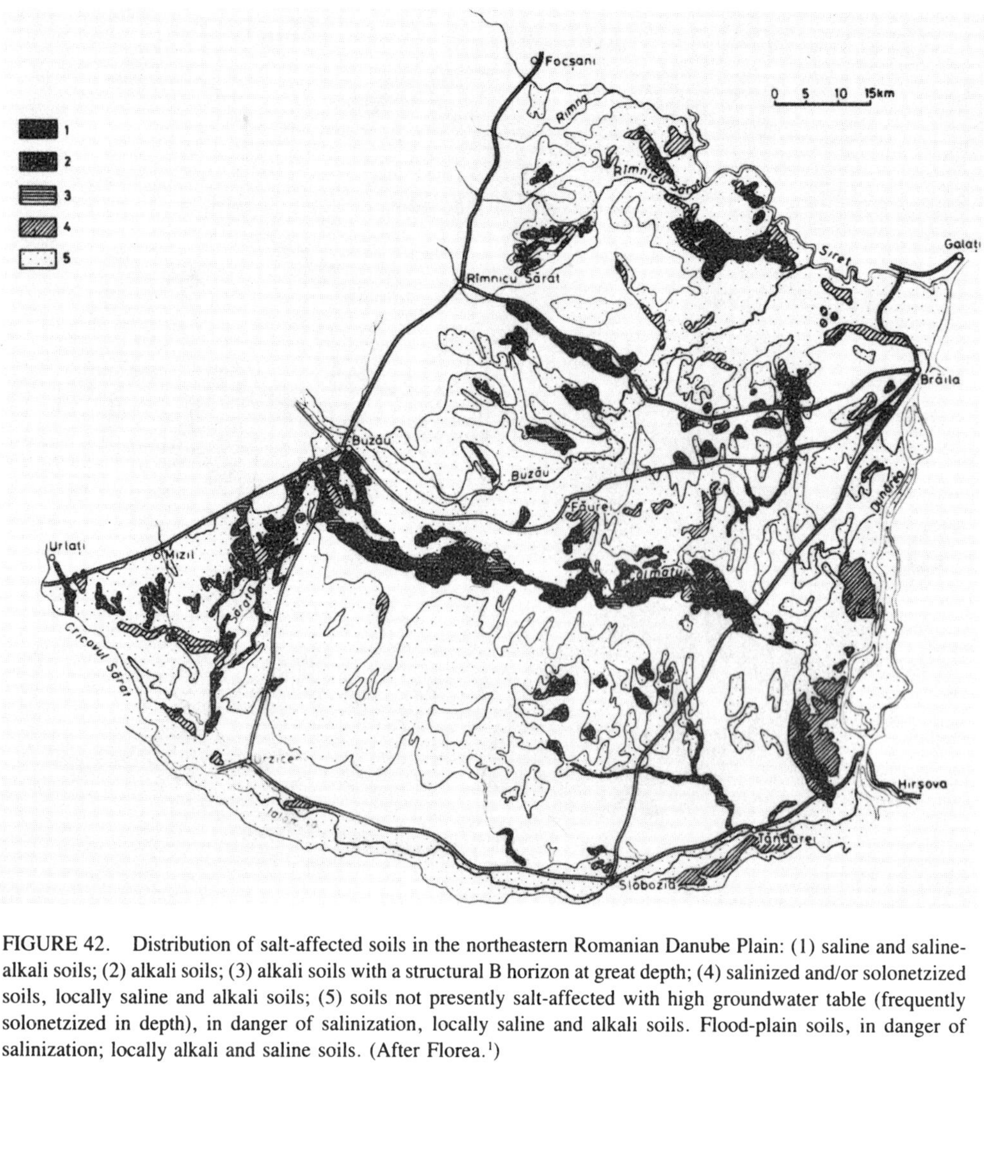

FIGURE 42. Distribution of salt-affected soils in the northeastern Romanian Danube Plain: (1) saline and saline-alkali soils; (2) alkali soils; (3) alkali soils with a structural B horizon at great depth; (4) salinized and/or solonetzized soils, locally saline and alkali soils; (5) soils not presently salt-affected with high groundwater table (frequently solonetzized in depth), in danger of salinization, locally saline and alkali soils. Flood-plain soils, in danger of salinization; locally alkali and saline soils. (After Florea.[1])

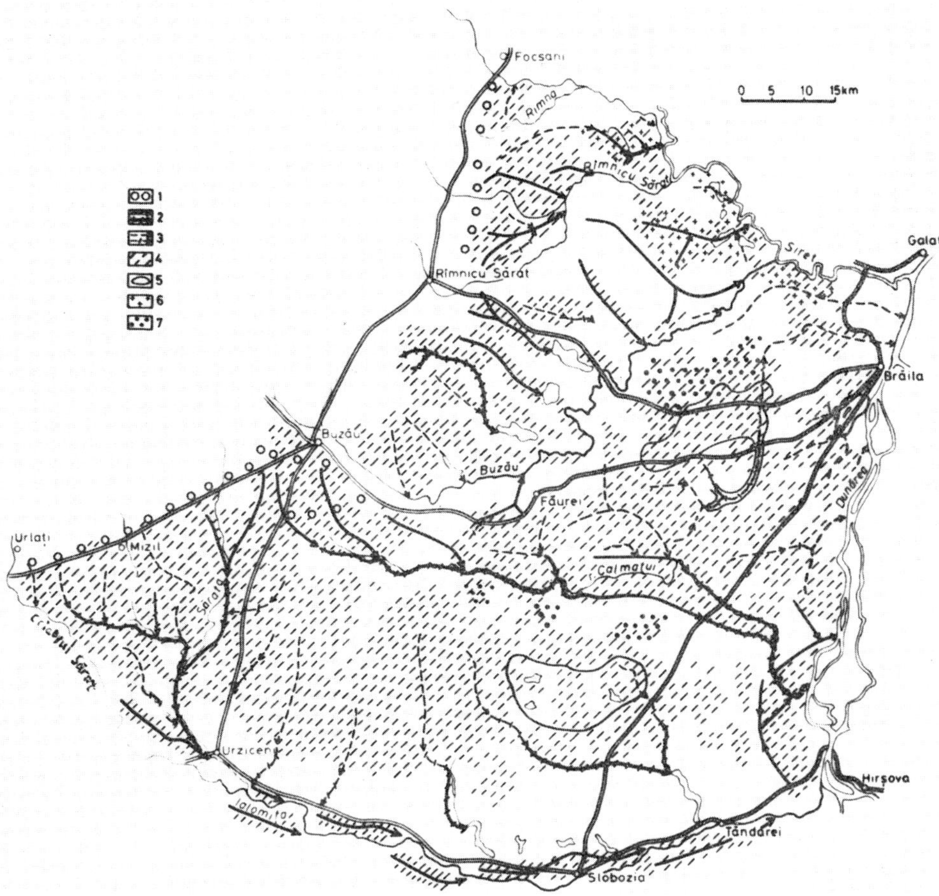

FIGURE 43. Measures proclaimed for the improvement of the hydrosaline regimes of soils in the northeastern Romanian Danube Plain: (1) drainage well (vertical drainage); (2) main drainage channels, in cases where irrigation is not employed; (3) main drainage channels in cases where irrigation is employed. (4) areas requiring drainage through secondary channels; (5) areas with saline lakes and closed valleys acting as natural drains; (6) areas which may be covered with alluvia (in the flood plains); (7) areas with spots of salt-affected soils which may be covered by leveling (through decapation of dune crests within a sandy relief). (After Florea.[1])

Chapter 3

THE BEHAVIOR AND EFFECT OF THE PRINCIPAL TYPES OF SALTS AND THEIR COMPOUNDS IN THE FORMATION OF SALT-AFFECTED SOILS

I. INTRODUCTION

As was mentioned in the previous chapters, several water-soluble salts and compounds may determine the processes leading to the formation of different groups and types of salt-affected soils. These salts and compounds are closely related to the environmental conditions, but they occur in nature always in mixture, in spite of the dominant occurrence of one or another. The origin of these salts may be very different, e.g., magmatic, weathering, sedimentary, surface- or groundwaters, airborne, biological, etc.[3,228]

In Table 35 the main salts influencing the formation of salt-affected soils are listed together with the regions where they occur, their origin, their solubility in water, their chemical reaction, and their effects on plants and soils. As Table 35 clearly shows, different chemical compounds lead to the formation of different groups of salt-affected soils, but these compounds are associated with very few chemical elements.

Salts leading to the formation of salt-affected soils are transported both by air and water to the places where they accumulate. These areas are closely related to the environmental conditions, as is shown in Table 36, where accumulation processes in Euro-Asia are listed. The Table shows that the main areas of salt accumulation are arid ecosystems, such as deserts and semideserts, where the largest quantity of soluble salts accumulate.[3,8,229] It is also clear that different ecosystems are characterized by specific chemical processes. The salinization of irrigated soils in the given areas is also shown in the Table, which confirms the fact that in most arid areas the salinization caused by irrigation is the major hazard.

Only a part of the salts causing salinity and alkalinity are transported from other places; the majority of them form due to epigenetic processes.[213,230] Perelman grouped the epigenetic geochemical processes into types shown in Table 37.[231] Besides listing the aqueous migrants, the processes resulting in the different types are also indicated (mainly oxidation and reduction). Table 37 contains not only the salts included in Table 35, but also other compounds such as organic acids. It can be seen from the table that a substantial transformation of different substances is caused by oxidation and reduction, which are the major processes in ecosystems in general and in soil-forming processes in particular. Comparing Table 35 with Table 37, it also becomes clear that the roles of several salts in the formation of salt-affected soils may be quite different, depending on the redox potential of the medium. As will be discussed later, sulfuric compounds may play a role in both acid and alkali soils depending on local circumstances.

Perelman also studied the occurrence of basic epigenetic processes in vertical subzones of the environment.[231] Under the shallow soil layer of the earth's crust, deeper layers of weathered crusts, deposits, and a vast area of aquifers exist, which are not only linked with the soil profile but also constitute a huge reservoir of substances which influence the soil-forming processes, often decisively. These conditions are demonstrated in Table 38, which also includes the occurrence and behavior of the main chemical types of salt accumulation, not only in the soil zones but also in the crust deposits. It is worth mentioning that the chemical types of epigenetic processes distinguished by Perelman are nearly identical to the system for grouping salt-affected soils applied by the author of this book.

Table 38 includes not only pedological aspects, but also other environmental systems in which the epigenetic processes of salt accumulation are significant, e.g., hydrosystems, oil deposits, mines, etc.

Similar investigations to that of Perelman were carried out by Kovda in relation to the

Table 35
THE VARIOUS SALTS INFLUENCING THE FORMATION OF SALT-AFFECTED SOILS

Group of salts	Occurrence	Origin	Solubility	pH in solution	Toxicity to plants	Effect on soil
Carbonates	All regions. In: soil, subsoil, groundwater, surface water, marine deposits.	Predominantly weathering products	Varies	Alkaline	Varies, depending on solubility	Varies, depending on mobility and solubility
$CaCO_3$	In: fresh river water, groundwater, soils (5 — 80%). Mostly in steppe and desert soils.	Predominantly weathering products in sedimentary rocks and from groundwater	Low, depending on: CO_3 concentration, CO_2 concentration, pH values in solution $L = 9.3 \times 10^{-9}$ $C = 9.8 \times 10^{-3}$ g/ℓ	Alkaline	No toxicity	Different form of $CaCO_3$ medium concretions, hardpan
$MgCO_3$	As for $CaCO_3$.	Predominantly weathering products	Low, but higher than $CaCO_3$ $L_{MgCO3} = 2.10^{-4}$ 1.2 g/ℓ Dolomite dissolution is poor	Alkaline	Toxicity due to alkaline hydrolysis	Rare in free form, mainly dolomite or dolomitized $CaCO_3$ concretions
Na_2CO_3	In: surface and groundwaters at mineralization of 0.5 — 3 g/ℓ; soils (mainly in the absence of gypsum), deposits.	Weathering products (in CO_2-containing water), exchange processes, sulfate reduction, from plants	Highly soluble CO_3 HCO_3 reaction	Alkaline up to pH 12	Very toxic, due to high solubility and alkaline hydrolysis	Peptization of soil, low water permeability, poor water physical properties, nonleachable
K_2CO_3	Similar to Na_2CO_3. Its occurrence in soils is very rare.					
Bicarbonates	All regions. In: soil, subsoil, groundwater, surface water, deposits.	Predominantly weathering products, sometimes formed by biological processes	Varies, depending on temperature and partial pressure of CO_2	Alkaline	No toxicity	Varies, depending on saturation, concentration and mobility

$Ca(HCO_3)_2$	All regions. In: groundwater, surface water, soil solution.	Predominantly by the dissolution of $CaCO_3$-containing minerals	Low, depending on temperature and partial pressure of CO_2	Slightly alkaline	No toxicity	Influences the calcium regime
$Mg(HCO_3)_2$	All regions. In: groundwater, surface water, soil solution together with $Ca(HCO_3)_2$.	Predominantly by the dissolution of $MgCO_3$-containing minerals	Low, but higher than $Ca(HCO_3)_2$ Depends on temperature and partial pressure of CO_2	Slightly alkaline	No toxicity	Affects the magnesium regime
$Na(HCO_3)$	All regions. In: groundwater, surface water, soil, subsoil, deposits.	Dissolution of soda, trona, and natricolite	Trona: 13 g/ℓ, natricolite: 6.9 g/ℓ	Alkaline	No toxicity	Low degree of peptization
Sulfates	In: deserts and steppes, deposits, soil, groundwater, dessicated tidal marshes.	Sometimes product of sulfide oxidation, sometimes weathering products, sometimes magmatic origin	Varies, but higher than carbonates	From strongly acidic to neutral	Different, depending on solubility and compounds	Varies, depending on compounds
$CaSO_4$	In: deserts and semidesert regions, groundwater, deposits, soils.	Sometimes weathering products, partly formed secondarily from SO_4 of magmatic origin in sediments due to the reaction: $Na_2SO_4 + CaCl_2 = CaSO_4 + 2\ NaCl$	$L_{CaSO4} + 2H_2O = 1.3 \times 10^{-4}$; $C = 2$ g/ℓ $L_{CaSO4} = 6.1 \times 10^{-5}$; $C = 1$ g/ℓ	Slightly acidic	No toxicity	Forms transparent mottles; compact layer used for soil amelioration
$MgSO_4$	In: desert and semidesert regions, saline soils, saline groundwater.	As for $CaSO_4$	High; $C = 262$ g/ℓ	Slightly acidic	Very toxic	Always accumulates in combination with other soluble salts. Reclamation by leaching
Na_2SO_4	In: desert and semidesert regions, saline groundwater, saline lakes, saline soils.	Partly weathering products, partly magmatic origin	High solubility, 280 g/ℓ (25°C). Depends very much on temperature	Nearly neutral	Two or three times less than $MgSO_4$	Accumulates together with other easily soluble salts. In warm periods dehydration. Reclamation by leaching in dry season
Al-Sulfates	All regions. In: desiccated tidal	Interaction between oxidation products of sul-	Very good	Acidic	Very toxic	Development of soil acidity, basic unsaturated

Table 35 (continued)
THE VARIOUS SALTS INFLUENCING THE FORMATION OF SALT-AFFECTED SOILS

Group of salts	Occurrence	Origin	Solubility	pH in solution	Toxicity to plants	Effect on soil
	marshes in form of Na aluminum- and Mg aluminum sulfate and basic aluminum sulfate.	fides and dissolution of silicate minerals				organo mineral complexes increasing the aggregation of clay particles
Fe-Sulfates	All regions. In: desiccated tidal marshes in form of K-, Na-, H^+ jarosite.	Products of sulfide oxidation	From poor to medium, depending on pH and type of Fe-sulfate	Acidic	Toxic depending on Fe^{++}: Fe^{+++}ion ratio in solution and on the pH of the media	Indirect, due to the acid production during pyrite oxidation
Sulfides	All regions as minerals in igneous and metamorphic rocks, in sediments. In: tidal seashores, lake bottoms, waterlogged depressions.	Accessory minerals of igneous and metamorphic rocks forming in a reduction phase of geomorphic cycle of sedimentation	Varies	Alkaline	Toxic	Indirect, depending on pH and microbiological processes
Fe-Sulfides	All regions. In: some igneous and metamorphic rocks, in sediments at tidal seashores and lake bottoms.	Pyrite occurs as accessory mineral of igneous and metamorphic rocks; iron sulfides form during the chemical and microbiological reduction phase of sedimentation	Low solubility, 4.9 ppm in case of pyrite and marcasite; depends on forms of iron sulfides and pH of the media	Alkaline	Toxic	Indirect effect, depending on pH and microbiological processes
Na-Sulfide	All regions. In: slightly saline-alkaline moorlands and waterlogged depressions.	Chemical and microbiological reduction of S-bearing compounds	Good solubility; 154-475 g/ℓ	Alkaline	Very toxic	Takes part in the formation of Na-carbonates and of some kinds of alkali soils
Chlorides	In: deserts and semi-desert regions, sa-	Partly magmatic origin, partly weathering	High solubility	Nearly neutral or slightly acidic	High toxicity	Saline soil; Physiological effect

	line groundwater, saline lakes, saline soils, seawater, seashores, marine deposits.	products				
$CaCl_2$	Waters of saline lakes (at salinity 400—500 g/ℓ), deep-lying groundwater.	Partly magmatic origin, partly weathering products	High solubility	Slightly acidic	Toxic in high concentrations	Seldom present in soil (forms $CaCO_3$ or $CaSO_4$); only at very high salinity.
$MgCl_2$	Common in saline groundwaters, lakes, soils. Only at very high salinity.	Partly magmatic origin, partly weathering products	High solubility; 353 g/ℓ	Nearly neutral	Very toxic	Very hygroscopic when found together with $CaCl_2$. Saline soils with $CaCl_2$ and $MgCl_2$ remain humid for a long time after rain. Reclamation: intensive leaching
NaCl	Seawater, marine sediments, coastal areas, saline surface waters, groundwaters, and soils, desert and semidesert regions.	Magmatic. Only rarely weathering products	High solubility; 264 g/ℓ	Neutral	Very toxic from 1 g/ℓ	In saline soil together with Na_2SO_4 and $MgSO_4$. Amelioration: leaching if soil contains gypsum. In absence of gypsum, alkalization
Nitrates						
$NaNO_3$	In: extremely arid desert regions, extremely polluted groundwaters.	Biological	High solubility	Nearly neutral	No toxicity	Causes salinity if accumulates
$Mg(NO_3)_2$	In: extremely arid desert regions, extremely polluted groundwaters.	Biological	High solubility	Slightly acidic	No toxicity	Causes salinity if accumulates
Silicates						
Na_2SiO_3	In: surface and groundwaters mainly together with Na_2CO_3.	Weathering products (in CO_2-containing water from plants)	Highly soluble in absence of CO_2	Alkaline	Toxicity occurs at pH over 9	Peptization of soil, soil mineralogical effects

Table 36
CHARACTERISTICS OF ACCUMULATION PROCESSES IN EURASIA, IN RELATION TO NATURAL CONDITIONS

Conditions	Residual salinization of sedimentary rocks	Maximum mineralization of waters (g/ℓ)			Max. quantity of soluble salts in top horizons of solonchaks (%)	Typical compounds	Salinization of irrigated soils
		River	Ground	Lake			
Desert	Common	20—90	200—350	350—400	25—75	NaCl KNO_3 $NaNO_3$ $MgCl_2$ $MgSO_4$ $CaSO_4$ $CaCl_2$	Widespread
Semidesert	Frequent	10—30	100—150	300—350	5—8	NaCl Na_2SO_4 $CaSO_4$ $MgSO_4$	Often found
Steppes	Rare	3—7	50—100	100—250	2—3	Na_2SO_4 NaCl Na_2CO_3 $NaHCO_3$	Rarely found
Forest steppes	None	0.5—1.0	1—3	10—100	0.5—1.0	$NaHCO_3$ Na_2CO_3 Na_2SO_4 Na_2SiO_3	Unknown
Forests	None	0.1—0.2	0—1	None	None	R_2O_3 SiO_2	None

differentiation of various compounds during salt accumulation on continents.[3] This differentiation is depicted diagrammatically in Figure 44. The Figure shows that quaternary deposits and ancient residual rocks with underlying igneous rocks move down the slope in close relation to the solubility values of various salts and compounds. According to this rule, the most mobile ions (like chloride and nitrate) accumulate at the bottom of the relief, while less mobile compounds (like silicates and sesquioxides) accumulate at the top.

Evidently, the features shown in Figure 44 are only of general applicability. Under specific conditions, local processes and epigenesis will not only transform the compounds in consequence of changes in the local circumstances, but also alter the intensity of their migration.

In Table 39 the migration series for element release from silicious rocks in temperate climate are shown in oxidizing and strongly reducing environments.[231] It can be seen that the most mobile element, chloride, has very strong migrational intensity under both oxidizing and reducing conditions, while others (e.g., sulfur) behave differently in different types of environments.

Although the grouping is made for single elements or compounds, in nature several compounds always occur together. It is seldom possible to speak of an exclusive, but rather of a dominating role of one or another element or compounds in the processes of salt accumulation and in the formation of salt-affected soils. There are various grouping systems based on the ratio of different compounds in salt solutions, related to their dominating role in the soil-forming processes. One of such systems, compiled by Borovski,[186] is given in Table 40. This Table refers to the most frequent cations in salt-affected soils.

As can be seen in Table 40, the ratio between the cations is very approximate when determining the type of salinization, and different authors recommend different numerical

Table 37
BASIC GEOCHEMICAL TYPES OF EPIGENETIC PROCESSES

	Aerial migrants			
	Oxidizing			
Aqueous migrants	In rocks containing reducers (O_2 oxidizing)	In rocks not containing reducers (O_2 oxidizing)	CO_2, CH_4 reducing without H_2S	Reducing with H_2S
H^+, SO_4^{2-}, Fe^{3+}, Zn^{2+}, Cu^{2+}	Sulfatic	—	—	Sulfatic, sulfidic
H^+, HCO_3^-, organic acid	Oxidation by acid waters	Acid	Carbonate-free colloidal	—
Ca^{2+}, HCO_3^-, Mg^{2+}	Oxidation by neutral weakly mineralized waters	Neutral, calcareous	Carbonatic, colloidal	—
Cl^-, SO_4^{2-}, Na^+	Oxidation by neutral strongly mineralized waters	Chlor-sulfatic	Oxysalt-bearing colloidal	Oxysalt-sulfide-bearing
Ca^{2+}, SO_4^{2-}	—	Gypsiferous	Oxysalt-bearing colloidal	
Na^+, HCO_3^-, OH^-, SiO_2	Oxidation by soda waters	Sodic	Sodic, colloidal	Soda and hydrogen sulfide

After Perelman.[231]

Table 38
THE OCCURRENCE OF BASIC TYPES OF EPIGENETIC PROCESSES

	Vertical subzones			
Epigenetic process	**Soil**	**Weathered crust**	**Continental deposits**	**Subzone of catagenesis (aquifers)**
Sulfatic	Over sulfide ore bodies	Oxidation zone of sulfur and sulfide ore deposits; over pyritic shales and clays	—	Sulfide-bearing rocks, flushed by oxygenated waters
Acidic	Meadow podzols; red, gray, brown soils of forested steppes; salt flats	Carbonate-free rocks in temperate and tropical humid climates	Deluvium in humid climates	Carbonate-free rocks in humid climates
Neutral carbonatic	In steppes; chernozems, chestnut-brown, or gray soils	In steppes and deserts	Deluvium and proluvium in arid climates	Carbonate-rich rocks (limestones, marts, loesses)
Chlor-sulfatic	Upper layers of some salt marshes in steppes and deserts	Over salt-bearing rocks, and salt deposits	Saline sediments in deserts	Salt-bearing rocks; aquifers in deserts
Gypsiferous	Ancient salt marshes on terraces	Gypsum cappings over salt bodies	—	Places where deep chlor-calcic waters mix with surficial sulfatic waters
Sodic	Salt marshes	—	Red sediments	Wastes of steppes: deep-lying sand with artesian waters
Carbonate-free colloidal	In swamps of taiga, tundra, or tropics	Plains of northern taiga and tundra	Alluvium in humid climates	Carbonate-free rocks in predominantly humid climates
Carbonate-rich colloidal	Meadow and swamp soils of northern steppes; carbonate-rich on meadows and in swamps of forest and tundra zones	—	Alluvium in forested and northern steppes	Carbonate-rich and sulfate-poor rocks
Saline colloidal	Salt marshes with weakly reducing quality	—	Muds in saline lakes with weak reducing media	Rocks in arid climates
Gypsiferous colloidal	Gypsum horizons of meadows	—	—	Aquifers in gypsum strata
Sodic colloidal	Soda-salt marshes on meadows	—	Muds of soda lakes	—
Oxysalt-sulfidic	Lower horizons of salt marshes	—	Muds of saline lakes with strong reducing media	Deep oil deposits
Sodic with H_2S	Salt marshes	—	—	Oil deposits, bituminous limestones, and shales in process of destruction flushed by weakly mineralized waters

values. This is hardly surprising, because the ratio between the cations may differ depending on the environmental conditions, even in the case of similar soil properties.

As was mentioned above, the main carrier of salts in the ecosystems is water, — its surface flow, as well as the effect of rising groundwater. However, particularly in dry areas, airborne salt accumulation can also be considerable. Furthermore, in several districts the

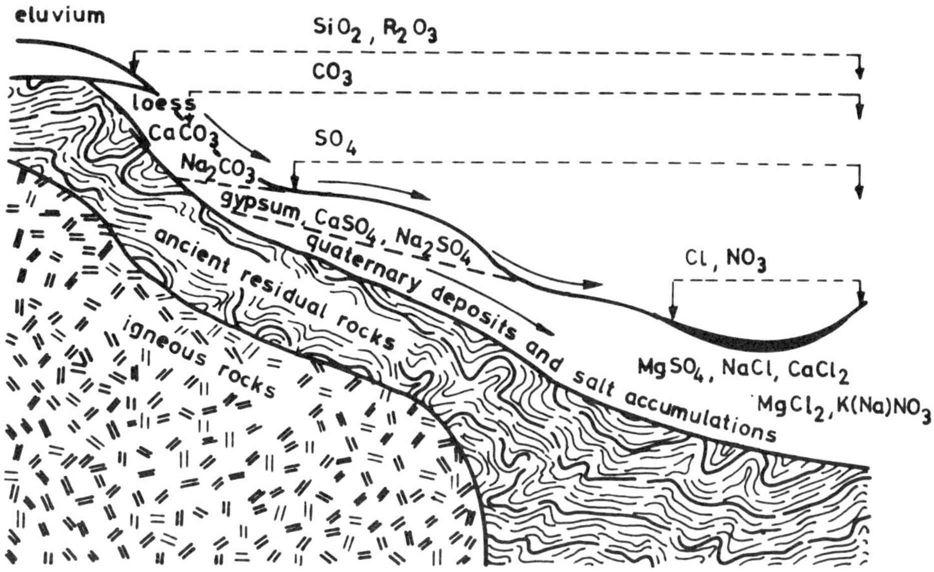

FIGURE 44. Diagram of the differentiation of compounds during salt accumulation on continents. (After Kovda.[3])

Table 39
MIGRATION SERIES FOR ELEMENTS RELEASE FROM SILICIOUS ROCKS IN TEMPERATE CLIMATES

Migrational intensity	Oxidizing environment	Strongly reducing environment
Very strong	Cl, I, Br, S	Cl, Br
Strong	Ca, Mg, Na, F, Sr, Zn, U	Ca, Mg, Na, F, Br
Medium	Co, Si, P, Cu, Ni, Mn, K	Sr, P, K
Weak and very weak	Fe, Al, Ti, Y, Th, Zr, Hf, Nb, Ta, Ru, Rh, Pd, Os, Pt, Sn	Al, Ti, Se, V, Cu, Ni, Co, Mn, Th, Zr, Hf, Nb, Ta, Ru, Rh, Pd, Os, Zn, U, Pt

After Perelman.[231]

precipitation also contains a considerable amount of water-soluble salts. Table 41 illustrates the salt input in the soils of Kazakhstan due to precipitation.[232]

Table 41 demonstrates that in areas with both high and low elevation, the salt input from precipitation is quite considerable and contributes to the salt accumulation in soils and waters. It is evident that the accumulation is more pronounced on lower parts of the slope because the salt balance of the territory is influenced strongly by surface flow and also by leaching. These features are demonstrated in Figure 45, where the joint effect of weathering and leaching is demonstrated. Figure 45 shows that the accumulation of soluble salts is directly proportional to the intensity of the weathering processes and inversely proportional to the intensity of leaching. The Figure also shows the distribution of different water-soluble products according to their solubility. Besides water-soluble salts it also contains some clay minerals and hydrous oxides.

II. BRIEF CHARACTERISTICS OF SEVERAL PRINCIPAL COMPOUNDS OF SALT-AFFECTED SOILS

A. Carbonates

Carbonic acid salts are widely found in the soils, subsoils and groundwaters of deserts,

Table 40
TYPE OF SALINIZATION DEPENDING ON CATIONS

Type	Ratio in meq of cations in 1:5 aqueous						
	Na + K/ Ca + Mg[a]	Ca/Mg[a]	Na + K/ Ca + Mg[b]	Ca/Mg[b]	Na/Mg[c]	Na/Ca[c]	Mg/Ca[c]
Sodic	>4	—	>2	—	>1	>1	—
Sodium-magnesium	1—4	<1	—	—	<1	>1	>1
Sodium-calcium	1—4	>1	—	—	>1	<1	<1
Magnesium	<1	<1	—	—	<1	—	>1
Calcium	1	1	—	—	—	—	—
Magnesium-sodium	—	—	1—2	<1	>1	>1	>1
Calcium-sodium	—	—	1—2	>1	>1	>1	<1
Calcium-magnesium	—	—	<1	<1	<1	<1	>1
Magnesium-calcium	—	—	<1	>1	<1	<1	<1

[a] Ivanova and Rozanov, 1939.
[b] Lebedev, 1951
[c] Bazilevich and Pankova, 1968.

After Borovski.[186]

Table 41
ACCUMULATION OF SALTS FROM PRECIPITATION IN SEVERAL DISTRICTS OF KAZAKHSTAN

Location	Altitude[a]	Year	Average precipitation per year (mm)	Salt content of precipitation (ppm)	pH	Total salt input (kg/ha)
Alma Ata	800 m	1967-70	654	33.9	6.3	216.1
Min — Jilki	3600 m	1964-70	836	20.5	6.0	167.2
Topoli	300 m	1965-68	135.2	78.2	6.6	105.5

[a] Above sea level.

After Borovski.[186]

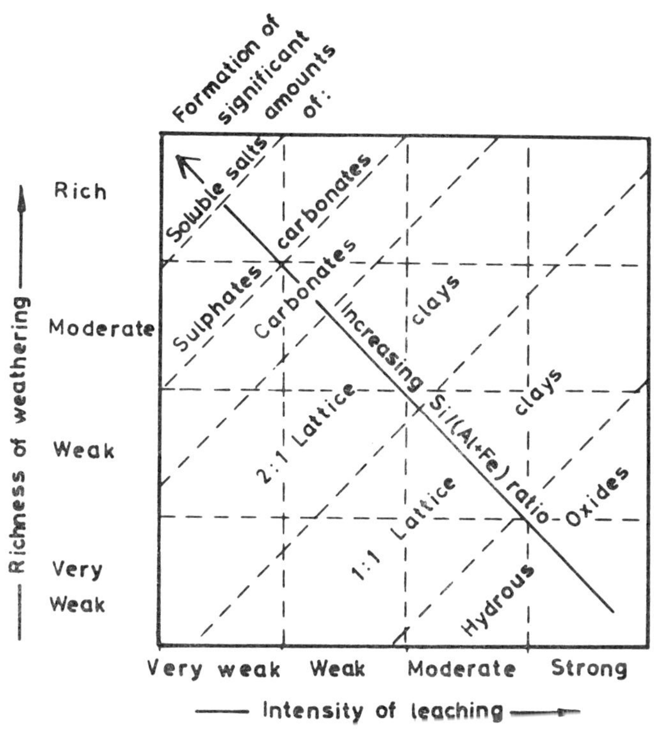

FIGURE 45. The joint action of weathering and leaching. (After Crompton.[231])

semideserts, steppes and even forest steppes. They have different effects on the soil, depending on:

- What cation the carbonate is bound to
- How much carbonate accumulates and at what depth
- The solubility of the accumulated carbonate
- What effect the carbonate has on the water balance and colloid chemical properties of the soils
- The concentration at which the carbonate becomes toxic to plants and the degree of toxicity

Table 42
CORRELATION BETWEEN CALCIUM CARBONATE SOLUBILITY, DISSOLVED CO_2, AND THE pH OF THE SOLUTION

Atmospheric CO_2 (%)	Dissolved CO_2 (mol/ℓ)	$CaCO_3$ conc. (g/ℓ)	Ca conc. (mol/ℓ)	pH of the solution
0.00	—	0.01	$1.3.\ 10^{-4}$	10.23
0.03	$1.22\ .10^{-5}$	0.06	$6.15.10^{-4}$	8.48
0.30	$1.22\ .10^{-4}$	0.14	$1.37.10^{-3}$	7.81
1.00	$4.10\ .10^{-4}$	0.21	$2.09.10^{-3}$	7.48
10.00	$4.10\ .10^{-3}$	0.47	$4.63.10^{-3}$	6.80

After Wiegner; cited in Darab.[290]

1. Calcium Carbonate

This is one of the most common inorganic mineral salts found in the soils. It may occur as pure calcium carbonate, but it is also found as calcium-magnesium carbonate. It has very low solubility in water: 9.8 ppm in the case of calcite and even lower in the case of dolomite. Since the salt is formed from a weak acid and a strong base, alkaline hydrolysis takes place in an aqueous medium, leading to the formation of $Ca(OH_2)$ and $Ca(HCO_2)$. A saturated aqueous solution of calcium carbonate has a pH of 8.3 to 8.4. The solubility depends on the carbon dioxide concentration and pH of the solution. The interaction between the three factors, CO_2 concentration, hydrogen ion concentration and calcium concentration, can be expressed by the following relationship:

$$[Ca^{++}] = \frac{K_{sp}}{K_1K_2} \cdot \frac{[H^+]^2}{[H_2CO_3]} \qquad (1)$$

where: K_{sp} = solubility product of $CaCO_3$, K_1, K_2 = first and second dissociation constants of carbonic acid, and [ ] = activity in mol/ℓ. On substituting the numerical values of the $CaCO_3$ solubility product and the first and second dissociation constants of carbonic acid, the relationship becomes:

$$\log Ca^{2+} = 8.63 - 2pH - \log(CO_2) \qquad (2)$$

It should be noted that the equation is only approximately valid, since the values of the first and second dissociation constants of carbonic acid are not constant but, as will be discussed later, depend on the ionic strength of the solution.[17,233]

The pH value and the carbon dioxide concentration, which influence the solubility of calcium carbonate, also have a mutual effect on each other. An increase in the CO_2 concentration of the solution leads to a decrease in the soil pH, thus increasing the solubility of calcium carbonate. The correlation between calcium carbonate solubility, dissolved CO_2, and the pH of the solution is illustrated in Table 42.

Dissolution is also influenced by the fact that the Ca^{++}, HCO_3^- and OH^- ions formed during dissolution are apt to form ion pairs, which means that $Ca(HCO_3)^+$ and $Ca(OH_2)^+$ ions are also present in the solution. The probability of ion pair formation can be expressed in terms of the negative logarithm of the association constant.[234]

Due to the low solubility, mobility is also low. Calcium carbonate is frequently found in soil-forming rocks and soils. In the case of soils its occurrence may be primary, i.e., the soil-forming rock itself contains calcium carbonate, or secondary, being formed during the soil-forming process.

Among the soil-forming rocks, the occurrence of calcium carbonate is chiefly associated with sedimentary rocks. The quantity found in sedimentary rocks varies according to whether chemical or biological reactions play the dominant part in the decomposition of calcium-containing rocks and in the precipitation of the $CaCO_3$ content of the solution. On dry land, when crystalline rocks of volcanic origin begin to decompose, chemical reactions and equilibria are dominant. In this case the saturating concentration of the solution flowing in the decomposed rocks is very low, and the $CaCO_3$ formed in the course of decomposition has low mobility. When plants appear, the CO_2 concentration of the solution in the decomposed rock increases, with a parallel increase in the saturation concentration of $CaCO_3$. The greater the moisture content and the more intensive the plant vegetation, the greater the dissolution and mobility of calcium carbonate.

Among the calcium carbonate-containing sedimentary rocks, those of marine origin form a special case, since the majority of the calcium carbonate in these rocks is made up of calcareous skeletons of biogenic, animal, or plant origin.

The calcium carbonate accumulating in the course of soil formation may originate from:

- The decomposition of crystalline volcanic rocks
- The calcium carbonate content of sedimentary rocks
- The calcium carbonate dissolved in natural waters
- The calcium carbonate dissolved in the groundwater

In the first two cases the process is characterized by the rearrangement of $CaCO_3$ formed *in situ* or present in the sedimentary rock, while the other two cases are characterized by the precipitation of $CaCO_3$ transported into the soil profile by water. Calcium carbonate precipitated in the course of soil formation generally has a micro-crystalline structure.[235] Due to the poor solubility, the soil solution soon becomes saturated with $CaCO_3$, so mobility is relatively low and very little is leached out of the soil.[236] Calcium carbonate is to be found in soils in every climatic zone, but it is characteristic of steppe, semiarid and arid soils.[237,238] It is present in varying quantities within the soil profile, the distribution being determined by the moisture balance of the profile.[239] In the solonetz soils of the Hungarian Plain, where current soil-forming processes are characterized by slow and usually noncontinuous leaching, the $CaCO_3$ from the relatively dilute soil solution accumulates at the bottom of the B_2 level, where the capillary zone of the groundwater gives way to the lower limit of the zone affected by rainwater (Figure 46). The calcium carbonate distribution curve for solonchak soils formed on the calcareous Danube alluvium is also indicative of the combined effects of groundwater and surface waters (Figure 47). In this case the intensive effect produced by the closeness of the groundwater and the calcareous nature of the soil-forming rock leads to an intense accumulation of $CaCO_3$ in the soil. The quantity of $CaCO_3$ in the accumulation level may approach or even exceed 40 to 50%.[240]

In zones where the low quantity of precipitation, the uneven distribution of the rainfall, the warmer climate and the consequent intense evaporation result in a substantial reduction in the leaching of salts, or even to an accumulation of salts if there is a salt source nearby, the lime accumulation associated with soil formation may be significant even on the surface.

Calcium carbonate itself is not toxic to plants. If there is intensive accumulation, however (particularly if a dense, continuous layer is formed in the soil), moisture mobility is hindered, causing a deterioration in the water management properties of the soil. If the lime accumulation layer is on or near the surface, it physically inhibits the development of the root system. The intensive accumulation of $CaCO_3$ in the soil leads to changes in the element ratios and affects the dynamics and availability of the nutrients. A lime accumulation layer close to the surface thus has an unfavorable effect on the nutrient uptake of the plants.[241]

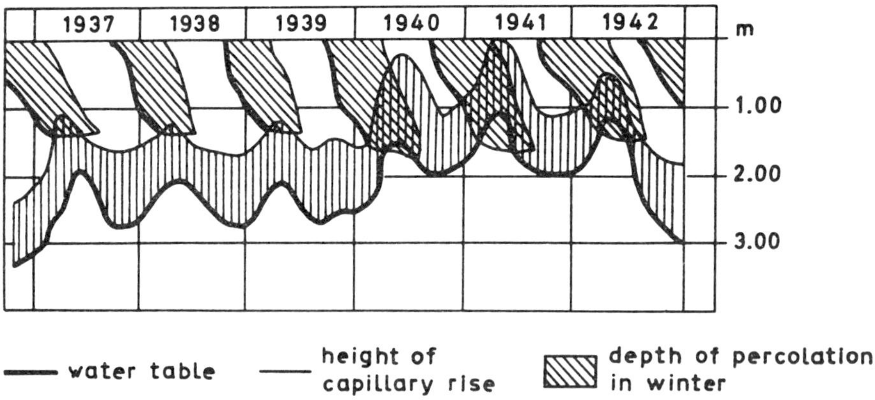

FIGURE 46. The groundwater and wetting conditions in a solonetz area in Hortobágy (Hungary), over a six-year period. (After Mados.[1])

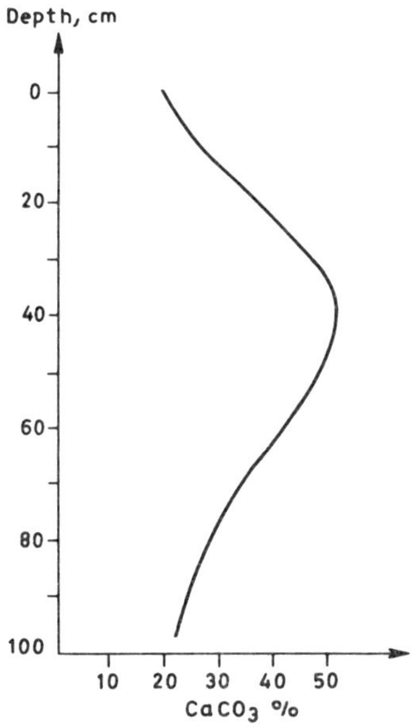

FIGURE 47. The calcium carbonate distribution curve for solonchak soils formed on the calcareous Danube alluvium (Hungary).

2. Magnesium Carbonate

This mineral is found in soil-forming rocks and soils in the form of magnesite ($MgCO_3$), dolomitized calcite and dolomite. The solubility of magnesite is moderate; that of dolomitized calcium carbonate is better than that of calcite but poorer than that of magnesite; whereas dolomite dissolves in water with great difficulty. In aqueous solution, alkaline hydrolysis takes place. A saturated aqueous solution of magnesium carbonate has a pH of around 10. The Mg^{++}, CO_3^{--}, HCO_3^-, and OH^- ions formed in the course of dissolution are capable of forming ion pairs.

Magnesium carbonates are components of sedimentary rocks and soils. The primary formation of magnesium carbonate stems from the chemical decomposition of crystalline rocks. As a decomposition product, it is transported by the solution from the decomposed rock into the sedimentary rock, where it accumulates.

As a lattice-forming element, magnesium plays an active role in clay formation, being bound to clays to a significant extent. For this reason, very little magnesium carbonate, only a few percent, is to be found in sedimentary rocks and soils. Magnesite is the rarest, while dolomite and dolomitized calcium carbonate are more frequent.

Since magnesite dissolves in water fairly, it plays little part in building up the solid skeletons of living organisms. Abiotic factors have more influence on its accumulation.

Among the various forms of magnesium carbonate, dolomite is a more frequent component of rocks and soils than magnesite. Abiotic factors play the major role in its accumulation. It precipitated in marine sediments mainly in the Pre-Cambrian period, when the biological activity of the sea was very slight and the accumulation of alkaline earth carbonates took place as a chemical process in quantities depending on the proportions of the elements in the solution and on their solubility.

As the biological activity of the sea increased, living organisms accumulated more and more of the calcium in the seawater in the form of calcium carbonate, thus giving prominence to the biotic accumulation of calcite. This in turn lead to a decrease in dolomite formation, as a result of which the magnesium concentration of the seawater increased.

Under the pressure and temperature conditions at present characteristic of the earth's surface, the formation of dolomite is only conceivable abiotically, in alkaline waters or solutions with a high salt content, e.g., in the water of Lake Balkhas (U.S.S.R.). Theoretically, the formation of dolomite and dolomitized calcite is also conceivable in the case of sodic solonchaks. A more detailed study of these soils, however, including the distribution of dolomite and calcite according to particle size, indicates that dolomite is more likely to occur in sedimentary rock than to be formed in the course of salinization. Nevertheless, in sodic soils with a strongly alkaline reaction and a high salt concentration, the possibility that dolomitized calcite or even dolomite may be formed cannot be completely excluded.

Due to the alkaline reaction of the salts, solutions of magnesium carbonates are toxic to plants. The degree of toxicity depends on the solubility of the compound.[242-244]

3. Sodium Carbonate

Sodium carbonate can be formed in nature in numerous ways. It dissolves readily in water.[245] The aqueous solution is subject to alkaline hydrolysis. The pH of the aqueous solution is alkaline, being between 9 and 11, depending on the concentration of the solution. The Na^+, CO_3^{2-}, HCO_3^-, and OH^- ions formed during dissolution are able to form ion pairs.

Due to its excellent solubility, it is mobile, so the Na_2CO_3 concentration of the solution flowing in decomposed rocks may be considerable. According to Kovda,[3] in nature Na_2CO_3 accumulates in preference to the other salts in the solution if the solution has a total salt concentration of 0.3 to 0.5 g/ℓ (Figure 48). In this range, when the total salt concentration in natural waters rises, the concentration of $NaHCO_3$ increases proportionately and its quantity exceeds that of other salts (Figure 49). If the total salt concentration is above 5 g/ℓ, the proportion and even the absolute quantity of $NaHCO_3$ declines as the concentration rises. As can be seen from Figure 49, at a total salt concentration of 6 to 14 g/ℓ the $NaHCO_3$ concentration decreases. In the groundwater this decrease is associated with a pronounced rise in the concentration of $CaSO_4$, indicating that the joint occurrence of $NaHCO_3$ — Na_2CO_3 and $CaSO_4$ is rare in moderately or strongly mineralized soils.

The soda accumulating on the earth's surface in rocks or soils may be formed due to geological, chemical, or biological factors.

Many authors have studied the processes and conditions leading to the formation of sodium carbonate in soils. Various theories have been elaborated and disputed, and a number of scientists have discussed these questions in the literature (Hilgard,[10] Treitz,[14] 'Sigmond,[12] Muraközy,[1] Kelley,[18] Arany,[114] Scherf,[21] Verner and Orlovski,[246] Kovda,[51,52] Szabolcs,[113,188] Cheverry,[71] and others).

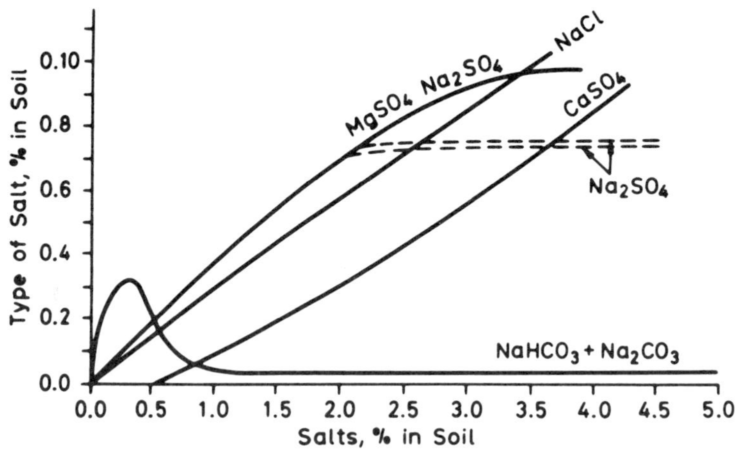

FIGURE 48. Relationship between the degree of alkalization of the soil and the accumulation of alkaline salts. (After Kovda.[3])

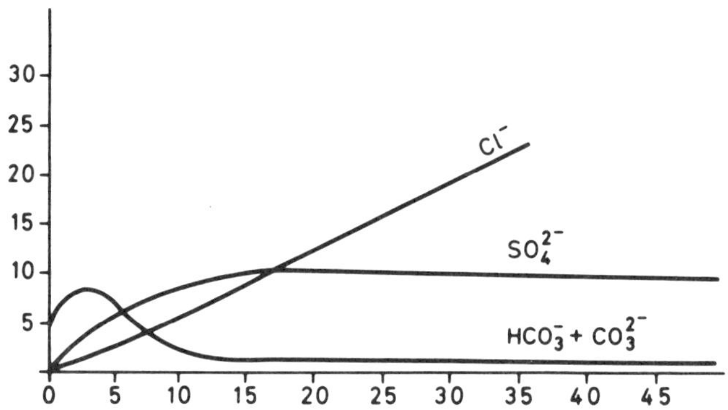

FIGURE 49. Relationship between alkalinity and salt concentration in the groundwater. (After Kovda.[3])

It may be stated that — although some of these authors and their followers have strived to verify the exclusiveness of one theory of sodium carbonate formation in soils and to disprove or belittle the importance of the processes described by other authors — virtually every theory has its justification, or even its importance, under certain circumstances and at certain times. It is clear that sodium carbonate may be formed in soils in several ways, by several different processes. Arany,[114] Kelley,[18] and Kovda[3] have described these sequences as follows:

First — Sodium carbonate may be formed by the ordinary weathering processes of the earth's crust, that is, by the interaction of silicates and CO_2-containing water. During this process bicarbonates of Ca, Mg, Na, and K are formed. Upon evaporation of the resulting solutions, Ca and Mg carbonates are precipitated, while $NaHCO_3$ gradually loses CO_2 and is transformed into Na_2CO_3. According to Kovda,[213] this kind of sodium carbonate formation, the transport of the products evolved by these processes and their geochemical distribution in waters, soils, and subsoils, constitutes the main source of sodium compounds capable of alkaline hydrolysis, and plays a very important role in soil-forming processes. The other processes by which sodium carbonate formation may develop — either partially or completely — are conditioned by the presence of these products.

Second — According to Hilgard's theory,[10] sodium carbonate may be formed by the action of NaCl or Na_2SO_4 on $CaCO_3$:

$$2NaCl + CaCO_3 = Na_2CO_3 + CaCl_2 \quad (3)$$

$$Na_2SO_4 + CaCO_3 = Na_2CO_3 + CaSO_4 \quad (4)$$

Hilgard and many of his followers maintained — in their time — that this was the primary way in which sodium carbonate was formed in soils. Kovda stipulates the conditions under which this procedure plays a decisive role.

Third — Gedroitz,[237,238] Kelley,[18] and 'Sigmond[12] associate sodium carbonate formation with the fine structure of soils, and with processes taking place in the colloid fraction, according to the following equations:

$$\text{(absorbing complex)} - 2Na^+ + H_2CO_3 = \text{(absorbing complex)} - 2H^+ + Na_2CO_3 \quad (5)$$

or

$$\text{(absorbing complex)} - 2Na^+ + CaCO_3 = \text{(absorbing complex)} - Ca^{2+} + Na_2CO_3 \quad (6)$$

Many soil scientists agree that this scheme represents the basic process of sodium carbonate formation in soils. It must be realized, however, that it can be decisive or important only under certain conditions. Gedroitz's scheme is based on exchange reactions taking place in highly dispersed materials; thus, the process can occur only in the presence of sufficient amounts of colloid materials. The equilibrium and reversibility of ion-exchange processes are also very important because they explain why only a limited sodium carbonate concentration can be formed in the soil in this way. It follows that the Gedroitz scheme only plays a significant role in soil formation processes in soils where the sodium carbonate concentration is not too high. It is obvious how wrong one would be to draw the conclusion that this is the only way, or the most important way, in which sodium carbonate is formed, because the compound often is present in such large amounts that it cannot possibly be explained by the scheme of Gedroitz. Naturally, in such cases, the nature of the medium, the concentration of the solution and the ion-exchange equilibrium determine both the tendency and the intensity of soil-forming processes. Therefore, regardless of the way in which the majority of the sodium compounds capable of alkaline hydrolysis have formed in the soil, and regardless of how the finely dispersed materials in equilibrium with the soil solution have formed under certain conditions, the processes determining soil formation and leading to the development of particular salt-affected soils may be correctly described by the above and other equations formulated by Gedroitz.[237,238]

Fourth — In the literature the biological soda forming process is also discussed by numerous authors. Sodium carbonate may be formed through the biological reduction of sulfates in the soil, according to the following schematic equations:

$$Na_2SO_4 + 2C = Na_2S + 2CO_2 \quad (7)$$

$$Na_2S + CO_2 + H_2O = Na_2CO_3 + H_2S \quad (8)$$

According to Kovda,[3] Szabolcs,[1] Timár,[21] Orlovski,[246] and Cheverry,[71] this sequence occurs only under certain conditions. The first and most important of these is that the anaerobic conditions be continuously and permanently maintained throughout the course of the process. Under aerobic conditions, the oxidation of sulfur-containing materials provokes a shift in the pH of the soil towards the acid range, and not towards the alkaline (as pointed out by Kovda and Szabolcs[39]). In addition to the effect of the reduction, the quantity and quality of sulfur-containing compounds present in the soil also influence this process. Microbiological processes leading to sodium carbonate formation cannot take place above a certain concentration of water-soluble sulfates in the soil. This process of sodium carbonate formation is by no means common and decisive in salt-affected soils; nevertheless, in certain areas it does take place, and its extent can be considerable. It may occur, and may be significant in swampy or boggy areas, in a slightly alkaline medium where diluted salt solutions are present, and in the presence of organic matter of adequate quantity and quality (which is also necessary for the development of these processes). Considerable amounts of sodium carbonate have formed through these processes in several alkaline areas, e.g., in the Baraba and Kulunda Plains in West Siberia,[246] in the environs of Lake Chad in Africa,[72] and (especially in the past) in some places on the Hungarian Plain.

Fifth — Numerous authors claim that other biological processes may lead to the decay of sodium-containing plant materials which in turn produce sodium compounds capable of alkaline hydrolysis. When discussing the possibilities of sodium carbonate formation in steppes and deserts, Kelley, Kovda, and Bazilevich described biological processes brought about by the growth and decomposition of various plant materials.[3,18,214]

It may be seen that each process of sodium carbonate formation in soils is associated with certain conditions. It is evident that in one place one process may have a decisive influence on soil-forming processes, while a quite different process may be decisive in another.[247]

If the salt concentration is relatively low in a soil, and the percentage of clay high, then the scheme of Gedroitz represents the situation. But where the electrolyte concentration is high in sandy soils, it cannot be applied.[248] It must be pointed out that the role and significance of the various processes in a given place may alter as a consequence of changes in the local conditions. In the 19th century in the Hungarian Plain, waterlogging prevailed over far more extensive areas than today — in the presence of large amounts of organic matter in the upper soil layers the salinization and alkalization processes were intertwined with swamp- and bog-forming processes. At that time the importance of biological sodium carbonate formation brought about by reduction processes was much greater than it is today. As a consequence of the regulation of the river system, and of soil amelioration, the extent of waterlogged soils has sharply decreased, and a slow sinking of the water table has taken place. The formation of sodium carbonate on the soil surface has ceased, and now in most places the concentration and dynamics of water-soluble salts in the soil profile have decreased to such an extent that the reactions between the colloid fraction and the soluble salts, and the resulting dynamics of sodium carbonate formation, correspond more closely to the theory of Gedroitz than they did in the 19th century.

Sodium carbonate is a highly soluble compound. As a result of hydrolysis, the sodium carbonate-containing solution is always alkaline, and can be as high as pH 12. Owing to its high solubility and alkalinity, the sodium carbonate solution has a very toxic effect on the majority of plants, causes peptization of the soil colloids and reduces permeability, resulting in poor water-physical properties in the soils.[249]

Sodium carbonate can easily be transformed in the presence of carbon dioxide, i.e., under conditions of intensive decomposition of organic matter and at low temperatures. Due to the formation of bicarbonate, the alkalinity of the solution decreases. The transformation of carbonate into bicarbonate is a reversible process, and a decrease in carbon dioxide concentration in the solution provokes the formation of carbonate from bicarbonate. This occurs where the activity of micro-organisms is weak and the content of organic matter low. It can also occur if the solution is warmed up. Most natural waters with a concentration of 0.5 to 3 g/ℓ contain substantial quantities of carbonates and bicarbonates of sodium.

A precondition for the formation of soda-saline soils is that, where soda is present in the water and/or in the deposits with high sodium carbonate and bicarbonate contents, the sodium carbonate should be able to accumulate in sufficient quantities. Another precondition, besides the presence of soda in the water and/or parent rock, is the absence of gypsum. These

Table 43
CHEMISTRY OF GNEISS, THE WEATHERED SEDIMENTS OF GNEISS, AND THE DRY RESIDUE OF THE LOCAL RIVER WATER

	SiO_2	SO_4^{2-}	Cl^-	Ca^{2+}	Mg^{2+}	Na^+	K^+	Al_2O_3	Fe_2O_3
Primary rock (gneiss)	64.1	0.26	0.04	0.64	1.10	1.20	2.67	20.35	6.97
Weathered sediment of primary rock	60.1	0.22	0.04	0.38	1.31	1.12	2.65	20.17	8.55
Dry residual of river waters	21.3	36.7	Tr.	13.00	2.80	7.8	—	3.6	—

Note: Values given are percent, Tr., trace.

After Polynov.[193]

conditions occur in some parts of the prairies in the U.S., in Canada, in the chernozem regions of the U.S.S.R., in Hungary, in the monsoon areas of Pakistan and India, and in low-lying river plains in Argentina, Chile, Chad, and the Armenian SSR. In arid and semiarid regions, where the accumulation of calcium sulfate is intensive, soda-saline or saline-alkaline soils are rare.[250,251]

4. Potassium Carbonate

With regard to its chemical properties, potassium carbonate is very similar to sodium carbonate. Soils having high quantities of potassium carbonate are very rare.

B. Sulfates and Sulfides

1. Calcium, Magnesium, and Sodium Sulfates

The sulfate compounds in natural waters, sediments, and soils are mainly weathering products of the minerals of volcanic and sedimentary rocks, as can be seen in Table 43.

Calcium sulfate occurs in natural waters, sediments, and soils in all regions under different climatic conditions as a product of the weathering of sulfide minerals, due to the reaction of sulfate ions with the calcium removed from calcium-bearing minerals. Calcium sulfate also accumulates secondarily by the reaction of sodium sulfate and calcium chloride in sediments and soils. In semiarid and arid regions, the sediments and rocks contain large quantities of calcium sulfate, precipitated during the evaporation of saline lacustrine waters and groundwaters.

Calcium sulfate accumulates in the soil under various climatic conditions, in various soil horizons, and in various quantities and forms. Usually, in soils in the moderate climatic zone, calcium sulfate accumulates in the B_2 horizon of hydromorphic soils, as for instance in meadow solonetz, solonetz-like meadow soils and meadow soils. In Table 44, this phenomenon is demonstrated for the hydromorphic soils of the Hungarian Plain. Calcium sulfate often occurs in the deeper layers of the soils in the steppe region and it may be found in high amounts in the gypsiferous soils of deserts and semideserts.

In Table 45, the chemical composition of the soluble salts in a meadow solonchak in the Kysil-Kum Desert is shown.

Calcium sulfate usually occurs in soils as gypsum. Gypsum crystallizes in soils in a great variety of forms, from the transparents to large nodules, concretions, or regularly shaped slabs. Gypsum sometimes forms a spongy porous mass in soils, causing cementation in the horizon of accumulation on the surface, or in the entire soil profile. Under the very dry climate of deserts the gypsum may be dehydrated and turn into a dry powdery mass of calcium sulfate hemihydrate.

Magnesium sulfate, which has high solubility, is mainly a product of weathering. It is a typical component of seawater, occurs in saline groundwaters, in many saline lakes, and

Table 44
THE CHEMICAL COMPOSITON OF SATURATION EXTRACTS OF HYDROMORPHIC SOILS IN HUNGARY

Soil	Depth of sampling (cm)	Soluble salts (%)	Ca^{2+} (meq/ℓ)	Mg^{2+} (meq/ℓ)	Na^+ (meq/ℓ)	HCO_2^- (meq/ℓ)	Cl^- (meq/ℓ)	SO_4^{2-} (meq/ℓ)
Meadow soil	7—17	0.10	5.25	0.98	0.87	0.64	3.90	0.55
	58—72	0.30	14.80	12.60	6.10	1.46	12.80	6.60
	90—110	0.35	22.50	12.10	4.80	0.88	5.30	27.00
	126—136	0.22	17.80	9.80	7.60	1.00	5.60	22.90
Solonetz-like meadow soil	5—17	0.01	2.40	2.80	0.97	0.25	2.00	0.50
	30—42	0.06	2.50	5.30	1.60	3.00	2.50	1.30
	56—76	0.10	3.75	3.60	1.60	1.60	0.72	2.60
	89—100	0.55	17.40	16.00	21.30	1.40	1.10	31.40
Meadow solonetz	8—16	0.06	4.30	6.00	6.40	3.10	4.40	1.40
	26—37	0.12	4.50	3.70	10.00	3.50	3.40	4.30
	45—55	0.20	40.0	3.40	18.20	4.80	4.60	4.10
	75—85	0.75	15.10	41.70	37.70	1.50	1.70	37.80
	97—105	0.75	15.50	34.80	34.30	1.40	1.30	43.50
	117—126	0.25	3.10	8.30	24.50	3.30	2.70	8.30

Table 45
CHEMICAL COMPOSITION OF THE SOLUBLE SALTS IN A MEADOW SOLONCHAK IN THE NORTHERN KYSIL — KUM DESERT, DETERMINED FROM 1:5 AQUEOUS EXTRACT

Depth of sampling (cm)	Dry residue (%)	HCO^-	CO_3^{2-}	Cl^-	SO_4^{2-}	Ca^{2+}	Mg^{2+}	$Na^+ + K^+$
0—2	21.73	0.04	Tr.	1.85	11.36	2.63	0.36	5.68
2—6	16.00	0.04	Tr.	1.46	8.62	2.52	0.33	4.18
10—15	4.41	0.02	—	1.05	1.61	0.33	0.18	0.74
25—30	3.52	0.02	—	1.09	1.04	0.11	0.16	0.79
40—45	3.23	0.01	—	0.89	1.04	0.13	0.12	0.71
120—125	1.24	0.01	—	0.20	0.58	0.12	0.04	0.19
190—195	1.94	0.02	—	0.18	1.04	0.18	0.05	0.32

Note: Chemical values are percent composition. Tr., trace.

After Borovski.[186]

accumulates in the form of epsomite ($MgSO_4 \cdot 7H_2O$) in the saline soils of desert and semidesert regions. Magnesium sulfate never accumulates in soils in the pure form, but is always found together with other readily soluble salts, such as sodium sulfate, sodium chloride, and magnesium chloride.

Sodium sulfate is mainly a weathering product, with high solubility and high mobility. High concentrations of sodium sulfate may be present in saline groundwaters, in the water of saline lakes, and seawater. Sodium sulfate can be found in the salt-affected soils of the steppe region, but it is a typical compound in the saline soils of desert and semidesert regions. Depending on the temperature, sodium sulfate crystallizes with a different number of water molecules. On the surface of solonchak soils, at relatively low temperature, it precipitates in the form of large transparent mirabilite ($Na_2SO_4 \cdot 10\ H_2O$), which, with an increase in temperature and a decrease in humidity, dehydrates and turns into white powdery thenardite (Na_2SO_4).[252]

The sulfate compounds taking part in the formation of salt-affected soils have very different solubilities. The concentrations of saturated solutions of gypsum, magnesium sulfate, and sodium sulfate at 25°C are 2 g/ℓ, 262 g/ℓ, and 280 g/ℓ, respectively.

Solubility of calcium sulfate: the saturation concentration of gypsum in water is fairly low. In the case of a mixed salt solution, the solubility of calcium sulfate depends on the ionic concentration and the chemistry. The amount of gypsum dissolved at saturation concentration can be calculated from:

1. The thermodynamic solubility product of $CaSO_4 \cdot H_2O$
2. The total ionic concentration of the solution to be saturated for gypsum
3. The ionic composition of the solution before saturation with gypsum

The ion concentration of a salt calculated from the concentrations of the related ions depends on the ionic concentration and ion composition of the electrolyte. The thermodynamic solubility product (which is calculated from the activities of the ions and takes into account the short and long range electrostatic interaction of particles with opposite charges) remains constant in electrolytes having different concentrations and different compositions.

Saturation extracts from solonetz soils were prepared and analyzed before and after saturation with calcium sulfate (Table 46). The solubility products (thermodynamic solubility products) were calculated with and without subtracting the sum of the ion-pair concentrations from the analytical concentrations of calcium and sulfate ions.

It is evident that ion concentration products of gypsum (Table 47) depend on the ionic concentration of the solution calculated either from the analytical concentrations of calcium and sulfate or by reducing the ion concentrations by the sum of the ion-pair concentrations. The thermodynamic solubility product changes only slightly with the change in the ionic concentration of the solution and is independent of the electrolyte concentration if the thermodynamic solubility product is calculated from the reduced activities of calcium and sulfate ions.

The solubility of calcium sulfate is higher in the soil solution when the concentration of nonsulfate electrolytes increases. In a solution containing sodium sulfate and/or magnesium sulfate the dissolution of calcium sulfate is limited by the concentration of sulfates.

In systems containing solid calcium carbonate and calcium sulfate, the ratio of the amounts of $CaSO_4$ and $CaCO_3$ dissolved must be determined by the ratio of their thermodynamic solubility products, according to the following equation:

$$\frac{K_{SPCaSO_4}}{K_{SPCaCO_3}} \cong 5 \cdot 10^3 \quad \text{(at 25°C)} \tag{9}$$

where: K_{SP} = thermodynamic solubility product.

For the sulfate type of salinization, the solubility of calcium sulfate decreases with an increase in the sulfate concentration of the soil solution, and the dissolution of $CaCO_3$ increases due to the increasing ionic concentration. The dissolution of calcium carbonate shifts the pH value of the media towards the alkaline range.

Saturation extracts prepared from solonetz soils containing $CaCO_3$ were saturated with calcium sulfate and analyzed as to their chemical composition (Table 48). The data of analyses indicate clearly that the carbonate concentration, the total alkalinity and the pH value of the extract increase parallel with the increasing sulfate concentration. The increase in the sulfate concentration stabilizes the calcium concentration of a solution saturated with calcium sulfate at a certain level. The thermodynamic solubility product of $CaSO_4$ was calculated from the reduced activities of calcium and sulfate ions. The reduced activities of the ions were calculated by subtracting the sum of the ion-pair concentrations from the

Table 46
THE CHEMICAL COMPOSITION OF SATURATION EXTRACTS OF A SOLONETZ SOIL BEFORE AND AFTER SATURATION WITH GYPSUM

Depth of sampling (cm)	Ca^{2+}		Mg^{2+}		Na^{+}		SO_4^{2-}		HCO_3^-		Cl^-		I	
	1	2	1	2	1	2	1	2	1	2	1	2	1	2
40—60	8.42	11.5	16.7	13.5	97.8	90.6	65.8	63.9	0.87	1.33	14.7	13.8	238.6	230.5
60—80	1.80	12.1	4.1	4.1	63.0	59.3	31.7	39.6	1.71	1.48	12.4	11.4	113.7	147.6
80—100	1.90	12.3	4.2	4.5	63.0	49.8	31.6	38.9	0.95	2.14	13.8	12.7	114.2	143.9
100—120	1.40	12.5	2.20	2.1	43.5	39.5	17.7	29.5	1.1	1.62	12.2	11.2	71.0	115.0
120—140	1.40	12.0	·1.60	2.7	37.0	33.3	15.3	25.8	0.67	1.08	11.4	10.7	51.0	103.5

Note: 1, Before saturation with gypsum; 2, After saturation with gypsum. Chemical values are 10^3 mol/ℓ.

Table 47
THE SOLUBILITY PRODUCT OF GYPSUM CALCULATED FROM THE CONCENTRATION AND IONIC ACTIVITIES OF SOIL SATURATION EXTRACTS SATURATED WITH $CaSO_4$

Ionic strength of the solution		Solubility product calculated from		Thermodynamic solubility product calculated from		
Total 10^3 mol/ℓ)	Reduced (10^3 mol/ℓ)	Total ion conc.	Reduced ion conc.		Total ion activities	Reduced ion activities
230.5	170.7	$7.33.10^{-4}$	$2.92.10^{-4}$		$5.81.10^{-5}$	$2.85.10^{-5}$
147.6	114.0	$4.78.10^{-4}$	$2.12.10^{-4}$		$5.19.10^{-5}$	$2.74.10^{-5}$
143.9	109.9	$4.80.10^{-4}$	$2.12.10^{-4}$		$5.39.10^{-5}$	$2.81.10^{-5}$
114.5	89.3	$3.69.10^{-4}$	$1.72.10^{-4}$		$4.76.10^{-5}$	$2.62.10^{-5}$
103.5	78.9	$3.10.10^{-4}$	$1.46.10^{-4}$		$4.28.10^{-5}$	$2.39.10^{-5}$
				$\overline{K}_{sp}$ =	$5.07.10^{-5}$	$2.68.10^{-5}$
				$\overline{S}$ =	$2.59.10^{-6}$	$8.24.10^{-7}$
				CV =	11.4%	6.87%

Table 48
CHEMICAL COMPOSITION OF SATURATION EXTRACTS AFTER TREATMENT WITH $CaSO_4$ ANHYDRITE

Depth of sampling (cm)	pH	Ca^{2+} meq/ℓ	CO_3^{2-} meq/ℓ	HCO_3^- meq/ℓ	SO_4^{2-} meq/ℓ	I (10^3 mol/ℓ)	Thermodynamic solubility product $CaSO_4$	Thermodynamic solubility product $CaCO_3$
2—8	8.25	39.75	2.06	6.54	141.30	266.3	$4.635.10^{-5}$	$2.2346.10^{-8}$
8—17	8.10	29.01	2.16	3.79	139.30	251.0	$3.582.10^{-5}$	$0.7867.10^{-8}$
17—32	8.13	29.68	1.26	3.64	154.67	272.9	$3.743.10^{-5}$	$0.8906.10^{-8}$
32—42	7.85	29.42	1.34	3.52	140.24	250.9	$3.631.10^{-5}$	$0.2544.10^{-8}$
42—63	7.75	30.19	1.18	3.34	100.24	180.1	$3.520.10^{-5}$	$0.1916.10^{-8}$
63—79	7.80	34.42	0.98	1.99	70.90	136.3	$3.703.10^{-5}$	$0.3251.10^{-8}$
79—90	7.75	35.65	0.59	2.21	63.40	124.8	$3.725.10^{-5}$	$0.2839.10^{-8}$
							$\overline{X} = 3.791.10^{-5}$	$\overline{X} = 0.7096.10^{-8}$
							$S = 3.81.10^{-6}$	$S = 0.2747.10^{-8}$
							$\overline{S} = 1.44.10^{-6}$	
							CV = 9.74%	

$$\lg K_{CaCO3} = \lg (Ca^{2+}) + 2pH + \lg[K_1 \cdot K_2(H_2CO_3)]$$
$$\lg K_{CaSO4} = \lg (Ca^{2+}) + \lg (SO_4^{2-})$$

analytical concentrations of the relevant ions and multiplying by the coefficients of activities from the enlarged equation of Debye-Hückel:

$$-\log \gamma = \frac{Az^2 \sqrt{I}}{1 + Bv \sqrt{I}} \tag{10}$$

where: A and B = constants; I = ionic concentration of the solution, mol/ℓ; z = valence of the ion; v = diameter of the hydrated ion, Å; and γ = coefficient of activity.

The average thermodynamic solubility product of calcium sulfate anhydrite is 3.79×10^{-5}, the standard deviation of the mean is 1.44×10^{-6} and the coefficient of variation amounts to 9.74%. The thermodynamic solubility product in the extracts investigated does not show any dependence on the concentration of the solution. The thermodynamic solubility product of calcium carbonate was calculated from the reduced activity of calcium ions and the pH value of the solution:

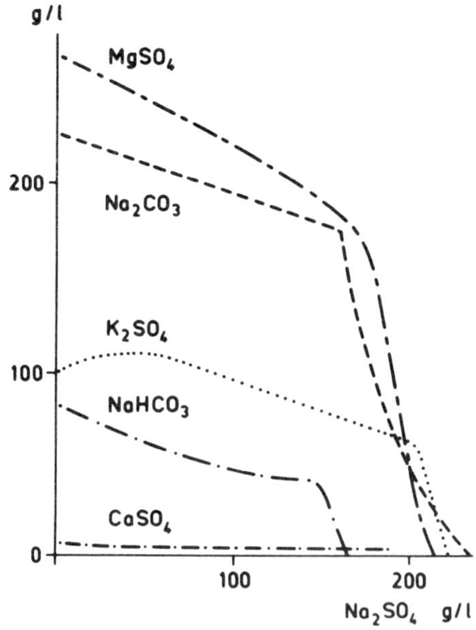

FIGURE 50. Solubility of salts depending on Na_2SO_4 concentration.

$$\lg K_{SP} = \lg K_1K_2(H_2CO_3) + 2pH + \lg(Ca^{2+}) \tag{11}$$

where: K_1 and K_2 = the first and second dissociation constants of H_2CO_3; K_{SP} = thermodynamic solubility product of $CaCO_3$; and (Ca^{2+}) = activity of related ion compound.

The average thermodynamic solubility product of calcium carbonate is 0.7096×10^{-8}. The standard deviation of the mean amounts to 0.2747×10^{-8}.

Solubility of magnesium sulfate — This is very good in water and depends only slightly on the temperature. The saturation concentration of magnesium sulfate decreases with increasing sodium sulfate concentration (Figure 50). Magnesium sulfate has high mobility as a result of its high solubility.

Solubility of sodium sulfate — Sodium sulfate has high solubility in water. The dissolution of this salt depends very much on the temperature, and increases considerably if the temperature rises. In the case of sodium sulfate salinization the situation is rather complex. In warm dry seasons of the year sodium sulfate accumulates together with other easily soluble salts on the surface of soils. In cold rainy periods the solubility of sodium sulfate decreases and it is not leached.

The influences of various sulfate compounds on the soil properties and plant growth are quite different and are determined by the solubility of the salts, and by the valence and size of the cation of the sulfates.

Calcium sulfate may accumulate in large amounts in soils, but it has a slight toxic effect on the plants due to its low solubility. The compact layer of gypsum accumulation on the surface or in the root zone of soils may be impenetrable by water, air, and plant roots, and may thus have an adverse effect on the plants.

The calcium ion of calcium sulfate reacts with sodium salts capable of alkaline hydrolysis and with the exchangeable sodium in the soil. If gypsum is present in the soil, it neutralizes the sodium salts capable of alkaline hydrolysis, forming poorly soluble salts with the anion of the weak base, while the cation forms salts dissociating with a neutral reaction.

The calcium ion of calcium sulfate replaces the exchangeable sodium according to the extent to which the sodium adsorption ratio decreases in the equilibrium solution (Table

Table 49
SAR VALUES AND SODIUM SATURATION PERCENTAGES OF SOILS BEFORE AND AFTER TREATMENT WITH CALCIUM SULFATE

Depth of sampling (cm)	SAR		ESP		Δ SAR	Δ ESP
	1	2	1	2		
2—8	31.6	23.7	31.2	25.2	7.9	6.0
8—17	53.4	27.5	43.7	28.2	25.9	15.5
17—32	69.6	35.0	50.3	33.5	34.5	16.8
32—42	65.6	32.4	48.8	31.7	33.2	17.1
42—63	73.0	20.9	51.6	22.9	52.1	28.7
63—79	39.5	10.4	36.3	12.3	29.1	24.0
79—90	38.4	7.1	35.6	8.5	31.2	27.2

Note: 1, Before treatment with calcium sulfate; 2, after treatment with calcium sulfate.

49). If the sulfate type of salinization occurs, the leaching of sodium sulfate causes an increase in the solubility of calcium sulfate, and this phenomenon promotes the ameliorative effect. Gypsum and other materials containing calcium sulfate are widely used for the reclamation of sodic and alkali soils.

Magnesium sulfate may accumulate in soils in large quantities. A solution with a high concentration of magnesium ions is one of the most toxic and harmful agents for plants. An increase in the magnesium ion concentration in the soil solution during the accumulation period leads to an increase in the magnesium saturation of the soil and to an increase in the magnesium adsorbed by soil minerals. The cations of the soil solution among the magnesium ions may be adsorbed on the clay minerals by ion exchange, or they may enter the crystal lattice of the layer silicates or possibly combine with other products of weathering to form clay minerals. The degree of magnesium adsorption and desorption is determined by the equilibrium between the cations of the intermicellar and micellar solutions and the equilibrium between the cations of the crystal lattice in the mineral and the micellar solution.

During the accumulation period, magnesium sulfate accumulates together with other easily soluble salts, mainly with sodium salts. An increase in the sodium and magnesium ion concentration in the solution raises the sodium and magnesium saturation of the soil and increases the ratio of layer silicates rich in magnesium. If leaching takes place, the soluble salts are leached out according to their mobility. The degree of sodium saturation decreases proportionally with changes in the concentration and chemistry of the soil solution. Some of the exchangeable magnesium may enter the liquid phase too, but it is recharged by the magnesium ions entering from the crystal lattice in exchangeable positions.

Sodium sulfate accumulates in soils under different climates in different quantities. In the desert and semi desert regions it is a typical component of the salts accumulated in saline soils, as might be expected from its high mobility. The toxicity of sodium sulfate is two or three times less than that of magnesium sulfate, and it is two or four times less than the toxicity of sodium chloride. In a sodium sulfate type of salt accumulation, an increase in the salt concentration and in the sodium adsorption ratio of the soil solution shifts the ion exchange equilibria in favor of the sodium ions, and the degree of sodium saturation of the soil increases with increasing salt accumulation.

Comparing the equilibrium conditions in systems containing sodium chloride and sodium sulfate, the difference in the degree of sodium saturation of the adsorbent is as great as that of the activity coefficients of the chloride and sulfate ions under the same conditions. In soils containing sodium sulfate and calcium sulfate, the degree of sodium saturation increases parallel with the increase in the sodium sulfate concentration.

Sulfate solonchaks usually contain Na_2SO_4, $CaSO_4$ and $MgSO_4$. They occur mainly in continental depressions in semidesert and desert regions. They have favorable water physical properties. Sulfate solonchaks can be ameliorated easily and rapidly by leaching and drainage, as a result of their natural gypsum content and high water permeability. If the main compound among the sulfates is sodium sulfate, leaching must be carried out during the warm period to avoid a decrease in salt solubility following the decrease in temperature.

In chloride-sulfate and sulfate-chloride types of salinization, chloride and sulfate salts accumulate together. The method of amelioration is nearly the same as that for sulfate solonchaks. The maximum permissible level of salt content and the desired degree of leaching are different because of the different toxicities of chloride and sulfate ions (Table 50).

In the case of sulfate-carbonate salinization, the sulfate ions dominate in the soil solution but the presence of carbonate ions shifts the reaction of the medium to the alkaline range, enhances the toxic effect of the salts on the plants, and increases the dispersion of soil particles, causing poor hydrophysical properties. The amelioration of soils formed by the influence of sulfate soda-salinization is carried out by the use of chemicals in combination with drainage and leaching.

2. Iron Sulfates and Sulfides and Aluminum Sulfates

Iron and aluminum sulfates occur in acid sulfate soils and result in the extremely acidic pH of this type. Both ferric and aluminum sulfates are rare in their pure form during the soil-forming processes, and both their geochemistry and their influence on soil-forming processes must be interpreted in close relation to the sulfur cycle in the formation of acid sulfate soils. Therefore, the behavior of ferric and aluminum sulfates in relation to the formation of salt-affected soils will be discussed below, together with the role of the sulfides.

In the earth's crust, aluminum is the second most abundant element after silicon, and is the most frequent metallic element. However, the majority of aluminum compounds are found in the lattice layers of the mineral part of the soils and only few aluminum compounds are mobile during soil-forming processes. One of these is aluminum sulfate, which is formed parallel to and in close correlation with the oxidation of iron sulfides and the acidification of the medium during soil-forming processes. According to numerous authors, the formation of aluminum sulfate during the genesis of acid sulfate soils is a secondary process following pyrite oxidation.[20,29]

The chemical hydrolysis of aluminum can be characterized schematically as follows:

$$Al^{3+} + H_2O = AlOH^{2+} + H^+ \tag{12}$$

In spite of the occurrence of aluminum sulfate in acid sulfate soils, the geochemistry and chemistry of the hydration of aluminum products constitutes a rather complex phenomenon, and by no means all the major aluminum ions in soil solutions with a strongly acidic reaction are associated with sulfate ions. This can be explained by the chemical behavior of aluminum and by the interaction of the soil solution with soil minerals, where aluminum is the major compound in the lattice layers. A great number of publications on this subject are available in the literature, so the problem will not be discussed in detail here. Due to their nature, mentioned by different authors, the various polynuclear hydroxy aluminum cations have an even more uncertain composition in the soil solution than in the solid phase. From the practical point of view, however, it can be agreed that aluminum sulfate and other acidic aluminum compounds play a substantial role in the processes involved in the formation of acid sulfate soils.[29,253]

The relationship between iron and aluminum sulfates in soil solutions may vary. In most cases, iron sulfate dominates as a primary agent in the acidification of the soil medium and in the activation of the mobility of aluminum compounds. However, authors have described

Table 50
CLASSIFICATION OF SALINE SOILS BY DEGREE AND TYPE OF SALINITY IN RELATION TO FIELD CROPS

Condition of agricultural crops with medium salt resistance	Degree of soil salinity	Type of salts dominating in soils						
		Content of salts soluble in water in the horizon with maximum salt accumulation within the 0—60 cm stratum (as %)			Content of salts soluble in water on average, for the 0—100 cm stratum (as %)			
		Soda	Chloridic soda and soda-chloridic	Sulfatic soda and soda sulfatic	Chloridic	Sulfatic chloridic	Chloridic sulfatic	Sulfatic
Good growth and development (no bare patches normal yield)	Practically non-saline (or only very slightly saline)	0.10	0.15	0.15	0.15	0.20	0.25	0.30
Slight withering (bare patches and reduction of yield by 10—20%)	Slightly saline	0.10—0.20	0.15—0.25	0.15—0.30	0.15—0.30	0.20—0.30	0.25—0.40	0.30—0.60
Medium withering (bare patches and yield reduction of 20—50%)	Medium saline	0.20—0.30	0.25—0.40	0.30—0.50	0.30—0.50	0.30—0.60	0.40—0.70	0.60—1.0
Marked withering (bare patches and yield reduction of 50—80%)	Strongly saline	0.30—0.50	0.40—0.60	0.50—0.70	0.50—0.80	0.60—1.0	0.70—1.20	1.0—2.0
A few scattered plants survive (virtually no yield)	Solonchaks	>0.50	>0.60	>0.70	>0.80	>1.0	>1.20	>2

cases where the quantity of aluminum sulfate in the soil was greater than that of iron sulfate. It must be repeated that when examining these substances in soil-forming processes, the main factor is the total acidity, and the role of free sulfuric acid must not be forgotten.

In spite of all the material available in the literature, it is difficult to distinguish the presence and influence of iron and aluminum sulfates in soils from that of sulfuric acid, which is the main reason for the peculiar characteristics of acid sulfate soils. Due to the oxidation of reduced sulfur compounds, all these products may form during the genesis of acid sulfate soils. In the muds of tidal marshes the most common mineral is pyrite, the percentage of which may exceed 5% of the soil-forming material. A large number of sulfate minerals can be found in association with oxidizing pyritic material, but most of these compounds are very soluble and are formed only in the absence of leaching. During dry periods, gyttya soils sometimes have crusts of water-soluble sulfates of Al, Fe^{III}, Ca, Mg, K, and Na. Under nonarid conditions, most of the SO_4^{2-} produced by the oxidation of pyrite in low pH environments at normal temperature is bound in the minerals.

Figure 51 illustrates the pe-pH diagram of those minerals which play a dominant role in the dynamics of sulfides and sulfates, as well as of iron compounds in certain processes of the formation of salt-affected soils. The figure shows that over a fairly wide range of pH, the dominant mineral under reduced conditions is pyrite. Elemental S also occurs, but not over the whole range of the pe-pH zone. When the medium is oxidized to give an acid reaction, ferric oxide occurs. Other iron sulfate materials have the stability ranges, indicated in Table 51.

Both nonbiotic and biological processes contribute to the formation of ferric and aluminum sulfates when the sulfide minerals oxidize. Sato characterized the initial processes during pyrite oxidation as follows:[29]

$$2FeS + 3/2\ O_2 + n\ H_2O \qquad Fe_2O_3 \cdot nH_2O + 2S^o \tag{13}$$

$$2S^o + 3/2\ O_2 + H_2O \qquad 2\ H^+ + SO_4^{2-} \tag{14}$$

$$2FeS + 3\ O_2 + (n + 1)H_2O \qquad Fe_2O_3 \cdot nH_2O + 2H^+ + SO_4^{2-} \tag{15}$$

Based on the measurements of pyrite electrode potentials in aqueous ferrous sulfate solutions of different pH, pyrite oxidation can be represented as:

$$FeS_2 = Fe^{2+} + S_2^o + 2e \tag{16}$$

in agreement with potentials according to:

$$E_H = 0.751 + 0.0295 \log[Fe^{2+}] \quad (\text{at pH} < 2) \tag{17}$$

and as:

$$FeS_2 + 3H_2O = Fe(OH)_3 + S_2^o + 3H^+ + 3e \tag{18}$$

in agreement with:

$$E_H = 0.857 - 0.059\ \text{pH} \quad (\text{at pH} > 3) \tag{19}$$

If the purely chemical oxidation of pyrite is considered as a slow process, the following steps can be derived for the pyrite oxidation process:

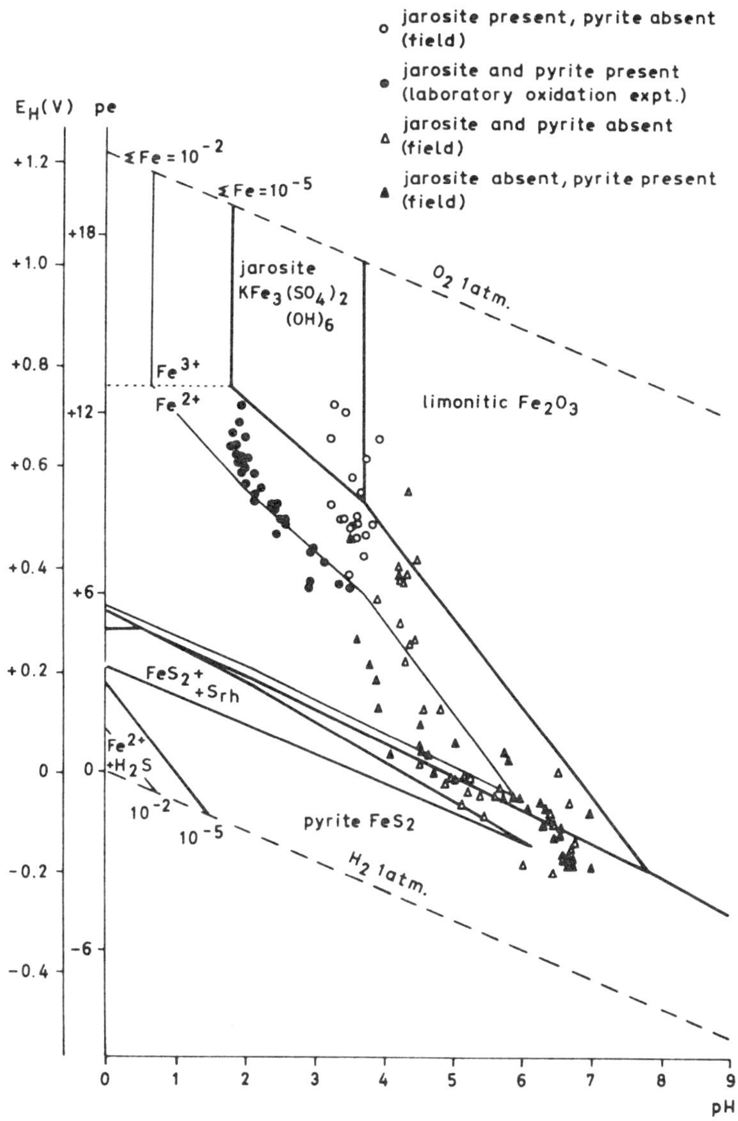

FIGURE 51. Diagram (pe-pH) of pyrite, rhombic sulfur, limonitic Fe_2O_3, jarosite, and dissolved K^+, SO_4^{2-}, Fe^{2+}, and Fe^{3+} at log $[SO_4^{2-}]$ = −2.3 log $[K^+]$ = −3.3, 25°C and 1 atm total pressure. Symbols refer to E_H-pH data measured in coastal plain soils in Thailand.[29]

$$FeS_2 + 1/2O_2 + 2H^+ \rightarrow Fe^{2+} + 2S^o + H_2O \quad (20)$$

$$Fe^{2+} + 1/4O_2 + H^+ \rightarrow Fe^{3+} + 1/2H_2O \quad (21)$$

$$Fe^{3+} + 3H_2O \rightarrow Fe(OH)_3 + 3H^+ \quad (22)$$

$$FeS_2 + 2Fe^{3+} \rightarrow 3Fe^{2+} + 2S^o \quad (23)$$

$$2S^o + 12Fe^{3+} + 8H_2O \rightarrow 12Fe^{2+} + 2SO_4^{2-} + 16H^+ \quad (24)$$

$$2S^o + 3O_2 + 2H_2O \rightarrow 2SO_4^{2-} + 4H^+ \quad (25)$$

Table 51
ΔG_f° VALUES FOR SEVERAL Fe AND Al SULFATES (25°C, 1 ATM)

Chemical formula	Mineral	ΔG_f° (kcal/F.W.)
$FeSO_4 \cdot H_2O$	Szomolnokite	−257.5
$FeSO_4 \cdot 7H_2O$	Melanterite	−599.4
$Fe_2(SO_4)_3$	—	−535.8
$Fe_3(OH)_4SO_4$	(Basic Fe^{II} sulfate)	−435.8
$Fe_4(OH)_2(SO_4)_3$	—	−1033.3
$Fe_3(OH)_3(SO_4)_2$	Butlerite?	−725.4
$Al_2(SO_4)_3 \cdot 6H_2O$	—	−1105.14
$(H_3O)Fe_3(SO_4)_2(OH)_6$	Hydronium jarosite	−773
$KFe_3(SO_4)_2(OH)_6$	Jarosite	−788.5
$KAl(SO_4)_2 \cdot 12H_2O$	K — alum	−1227.8
$KAl_3(SO_4)_2(OH)_6$	Alunite	−1112.95
$Al_4(OH)_{20}SO_6$	Basaluminite	−1104.13
$AlOHSO_4$	Amorphous?	−335.06

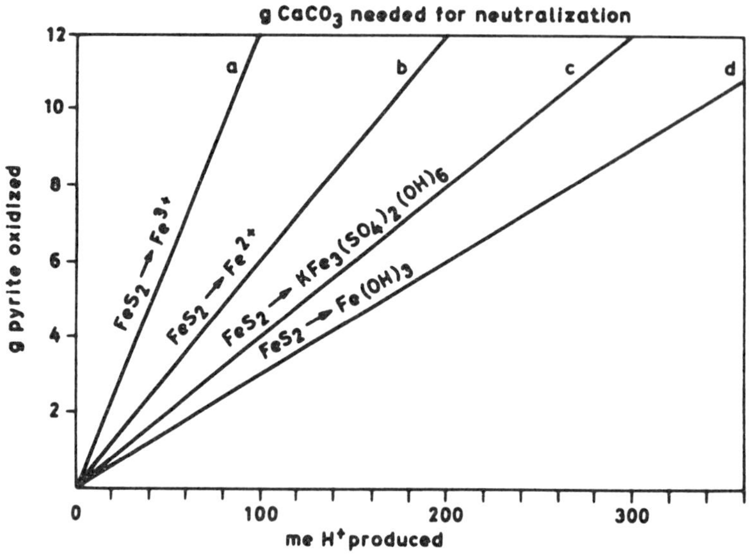

FIGURE 52. Acid production during pyrite oxidation.[29]

Considerable amounts of ferric ions can only remain in solution if the pH of the medium is below 3, which is quite common in acid sulfate soils.[29]

If sufficient buffering substances are available to prevent a rapid decrease in the pH below about 4, the Fe^{2+} produced will oxidize relatively fast and precipitate as $Fe(OH)_3$. At lower pH values jarosite may become the dominant solid Fe^{III} phase. Upon further acidification, these compounds may dissolve partly or wholly, and thus act as a source of Fe^{3+} for pyrite oxidation.

The oxidation of pyrite upon aeration is a cascade-like process because, after the first steps in the oxidation of FeS, the pH of the medium becomes acidic, and pyrite oxidation will continue at a higher rate. Sulfur compounds other than pyrite also participate in the process.

The production of acid during pyrite oxidation is demonstrated in Figure 52. The figure shows that the different routes of pyrite oxidation, resulting in different iron compounds, are associated with different degrees of hydrogen ion production, and consequently with different pH values.[29]

Besides pyrite, a considerable number of other iron sulfides also occur in tidal muds and take part in the formation of iron sulfates. Some characteristics of these minerals are demonstrated in Table 52.

Baas Becking et al. characterized the sulfate-reducing environments related to pH and E_H shown in Figure 53.[29] As the figure demonstrates, in the pH range 5 to 9, conditions are favorable for sulfate reduction, while below pH 5 — as was discussed earlier — acidic sulfur compounds prevail.

Sulfide ions contribute quite differently to the formation of salt-affected soils in another group of these soils. As was discussed when characterizing the behavior of carbonate ions, one of the schemes of soda formation in soils is closely related with the reduction of sulfates in the soils, parallel with an increase in the alkalinity of the medium and with the appearance of soda in the soil solution.

3. Sodium Sulfide

Both sodium sulfide and H_2S often occur under waterlogged conditions in the upper layers of alkali soils, and play a considerable role (due to reduction processes) in soda formation and in the development of some solonetz and soda-solonchak soils. This process was briefly discussed above in connection with sodium carbonate.

C. Silicates

Silicon, which is the major element both in the earth's crust and in the soil material, plays a substantial role in all soil-forming processes, but mainly as the main constituent of soil minerals and lattice layers. However, the compounds of silicic acid play an important part in the formation of alkali soils. The main compound in this respect is sodium silicate, which accumulates in considerable amounts on the surface of alkali soils and in the upper layers of their profiles.

The silicates include all the salts and esters of any silicic acids. The common feature of these compounds is that they contain SiO_4 tetrahedra in their crystal lattices, either isolated or linked through one or more of the oxygen atoms, forming groups, continuous chains, sheets, and three-dimensional structures.

Silicates are the most abundant compounds in rocks and soils, and the quantity of various silicon-bearing minerals surpasses that of all other minerals present in rocks of different geological origin. The average silicon content of the earth's crust amounts to 27.2%.

Silicic acid is present in the surface waters (i.e., in the waters of rivers, lakes, seas, and oceans) and is a component of groundwaters, layer- and artesian waters, and of pore waters of rocks and soils.

The silicic acid in the aqueous solution is the product of the hydrolysis of alumosilicates, the dissolution of quartz, calcedon, opal, and the disintegration of silicon-containing materials of biological origin. The degree of hydrolysis of alumosilicates is low, the solubility of quartz and other minerals consisting of silicon dioxide is poor, and the silicon content of biogenic materials is relatively small. The concentration of silicic acid in the aqueous solutions occurring in nature is usually not higher than 10 to 50 mg/ℓ SiO_2. The silicic acid presumably appears in the ionic and molecular forms of ortho- and meta-silicic acid. It may also be found as polymers of organo-silicic compounds or colloids of hydrosilicates, and it may be present as a suspended amorphic gel of silicates of the type $nSiO_2{\cdot}mH_2O$ in an aqueous medium.

The monomers, dimers, polymers, and organosilicates play a decisive role in the transport of "dissolved" silicic acid. In the transport processes, the mobility of silicic acid is limited by the solubilities of the salts formed with the cations in the solution. Among the mesosilicates, forsterite (Mg-orthosilicate) and fayalite (Fe-orthosilicate) have very low solubility. They occur in nature predominantly in the form of a solid solution (olivine), which is an

Table 52
MAJOR CHARACTERISTICS OF IRON SULFIDES

	Pyrite	Marcasite	Pyrrhotite	Mackinawite	Griecite	Smythite	Cubic FeS
Composition	FeS_2	FeS_2	$Fe_{(1-x)}S$	c.FeS	Fe_3S_4	Fe_3S_4	c.FeS
Structure	Cubic	Orthorhombic	(x 0.125) Hexagonal or monoclinic	Tetragonal	Cubic	Hexagonal, poss. monoclinic	Cubic
Stability	<743°C	Metastable below ca.400°C	<1083 °C	Metastable below ca. 130°C	Metastable below ca.200°C	Possibly stable <75°C	Metastable
Apparent H_2O requirement for formation	–	+	–	+	+	?	+
Occurrence	Most abundant crustal metal sulfide	Tends to form at acid pH. Absent or rare in rocks older than Palaeozoic	Poss. most abundant universal metal sulfides, Includes troilite	In "black FeS" of sediments, and certain ores. Often contains high Co, Ni and Cr	In "black FeS" of sediments and young sediments	In ores, usually associated with pyrrhotites	Not found naturally. Synthesized through the wet sulfidation of iron plates at 60°C
Heat	→$Fe_{(1-x)}S$	→Pyrite	—	→Pyrrhotite	→Pyrite + pyrrhotite/mackinawite	→Pyrite + pyrrhotite	→Mackinawite/pyrrhotite
+ S°	—	—	→Pyrite	→Marcasite/pyrite	→Marcasite pyrite	—	—
Reactions							
+ S_n^{2-}	—	—	—	→Pyrite	→Pyrite	—	—

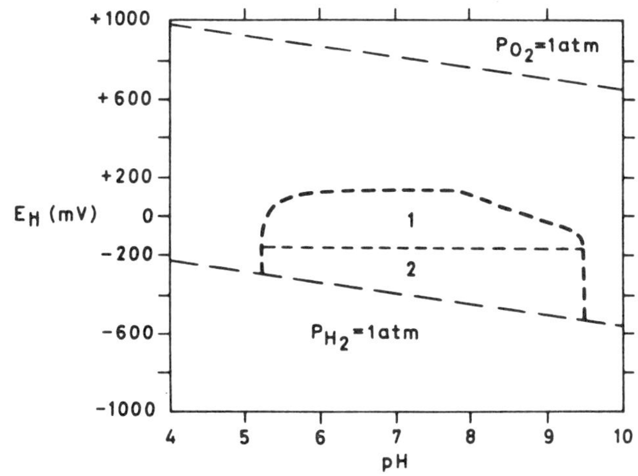

FIGURE 53. The sulfate-reducing environment in terms of pH and E_H from data collected by Baas Becking, et al.[29] 1. Environment of SO_4^{2-} reduction. 2. Max. E_H for intracellular SO_4^{2-} -reducing enzymes.

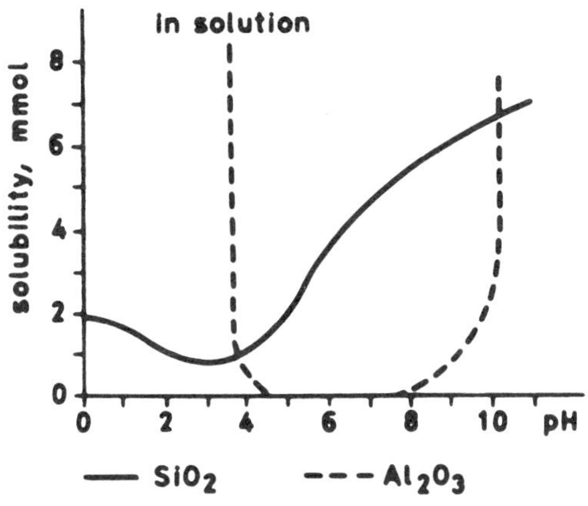

FIGURE 54. Solubility of Al and SiO_2 at various pH values.

important rock-forming mineral in mafic and ultramafic rocks. The saturation concentration of wolastonite (Ca-metasilicate) is 95 mg/ℓ. The different forms of sodium monosilicate (waterglass) dissolve in water, some of them having quite good solubility. The solubility of sodium monosilicates is influenced by the concentration of the Na^+ and K^+ ions and by the pH of the solution.

The saturation concentration of silicic acid depends on several factors:

Kind of rock and soil-forming material found in contact with the pore water — The equilibrium concentration of silicic acid in the solution of the pores becomes higher as the stability of the silicate minerals of the rocks decreases. 3.5 to 4 mg/ℓ SiO_2 dissolves out from quartz, the saturation concentration is 10 to 15 mg/ℓ in the case of biogenic opal, while 50 to 65 mg/ℓSiO_2 dissolves from amorphous silicon at pH 7.

Forms of silicic acid in the solution — The saturation concentrations are higher for monomers and dimers than in the case of silicate polymers. Organosilicates dissolve better than the inorganic salts of ortho- and metasilicic acids.

pH of the media — It is evident from Figure 54 that there is a slight increase in the silicic acid concentration in aqueous solution if the pH value drops below 3.5. The increase in solubility is very sharp if the pH of the medium increases to over 9. The saturation concentration of silicic acid depends even more on the pH value in the case of a soil-solution system. The increase in the H ion concentration of the solution, i.e., the acidic reaction of the medium, promotes the dissolution of metal-aluminum silicates through the desilification of rock and soil minerals, and the formation of orthosilicic acid:

$$\text{M-Al-silicate} + a\,H^+ + b\,H_2O = M^{2+} + cH_2SiO_4 + \text{Al (silicate residue)} \quad (26)$$

The sharp increase in the silicic acid concentration when the pH value is higher than 9 is due to the ionization of monosilicic acid:

$$Si(OH)_4 + OH^- \rightarrow Si(OH)_3O^- + H_2O \quad (27)$$

Concentration of sodium and potassium salts in the solution — An increase in the concentration of Na^+ and K^+ ions leads to an increase in the dissolved silicic acid due to a decrease in ion activity coefficients parallel to an increase in the ionic concentration of the solution, and the formation of readily soluble Na silicates. The chemistry of the groundwaters indicates the dependence of SiO_3^{2-} concentration on the total salt concentration of the water and on the chemistry of salinization. Differences in the dissolution of silicic acid are also caused by different geological conditions. The increased concentration of SiO_3^{2-} ions is caused in some cases by an increased concentration of Mg sulfates in the solution and a high carbonate content in the parent material. It should also be mentioned that the silicic acid in salinized water does not always play a decisive role in spite of an increase in the concentration. Water with a salinization of 150 to 350 mg/ℓ has a silicate type of chemistry, and as the salinization increases, the chemistry of the water is determined by the accumulation of carbonate, bicarbonate, sulfate, and chloride ions.

Size of silicate-containing rocks — With a decrease in the size of the particles, the specific surface of the ground and soil materials increases, as does the solubility of silicate minerals. Hydromorphic conditions promote the desilification of minerals and the destruction of organo-mineral complexes in the soil. The formation of the solodized "A" horizon of solonetz soils is a good example of the solubilization of soil organic matter, the destruction of silicate minerals, and the increase in amorphic silicate content. The destruction of silicates may also be noted in the surface layer of alkali soils under irrigation.[254]

Silicic acid is not toxic to plants. Plants grown under hydromorphic conditions, such as rice, assimilate part of the dissolved silicic acid and may contain a considerable quantity of silicic compounds.

D. Chlorides

The chlorides, together with the sulfates, are the compounds mainly responsible for the formation of saline soils. They are characterized by high solubility, and consequently, high toxicity.

1. Calcium Chloride

Calcium chloride is rarely present in soils. Reacting with sodium sulfate and sodium carbonate, the calcium precipitates in the form of calcium sulfate or calcium carbonate.

2. Magnesium Chloride

Magnesium chloride is more common in soils than calcium chloride, but it accumulates in large quantities only where there is high salinity. Its solubility is high, and as a result, it has high toxicity. Complex calcium and magnesium chlorides may occur in the soils of arid zones, e.g., in southern California, and the high plateaus of Algeria, where, because of their hygroscopic character, certain spots remain continually humid.

3. Sodium Chloride

Sodium chloride, together with sodium and magnesium sulfate, is the most common component of saline soils. It has high solubility and high toxicity.

The leaching of saline soils containing both sodium chloride and calcium sulfate is very easy. But if the soil does not contain calcium sulfate, the leaching of sodium chloride solonchaks can cause the alkalization of the soil.

E. Borates

Boric acid salts rarely occur in excess in soils. When this happens, it is usually due to specific geological conditions.

F. Nitrates

Nitrates (mainly $NaNO_3$) occur in extreme arid conditions in saline soils. Nitrate accumulation in non-salt-affected soils, due to the intensive application of mineral fertilizers or sludges, is a different problem not discussed in this book.

III. ION EXCHANGE PHENOMENON (by K. Darab)

The ion exchange phenomenon in soils results in pronounced changes in the soil colloidal status, and consequently in the physical, chemical, and biological properties of the soils.[18,249] Soil colloids and ion exchange are of very great importance in soil classification, soil utilization, and soil amelioration.[2] The ion exchange phenomenon is a function of the chemistry of both the liquid and solid phases of the soil as well as of the original colloidal and mineralogical media in the soil.[255-258]

Not only in soil science but in other branches of science, particularly in physical chemistry, mineralogy, etc., a large number of references are available on different aspects of the theory and practice of ion exchange.[259,260] Below, a short review will be given in order to facilitate an understanding of the contents of the following chapters.[261-263]

A. Ion Exchange in Soils

The term "ion exchange" indicates the reversible process by which cations and anions are exchanged between the solid and liquid phases, if these are in close contact with each other.[264-266] Theoretically, several approaches can be used to express the distribution of ions between solid and liquid phases at equilibrium.[261,268-269] In practice, most of the theories lead to equations identical with the mass action relationship:[270]

$$\frac{(A_{ad})^n}{(B_{ad})^m} = K_B^A \frac{(A^{m+})^n}{(B^{n+})^m} \tag{28}$$

where (A_{ad}) and (B_{ad}) = the activity of A^{m+} and B^{n+} ions on the adsorbent; (A^{m+}) and (B^{n+}) = the activity of the same ions in the intermicellar solution; and m and n = the valences of cations A and B, respectively.

It is evident from the equation applied to characterize the cation exchange equilibrium that the kind and quantity of ions adsorbed on the surface of soil particles depend on: the activity ratio of cations in the solution, the valences of the cations in the case of nonsymmetrical exchange, and the total concentration of the intermicellar solution.

It has been proved in several experiments that the relative replacing power of cations depends on the valences of the adsorbed and counter ions. In the case of nonsymmetrical exchange, the relative replacing power of the cations increases with the valence and follows the order $M^+ < M^{2+} < M^{3+}$ . . .[271,272]

The greater the dilution of soil solution systems containing cations with different valences, the greater the displacement of the equilibrium in such a way that the adsorption of cations

Table 53
THE AMOUNT OF EXCHANGEABLE SODIUM IN BENTONITE $NaCl$-$CaCl_2$ SOLUTION SYSTEMS[274,275]

I mol/ℓ	$[Na^+]$ (10^3 mol/ℓ)	$[Ca^{2+}]$ (10^3 mol/ℓ)	$\frac{[Na^+]}{[Ca^{2+}]}$	$\frac{[Na^+]^2}{[Ca^{2+}]}$	$\frac{[\gamma Na^+]^2}{[\gamma Ca^{2+}]}$	Exchangeable Na^+ (meq/100 g)	CEC (meq/100 g)	ESP
$5.0 \cdot 10^{-3}$	4.6	0.14	32.62	0.15	1.179	16.03	96	16.7
$7.5 \cdot 10^{-3}$	7.1	0.23	31.21	0.22	1.223	23.00		23.9
$1.0 \cdot 10^{-2}$	9.2	0.28	32.91	0.30	1.262	28.20		29.4
$2.5 \cdot 10^{-2}$	21.9	0.71	30.95	0.68	1.444	38.07		39.7
$5.0 \cdot 10^{-2}$	41.3	1.52	27.26	1.12	1.682	50.85		53.0
$2.5 \cdot 10^{-3}$	1.89	0.11	18.60	0.03	1.123	8.5	96	8.9
$5.0 \cdot 10^{-3}$	4.31	0.18	24.70	1.10	1.179	12.0		12.5
$7.5 \cdot 10^{-3}$	7.04	0.26	26.90	0.19	1.223	15.15		15.8
$1.0 \cdot 10^{-2}$	9.09	0.35	25.70	0.24	1.262	16.45		17.1
$2.5 \cdot 10^{-2}$	21.90	1.06	20.70	0.45	1.444	25.15		26.2
$2.5 \cdot 10^{-3}$	2.06	0.11	18.60	0.04	1.123	4.23	96	4.4
$5.0 \cdot 10^{-3}$	4.31	0.23	18.90	0.08	1.179	7.70		8.0
$7.5 \cdot 10^{-3}$	6.52	0.34	18.90	0.13	1.223	9.78		10.2
$1.0 \cdot 10^{-2}$	8.26	0.51	16.40	0.13	1.262	11.82		12.3

Note: The ratio of bentonite to solution is 1:100.

with higher valence increases, and the adsorption of cations with lower valence decreases. On increasing the ionic concentration of the intermicellar solution, the equilibrium shifts in favor of cations with lower valences.[265,273]

The effect of the valence of exchangeable and counter ions and the dilution of the intermicellar solution on the sodium saturation is reflected in the data obtained in bentonite-$CaCl_2$-$NaCl$ solution systems (Table 53).[274,275]

In the assymmetrical exchange of sodium and calcium ions, the adsorption shifts in favor of the calcium ions as the ionic concentration of the free solution decreases, because:

- The Na^+ concentration decreases much faster during dilution than that of Ca^{2+}, if the ratio of the exchanger to the solution is kept constant.
- The activity coefficients are influenced by the valences and sizes of the ions and by the total ionic concentration of the solution. The ratio of the activity coefficients of sodium and calcium ions decreases much faster in a free electrolyte solution during the dilution of the suspension than it does on the exchanger.
- The dilution effect can be achieved by decreasing the ionic concentration of the intermicellar solution,[256] by decreasing the ratio of the solution to the adsorbent, or by increasing the cation exchange capacity of the adsorbent (Table 54).[274]

In all these cases, the increase in the preference of calcium ion adsorption over that of sodium ions is due to an increase in the difference between the concentrations of the micellar and intermicellar solutions and to an increase in the potential differences between the surface of the adsorbent and the free solution. The preference of sodium ion adsorption comes to the fore as the concentration of free electrolytes increases, the solution-adsorbent ratio increases, or the cation exchange capacity of the adsorbent decreases.

The total ionic concentration of the soil solution varies over a wide range. In nonsaline soils it is relatively low and does not usually exceed 0.01 mol/ℓ. In salt-affected soils, depending on the chemistry and degree of salinization, and/or alkalization, the salt concentration can be as high as 0.1 to 5 mol/ℓ, or even more.

Table 54
THE AMOUNT OF EXCHANGEABLE SODIUM AT DIFFERENT CONCENTRATIONS OF BENTONITE AND DIFFERENT IONIC CONCENTRATIONS OF SOLUTION IN BENTONITE-NaCl-$CaCl_2$ SOLUTION SUSPENSIONS[274]

I (mol/ℓ)	Bentonite solution	$[Na^+]$ (10^3 mol/ℓ)	$[Ca^{2+}]$ (10^3 mol/ℓ)	$[Na^+]/[Ca^{2+}]$	Exchangeable Na^+ (meq/100 g)	Exchangeable Na^+ (meq/ℓ)	CEC (meq/ℓ)	ESP
	1:50	8.17	0.16	51.6	14.4	3.10	18.7	16.6
$1.14\ 10^{-2}$	1:75	8.34	0.16	52.2	18.1	2.30	11.8	19.5
	1:100	8.34	0.16	52.2	19.4	2.00	9.6	21.2
	1:200	8.34	0.15	57.6	29.8	1.30	4.0	32.9
$1.0 \cdot 10^{-2}$		6.95	0.69	10.1	5.98	0.60		6.2
$5.0 \cdot 10^{-2}$	1:10	34.30	3.69	9.3	16.83	1.68		17.5
$1.0 \cdot 10^{-1}$		86.90	7.88	11.0	40.35	4.03		42.0
$1.0 \cdot 10^{-2}$		7.83	0.39	20.1	11.03	0.44		11.5
$5.0 \cdot 10^{-2}$	1:25	38.70	1.50	25.8	31.36	1.25		32.7
$1.0 \cdot 10^{-1}$		100.01	4.45	22.5	66.81	2.67		69.6
$1.0 \cdot 10^{-2}$		8.17	0.16	51.1	14.30	0.29		14.9
$5.0 \cdot 10^{-2}$	1:50	40.40	0.79	51.1	50.00	1.00		52.1
$1.0 \cdot 10^{-1}$		105.20	2.30	45.7	97.17	1.94		100.0

Note: The average of CEC for bentonite: 90.5 meq/100 g.

In solutions with high concentrations of salts, a special electrolyte ''structure'' is formed, due to the electrostatic interaction of the ions.[276-279] This interaction has an effect on the concentration of ions with free valences, and as a result of the formation of the electrolyte ''structure'', it influences several properties of the electrolyte, including the degree of dissociation of the dissolved salts and the ratio of ion activities in the solution.[278,280-283] The ion-pair formation model is one approach used to describe the electrolyte ''structure'' formed during the interaction of ions with opposite charges.[280,284]

The decisive role of the sodium ions in the soil solution is strengthened by the different abilities of cations to form ion-pairs. The different degree of ion-pair formation in the case of sodium, calcium, and magnesium ions, raises the SAR value in the soil solution. The increase in the SAR value due to ion-pair formation becomes more important as the ionic concentration of the soil solution increases and shifts the balance of exchangeable cations in favor of sodium.

The ratio of exchangeable cations depends on the kind of anions present. The effect of the anion is due to the following:

- The ratios of ion activities in solution, a factor included in all the ion exchange equations, are different, and depend upon the kind of anions present.[285-287]
- In nonequilibrium systems, the solubility of calcium-sodium exchange products has an influence on the ionic composition of the solution and on the ratio of exchangeable cations. An inverse relation exists between the solubility of compounds formed in the course of ion exchange, the calcium concentration ratio, and the degree of sodium saturation of the adsorbent. Equilibrium is reached in a chloride-containing system at a relatively low degree of sodium saturation. The presence of calcium sulfate causes a relatively stable concentration of calcium ions in the free solution of sulfate-containing systems, and the ratio of sodium to calcium ions in the intermicellar and micellar solutions depends on changes in the total ionic concentration of free electrolyte solutions.

In carbonate-containing systems, the sodium ions dominate in the intermicellar solutions as a result of the poor solubility of calcium carbonate, and the adsorbent has a high degree of sodium saturation even if the ionic concentration of the liquid phase is low. In these systems the pH value of the medium determines the solubility of calcium carbonate, and consequently has an effect on the cation exchange equilibria of soil-solution systems. The activity of calcium ions in solution can be calculated from the thermodynamic solubility products of calcium carbonate:

$$\lg(Ca^{2+}) = \lg K_{CaCO_3} - \lg(CO_3^{2-}) \qquad (29)$$

The activity of the carbonate ions depends on the pH of the solution:

$$\lg(CO_3^{2-}) = \lg K_{d_1}K_{d_2}(H_2CO_3) + 2pH \qquad (30)$$

where K_{d_1} and K_{d_2} = the first and second dissociation constants of carbonic acid and (H_2CO_3) = the activity of carbonic acid in the solution.

From the data of the sodium carbonate and bicarbonate solution, it is found that the product of the two dissociation constants and the activity of carbonic acid is a linear logarithmic function of the square root of the ionic concentration in the solution:

$$\lg[K_{d_1}K_{d_2}(H_2CO_3)] = a + b\sqrt{I} \qquad (31)$$

The activity of carbonate ions depends on the pH value and ionic concentration of the solution, as follows:

$$\lg(CO_3^{2-}) = a + b\sqrt{I} + 2pH \qquad (32)$$

The calcium ion activity can be calculated by combining the equations, as follows:

$$\lg(Ca^{2+}) = a' - b\sqrt{I} - 2pH \qquad (33)$$

where $a' = -a + \lg K_{CaCO_3}$.

On substituting the activity of calcium ions into the cation exchange equation, it is found that the ratio of exchangeable sodium and calcium ion activities depends on the hydrogen-sodium potential and the ionic strength of the intermicellar solution:

$$\lg \frac{(RNa)^2}{(RCa)} = a'' + b\sqrt{I} + 2(pH\text{-}pNa) \qquad (34)$$

where $a'' = \lg K_{Ca}^{Na} - a'$.

The pH value of the media has an effect on the ion exchange equilibria of soil solution systems. Investigations carried out on phase-equilibrium systems show that the influence of the pH value on the soil-colloid systems not only manifests itself in changes in the solubility of poorly soluble salts, but also affects the degree of dispersion, the cation exchange capacity and the surface charge density of the adsorbent. The distribution quotient of ^{24}Na decreases as a rule with an increase in the concentration of the solution in the presence of chlorides and sulfates, but it increases and reaches a maximum value if sodium salts capable of alkaline hydrolysis are present in the free electrolyte solution (Figure 55). From the material balance of the system, it follows that with an increase in the isotope distribution quotients (in spite of the increase in the solutions), sodium ion concentrations indicate an increase in the cation exchange capacity of the adsorbent.

The increase in exchange capacity can be either apparent or real. Real changes in the CEC value originate from the colloid chemical and structural properties of the solid phase, while apparent increases are caused by changes in the exchangeable cation activity in swelling systems.[288]

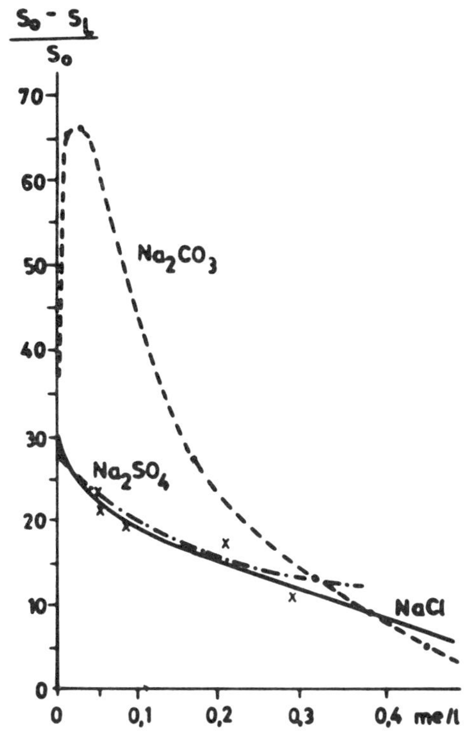

FIGURE 55. Dependence of bentonite Na-24 activity as a precentage of the total Na-24 activity on the equilibrium Na^+ concentration in bentonite – NaCl, Na_2SO_4 and Na_2CO_3 solution systems. Horizontal axis: Na^+ concentration of the equilibrium solution, meq/ℓ. Vertical axis: S_O = total Na-24 activity; S_L = activity of the equilibrium solution.

In the case of calcium-sodium ion exchange, the value of the selectivity coefficient (calculated from the Vanselow equation) depends on the molar fraction of exchangeable sodium and the total concentration of the intermicellar solution. The value of the Vanselow selectivity coefficient measured in bentonite-sodium chloride solution systems increases as the total concentration of the solution decreases.[289] The increase in the selectivity coefficient indicates that the system will approach the ideal state if the dilution of the solution increases.

In the case of constant ion concentration, the selectivity coefficient is nearly constant at low degrees of sodium saturation, but increases abruptly when a critical value for the molar fraction of exchangeable sodium is exceeded. The more diluted the external electrolyte, the lower the sodium saturation at which the selectivity of the adsorbent for sodium ions decreases. A change of this type indicates that the increase in exchangeable sodium is accompanied by certain alterations in the behavior of the adsorbent.

The swelling of the adsorbent is probably one of the most important changes. An increase in the swelling pressure changes the ion concentration close to the surface and reduces the differences between the concentrations of the micellar and intermicellar solutions. The increase in the value of the selectivity coefficient also indicates that the surface properties of the adsorbent may change parallel with the increase in its sodium saturation. The distribution of exchangeable cations on the surface is usually not homogeneous. At low degrees of sodium saturation, the sodium ions occupy sites with low energy and are more mobile than the calcium ions adsorbed on sites with high adsorption energy. As the sodium saturation increases, the alkalinity of the medium and the dispersion of colloid particles also increase, the degree of crystallization decreases, and sites with low mobility become accessible to exchangeable sodium. The occurrence of exchangeable cations with different energies and mobilities means that the cation exchange equilibrium is influenced by the interaction of the micellar and intermicellar solutions, the diffusibility of cations close to the surface, and the mobility of ions on the surface of the adsorbent.[290]

B. Irreversible Changes Connected with Cation Exchange Processes

Cation exchange is a reversible process. The state of equilibrium is determined by the concentration and chemistry of the intermicellar solution and the activity ratio of the exchangeable cations.

It is evident from the cation exchange equations that salinization favors the adsorption of sodium ions, even in cases where the ratio of sodium ions remains unchanged during the increase in the concentration of the soil solution. Salts usually accumulate in soils in reverse sequence of their solubility, and as the total concentration increases, the ratio of sodium ions also increases. On the other hand, the ion exchange equilibrium shifts in favor of cations with two valences during leaching, and not only soluble sodium, but also exchangeable sodium, is leached from the soil, which thus becomes nonsaline and nonalkaline. This evolution of salt-affected soils into fertile soils does not always take place. The decrease in exchangeable sodium during desalinization is often counteracted by the increasing alkalinity of the soil solution. The increase in pH value involves the precipitation of calcium ions in the form of calcium carbonate, an increase in the degree of dispersion, and the redistribution of exchange sites on the surface of the adsorbent. The result of these changes is the formation of alkali soils with a structural B horizon.

The structure of the adsorbent may also change during desalinization. Among these changes, mention should be made of the dissolution of sodium-saturated organic matter and the formation of swelling dioctahedral montomorillonite from its trioctahedral form.

The result of these changes in the structure of soil colloids is that the morphology and poor fertility of the soil remains, even if the degree of sodium saturation decreases below 5 to 10%.

Chapter 4

TYPES, FORMATION, AND PROPERTIES OF SALT-AFFECTED SOILS

I. INTRODUCTION

In the processes of salinization, the various salts accumulate in inverse proportion to their solubility. In nonsaline or slightly salinized soils, the soil solution is governed by the Ca^{2+}, Mg^{2+}, and HCO_3^- ions. As salinization increases, the more soluble salts become dominant. This also means that in the course of desalinization, the most soluble salts are leached out first.

The effect of water-soluble salts on soil-forming processes depends on environmental conditions as well as on the chemical types of electrolytes causing salinity and/or alkalinity (as discussed in Chapter 1). The mineral and organic parts of the soil material also influence the formation and properties of different groups of salt-affected soils.[1]

On the soil surfaces in the deserts of Asia, Africa, and Latin America, salt "crusts" often accumulate, which do not mix with fine earth or skeletons, and which are, chemically speaking, virtually pure. Their salt content can be as high as 80 to 100%, on a weight basis, and their thickness anywhere from a few centimeters to several meters. Some salt crusts are dense (monolithic) in structure, very nearly impermeable to solutions, difficult for roots and tilling instruments to penetrate, and devoid of natural fertility. They are nevertheless of economic value as sources of raw materials for the chemical or building material industries.[3,291]

Contemporary salt crusts, formed during the present period mostly through the evaporation of marine and lacustrine waters, or the effects of subsoil groundwaters (the mother solutions), are still in the process of growth, whereas residual salt crusts, whose mother solutions are not longer present, are not. According to their chemical compositions, they are classified as calcareous crusts, gypsum crusts, or soluble salt crusts.[292]

The different groups of salt-affected soils will be characterized below according to Table 2 (in Chapter 1).

II. SALINE SOILS

In arid regions, deserts, and semideserts, the majority of strongly saline soils (having water-soluble salts in their profile and a conductivity higher than 10 mmho) contain dominant quantities of NaCl and Na_2SO_4. In several soil classification systems these soils are referred to as chloride and/or sulfate solonchak. Most saline soils are solonchak. As a rule, they occur in large arid and semi-arid regions in Eurasia, North Africa, western North and South America, and in central Australia.

NaCl and Na_2SO_4 accumulate in varying degrees in the soil profile. The salinization process may be so intensive that a salt crust forms on the soil surface and the salt content increases appreciably in the various soil horizons of the soil profile. When referring to saline deserts, it is hardly possible to speak of soils, because these strongly saline formations are completely devoid of the most essential property of what is implied in the word fertility. It is evident that all other properties of saline soils are conditioned by their high salt content as well.

In the project for the preparation of the World Map of Salt-affected Soils, the following definition of the term *saline soils* was accepted:[1]

"Soils having a saline (salic) horizon within 125 cm below the surface (125 cm in the case of coarse texture, 90 cm in the case of medium texture, and 75 cm in the case of fine texture) or having an electrical conductivity of more than 4 mmho in at least some part of the soil within 25 cm below the surface; if the pH (H_2O 1:1) in this layer is 8.5 or less, an electrical conductivity of more than 15 mmho should occur within 125 cm in the case of coarse texture, 90 cm in the case of medium texture, and 75 cm in the case of fine texture below the surface.''

A similar definition was adopted in the course of the preparation of the FAO/UNESCO Soil Map of the World.[20-22]

In spite of the fact that the salt content of saline soils in most cases significantly surpasses the limit values given, the definition must be considered generalized, and may be modified according to local environmental conditions. For instance, in Europe or in other climates which are not arid or semi-arid, a salt accumulation exceeding 4 mmho EC on the surface of soils is only a precondition for salinization. In these soils the soil solution is nearly saturated with salts and, further the salts occur — in crystalline or non-crystalline forms — in the solid phase. If the soils have appreciable clay fractions, their colloidal properties are also determined by the high electrolyte concentration.

Salt-affected soils formed under the influence of neutral sodium salts do not always contain a large enough quantity of water-soluble salts to cause salt efflorescences on the surface.

According to the salt concentration, various types of saline or solonchak soils develop. As a rule, their profiles have no structural horizons, and no abrupt boundaries can be observed between the individual horizons. Morphologically they can hardly be distinguished.

Irrigation, drainage and salinity — An International Source Book described solonchaks in the USSR and other countries, where they occupy vast arid regions, in the following terms:[19]

The bottom part of solonchak profiles, soaked with capillary and groundwater solutions, is characterized by dark bluish, rust, and olive tints. The surface of solonchaks has not, as a rule, much plant cover, except for certain types of halophytes (pulpy, aliphatic saltworts; solonchak shrubs), which are adapted to saline soil conditions. It is only on the solonchaks of meadows that a plant cover is found and the profile of the soil is dark.

As regards their physical composition, solonchaks are frequently clayey, due to the fact that they are formed in lowlands, depressions, and hollows, and contain accumulations of clayey, soil-forming rocks; however, they are also occasionally formed from rocks of a light physical composition. They are characterized by the lack of a water-stable, cloddy structure. But at the same time, owing to the presence of a large number of electrolytes, the colloido-clayey solonchak substance is firmly coagulated, with the result that solonchaks, so long as they contain salts, are porous, permeable, and have some aggregation (pseudo-structure). Solonchaks as a rule contain only a very small quantity of humus (0.7 to 1.3%) in the top horizon. Only solonchaks of forest-steppe regions (such as the soda or sulfate solonchaks of Manchuria, Argentina, California, the Ukraine, and Siberia), which are formed from meadow soils, have a humus content as high as 2 to 3%. The soil solution reaction of solonchaks is slightly alkaline.

The content of readily soluble salts in solonchak profiles decreases in the lower horizons. It is in the top horizons that the accumulation of readily soluble salts (NaCl, $MgSO_4$, Na_2SO_4, Na_2CO_3, and $MgCl_2$) is greatest. Most solonchaks, however, also contain $CaSO_4$ and $CaCO_3$. The gypsum content in the top salt horizon may be as much as 3%, or even in some cases 10%. Under natural conditions, owing to the constant presence of capillary water, the readily soluble salts in solonchaks take the form of a saturated soil solution. The soil solution of solonchaks is characterized by a high concentration (100 to 200 g/ℓ or in some cases up to 400 g/ℓ) in the top horizons, dropping to 20 to 50 g/ℓ in the lower horizons.

Solonchaks are divided into various types on the basis of their hydrological, morphological, and chemical features. They vary greatly as regards economic potential. Before reclaiming solonchaks, it is important to assess them correctly with regard to hydrology and chemistry. From this point of view, it is essential to distinguish between the following five types:[3]

A. Active Solonchaks (Contemporary)

Active solonchaks are connected with groundwaters through a "capillary fringe" (or through a film-capillary flow) of salt solutions. In these solonchaks, the reserve of readily soluble salts, despite seasonal variations, gradually grows. The groundwater table of such soils lies at a depth of 0.5 to 3 m, according to the time of year. Solonchaks with film-capillary moisture have groundwaters at a depth of 5 to 9 m (i.e., not very great).

B. Residual (Dry) Solonchaks

Residual solonchaks are often to be found in the arid zones of North Africa, Central Asia, and Latin America.[294] Their groundwaters occur at depths of more than 20 to 25 m, but owing to the extreme aridity of the climate, the higher concentration of readily soluble salts is in the topsoil horizons. The formation of saline soils may be the result either of natural processes or of poor management. A distinction must therefore be drawn between natural or primary solonchaks, and secondary solonchaks, formed as a result of man's failure to control the salt regime of the soil. There may be great differences between the two types. (Secondary salinization and alkalization of soils will be treated later in this book.)

C. Wet Mineral Solonchaks (sabakh soils in Iraq, sebkhas soils in North Africa)

The term "wet mineral" is applied to solonchaks with a very high groundwater level, so that the soil is saturated with capillary water, which keeps it continually moistened. The term "wet" is also sometimes applied to solonchaks where the soil retains, in the upper salt crusts, quantities of magnesium chloride and calcium chloride, which, being highly hygroscopic, absorb atmospheric moisture, so keeping the soil surface very moist and making it dark in color.

D. Puffy (Fluffy) Solonchaks

The term "puffy" is applied to solonchaks whose surface consists of a fluffy, loose, dusty salt horizon. As a rule, the fluffy salt horizons are composed of a mixture of earthy substances derived from soil-forming rocks, a powdery, fine-crystalline mass with 10 to 20% $CaCO_3$ and $CaSO_4$, and a fine-crystalline mass of readily soluble salts (NaCl, Na_2SO_4). Fluffy salt horizons are often scattered by the wind and blown considerable distances to form salt dunes of pseudo-sand.

E. Crust Solonchaks

In some cases, due to periodical moistening, readily soluble salts (particularly solonchak gypsum through recrystallization) form a solid salt top crust which may sometimes be as much as 2 to 5 cm thick and have a high degree of mechanical stability.

Kovda divides solonchaks according to their salt composition as follows:[19]

Gypsiferous and Nongypsiferous	Nongypsiferous
Nitrate	Sulfato-soda
Nitrate-chloride	Soda
Chloride	Borate
Sulphato-chloride	
Chlorido-sulfate	
Sulfate	

F. Nitrate Solonchaks

Nitrate solonchaks contain $NaNO_3$, and KNO_3. As regards toxicity and the reclamation measures needed, nitrate solonchaks resemble chloride solonchaks. Nitrate solonchaks and accumulation of saltpeter are widespread in the deserts of the western U.S., Chile, and Peru, and also in the dry areas of Bolivia, Colombia, and Argentina. Saltpeter solonchaks are also common in the river deltas of eastern China. In the plains along the lower reaches of the rivers Syr Darya, Murghab, and Kashka Darya in Central Asia the solonchaks formed in some places contain as much as 5 to 10% NO_3 in the topmost soil horizons. In the deserts of Central Asia, Asia Minor, and America, nitrates accumulate in combination with chlorides. Nitrato-chloride solonchaks characterized by an exceptionally high content of readily soluble salts (up to 25%) are among the most harmful and barren solonchaks and are extremely difficult to reclaim and develop. When assessing the possibility of reclaiming these solonchaks, the question of whether or not they contain gypsum in the top horizons is of vital importance.[295]

G. Chloride Solonchaks

Chloride solonchaks contain NaCl, $MgCl_2$, and in some cases, $CaCl_2$. They exist mainly in areas of marine salinization, in areas rich in NaCl, or in regions where deep-lying salty groundwaters are forced up to the surface. In this last type of solonchak, the salt content and the degree of salinization of the groundwater are as a rule exceptionally high. The hydrophysical properties of chloride solonchaks during leaching depend upon whether they are gypsiferous or nongypsiferous. Where gypsum is present, as it very often is, reclamation is easy. Nongypsiferous chloride solonchaks, on the other hand, are difficult to leach because of their low permeability, their alkalinity, and the peptization of the soil (alkalization). In such cases, preliminary measures have to be taken in order to increase the effectiveness of leaching — deep loosening of the soil, subsoiling, installation of shallow temporary drains, and sometimes gypsuming.

Chloridic solonchaks may be hygroscopic, if Mg and Ca chlorides largely predominate over Na chlorides. They remain humid even during dry seasons and their surfaces are dark-colored. The profile described below is that of a chloride solonchak in Oued Melah (Tunisia) in a dissipating area of the Zeroud in the badly drained region of Bled Sesseb-Sidi Saidane (northern Kariouan), separated from the Sebkha Kelbia by the Draa Es-Souatir (sand bars). The profile represents the situation in February 1969 (see also Tables 55 and 56).

Vegetation:	*Halophytes, Suaeda fruticosa, Atriplex halimus.* Grain crops are grown after flooding; during rainy years the surface is cracked.
0 — 10	Gray-brown silty clay; compact surface structure; coarse blocky texture going to platy; dry; a few salt efflorescences.
10 — 30	Gray-brown clay; coarse blocky structure going to platy; compact; a few roots which go no deeper than this level.
30 — 55	Dark brown silty clay; massive structure with a few shrinkage cracks; matted with gypsum crystals and salt.
55 — 75	Gray-brown silty clay; massive structure; many gypsum crystals and salts on the walls; fresh.
75 — 135	Slight brownish-gray; massive structure; fresh; some rust-ochre mottles.
135 — 185	Brownish-yellow with some rust-ochre mottles and black ferro-manganese spots.

The soil is apparently two soils, one (recent alluvium) superposed on the other, older hydromorphic soil, is characterized by mobile iron and an accumulation of blackening organic matter. In the same zone there are soils presenting a hydromorphism with calcareous accumulation.

Chloride solonchaks are widespread on the seashores of practically all continents, on tidal marshes and coasts where, due to the permanent effect of seawater, temporary or permanent waterlogging results in the accumulation of sodium chloride at the surface. This ecosystem has various local names. On the Korean Peninsula, for example, it is known as "Kanzuchi", which means salty marshes. This formation is widely utilized for sodium chloride production in many countries of South and Southeast Asia as well as in Africa. In many places acid sulfate soils are often related to the saline soils at seashore marshes.[95,296]

In Figures 56 and 57 the environment and profile of a Kanzuchi soil (Korea) are demonstrated.

Table 55
CHEMICAL CHARACTERISTICS OF A CHLORIDE SOLONCHAK PROFILE IN TUNISIA

Depth (cm)	pH	Lime percentage Total	Lime percentage Active	Gypsum (%)	SP (%)	EC (mmho)	Cl^- (meq/100 g soil)	HCO_3^- (meq/100 g soil)	Ca^{2+} (meq/ℓ)	Mg^{2+} (meq/ℓ)	Na^+ (meq/ℓ)
10—30	8.7	10.6	7.5	—	80	29.5	203	3	35	25	362.5
30—55	8.7	13.1	12	3.6	60	25.5	165	2.5	30	32	307.5
55—75	8.9	25.1	30.5	10.1	80	33.7	265	—	—	—	—
75—135	8.95	37.4	15	7.1	70	41	337	—	—	—	—

Table 56
ABSORPTION COMPLEX OF A SOLONCHAK PROFILE IN TUNISIA

Depth (cm)	Ca^{2+}	Mg^{2+}	K^+	Na^+	ESP
0—30	8.96	1.14	1.1	13	53.7
30—55	15.63	0.55	0.55	8.55	49.5

Note: Values are meq/100 g soil.

FIGURE 56. Environment of a Kanzuchi soil on the Korean Peninsula.

In Table 57 the chemical composition of a coastal saline soil is given, which shows that the degree of salinity in the soil is close to that of seawater.

H. Sulfato-Chloride Solonchaks

Sulfato-chloride solonchaks are common in Central Asia (e.g., Tadzhikistan and Uzbekistan), in the Kura-Araks depression in Asia Minor, and around the lower reaches and deltas of the Rivers Volga, Ural, Emba, and Atrek on the Caspian Sea coast. They are also to be found in North Africa, Central Australia, the Indian subcontinent, and in other semideserts and dry steppes.

Sulfato-chloride solonchaks contain chlorides and sulfates in varying proportions, with the chlorides predominating, a large quantity of readily soluble salts in the top horizons (3 to 5%), and in some cases gypsum. In order to reclaim sulfato-chloride solonchaks and bring them under cultivation, a considerable outlay on leaching and salt removal is necessary; but there is less danger of alkalization during leaching than in nongypsiferous solonchaks. In some cases, chloride and even sulfato-chloride solonchaks, usually rich in clay, may undergo alkalization processes through a large increase in the exchangeable sodium in their absorbing complexes. Despite their salt contents, their permeability is low, because of the dispersion of the clay. These solonchaks are referred to as saline-alkali soils, and their reclamation is usually very difficult.

I. Chlorido-Sulfate Solonchaks

Chlorido-sulfate solonchaks are characterized by the predominance of sulfates (Na, Mg) over chlorides. These solonchaks are, as a rule, rich in gypsum. They are widespread in

FIGURE 57. Profile of a Kanzuchi soil (Korea).

Table 57
CHEMICAL CHARACTERISTICS OF A SALTY MARSH SOIL (KOREA)

Depth (cm)	% of salts	pH	Organic matter (%)
0—4	3.15	6.5	1.2
5—15	2.62	7.1	0.8
20—25	2.32	6.8	0.6
40—50	1.85	5.2	0.5
80—90	1.94	6.5	—
110—120	1.89	6.2	—

the Ferghana Valley, along the lower reaches of the Zeravshan (Bukhara and Karakum cases), along the lower reaches of the Kashka Darya, and also in southern Azerbaidzhan. They are also extensive in Indo-China, China, dry areas of Africa, etc.

Although the salinity of chlorido-sulfate solonchaks is as high as that of sulfato-chloride and nitrato-chloride solonchaks (2 to 5%), their salt toxicity is substantially lower because of the high content of Na_2SO_4, and $CaSO_4$. The reclamation of chlorido-sulfate solonchaks by leaching and drainage is comparatively simple. Alkalization, where it occurs, is neither marked nor lasting; there is virtually no drop in the permeability and little or no increase in alkalinity.[3,297,298]

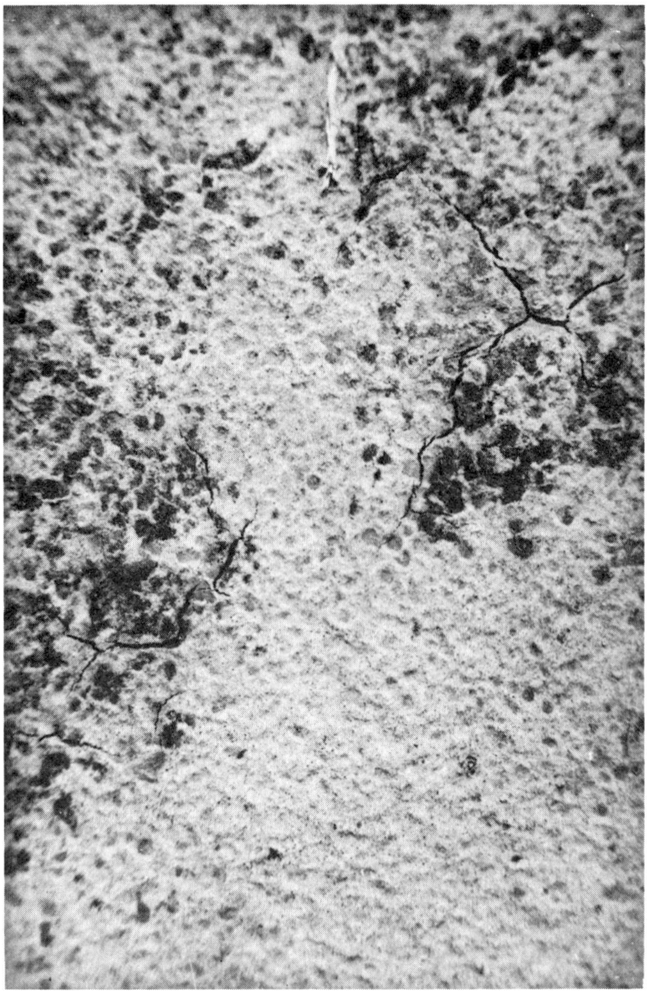

FIGURE 58. Surface of a saline soil developed on sand (Hungary).

Figure 58 shows the surfaces of saline soil developed on sand, while Figure 59 illustrates the surface of a saline soil with salt efflorescences developed on clay. The surface salt efflorescence and the profile of a saline soil with a heavy texture can be seen in Figure 60.

J. Sulfate Solonchaks

Sulfate solonchaks usually contain Na_2SO_4, $MgSO_4$, and $CaSO_4$. They occur most often in continental depressions. Sulfates are the least toxic among the readily soluble salts. The hydrophysical properties of these solonchaks are generally favorable, since, as a result of their natural gypsum content, they do not alkalize. Leaching and drainage are easy and rapidly effective in the reclamation and development of these soils.

K. Sulfato-Soda Solonchaks (Saline-Alkali Soils)

Sulfates predominate in the salt composition of these solonchaks; chlorides are as a rule also present, but the most characteristic feature is the presence of bicarbonates of alkali to a value of 0.1 to 0.15% HCO_3 and a pH between 8.7 and 9.5. In the U.S.S.R., sulfato-soda solonchaks occur in relatively northern saline soil areas, such as the Ukraine, Middle Volga, Kulunda Steppe, Siberia, Azerbaidzhan, Armenia, etc. But this type of solonchak

FIGURE 59. Surface of a saline soil developed on clay (Hungary).

is also common in Hungary, Manchuria, Mongolia, India, Pakistan, North and South America, as well as in Africa south of the Sahara. Their salinity is not as a rule very high (1.5 to 2%), but as a result of their intense alkalinity and the presence of soda, their hydrophysical properties are extremely unfavorable (poor water permeability, lack of structure), and they have severe toxic effects on plants.

L. Soda Solonchaks

In these same regions, pure soda solonchaks occur, containing mainly Na_2CO_3 and $NaHCO_3$ with an admixture of Na_2SO_4 and $MgCO_3$. The toxicity of soda is very great, owing not only to its direct influence but also to the high resultant alkalinity (pH 9.5 to 12). These soils are characterized by a residual humus content (mostly of meadow origin), a high exchangeable sodium percentage (70 to 80%), a prismoid-block structure, a very high density, and a relatively nonsaline but alkaline groundwater close to the surface. The readily soluble salt content is not very great (rarely more than 1.5%), but the properties of these soils are extremely unfavorable for reclamation and development because of their lack of structure, their impermeability to water, and the extreme toxicity of their free soda. Reclamation is mainly carried out through the use of chemicals such as gypsum, sulfur and industrial acid

FIGURE 60. Surface salt efflorescence and profile of a saline soil with a heavy texture (Hungary).

waste in combination with drainage and leaching. Rice cultivation is sometimes an effective method of amelioration. The reclamation of soda solonchaks is difficult and not always economically sound.

Soda solonchaks are classified among the alkali soils, according to the designations adopted by the Subcommission on Salt-Affected Soils. Figure 61 shows a soda solonchak profile from the Hungarian Plain.

M. Borate Solonchaks

The formation of these solonchaks is caused by the accumulation of borates contained in volcanic exhalations under very arid desert conditions. The high borate content causes plants to wither and ruins the fertility of the soil. Borate is usually accompanied by chlorides and sulfates which generally play an important role. Borate solonchaks are common around the lakes in Tibet and in the deserts of the U.S. and Latin America. In Bolivia there is an exceptionally large area of such solonchaks around Lake Ascota where the salt horizon is as much as 15 to 20 cm thick and the B_2O_3 content may reach 28 to 37%.

*   *   *

FIGURE 61. Profile of a soda solonchak soil (Hungarian Plain).

The salt content of saline soils and the distribution of salts in the various layers of the salt profile, may undergo seasonal changes; these may be climatic, or may be due to differences in irrigation and drainage connected with the seasons. These changes must be taken into consideration during any sampling and analytical procedures, and in making any authoritative description of the soil.

The basic morphological precondition for a saline is the lack of a structural B horizon (see Figure 62). Although several morphological systems use the letter "B", they do not distinguish it from the A horizon by its well-developed structural formation. Consequently, the profiles of saline soils are rather monotonous, from the surface down to the parent material, with the exception of some gypsiferous saline veins and nodules at deeper levels. In a few cases, where saline soils have formed under bog conditions, the top layers are very humic; but usually, where they have developed under arid conditions, these layers have a humus content lower than 1%. Low plant-nutrient content (particularly N and P_2O_5) is characteristic of most saline soils.[298,299]

In Figure 63 the profile of a saline soil is shown. In this case the groundwater table is close to the surface and the wetness caused by groundwater appears at a depth of 30 to 35 cm.

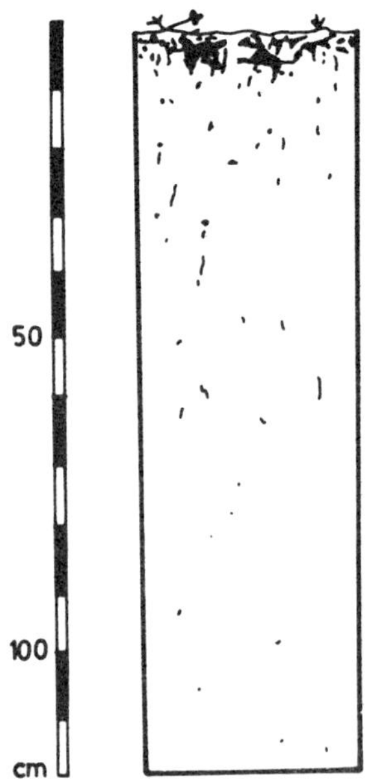

FIGURE 62. Schematic profile of a saline soil.

The high salinity determines virtually all the physical and chemical properties of saline soils; consequently, when these properties are evaluated, the consideration of the salt content and its influence should be given priority.

Many classification systems employ the term chloride and/or sulfate solonchak for saline soils.

Figure 64 demonstrates the salt profile of a saline, chloride-sulfate solonchak soil in the Caspian Lowlands, U.S.S.R.

The description of a saline soil (solonchak) profile, characteristic of the eastern part of the Romanian Danube Plain, is as follows:

A_{11}sa (0 — 3 cm)	Very dark gray to very dark grayish-brown (10YR 3/1.5) silt loam; gray (10YR 5/1) when dry; weak; thin platy structure; very friable; slightly hard, plentiful roots; salt efflorescences; calcareous; clear wavy boundary.
A_{12}sa (3 — 17 cm)	Texture and colors similar to the above horizon; weak, thin platy structure; very friable; slightly hard; many efflorescences, veins, and nests of soluble salts; grass roots; calcareous; gradual boundary.
$A_c$sa (17 — 27 cm)	Dark gray to grayish-brown (10YR 4/1.5) silt loam; gray to light brownish-gray (10YR 6/1.5) when dry; mottles of dark reddish-brown (5YR 3/4) when moist, and yellowish-red (5YR 4/6) when dry; massive; very friable; slightly hard; numerous thin veins of soluble salts; calcareous; abrupt boundary.
C_1sa (27 — 70 cm)	Alluvium layers; very dark grayish-brown to dark grayish-brown (10YR 3.5/2) silt loam, gray (10YR 5/1) when dry; coarse-textured layers are lighter colored; mottled with dark reddish-brown (5YR 3/4) when moist, and yellowish-red (5YR 4/6) when dry; massive; large amounts of soluble salts; calcareous; abrupt boundary.
IIC2sa (70 — 95 cm)	Dark gray (10YR 4/1) silt loam, gray (10YR 5/0.5) when dry, mottles of reddish-brown (5YR 3/4) when moist, and yellowish-red (5YR 4/6) when dry; massive; salt crystals; calcareous; abrupt boundary.
IIIABsa (95 — 113 cm)	Buried solonchak, black (10YR 2/1) silt loam, very dark gray to dark gray (10YR 3.5/1) when dry; few yellowish-red (5YR 4/6) mottles; massive; old roots; thin veins and nests of soluble salts; calcareous; gradual boundary.
IIIACb'sa (113 — 137 cm)	Very dark gray (10YR 3/1) silt loam, gray (10YR 5/1) when dry (colors become lighter in the lower part of the horizon); yellowish-red mottles; massive; crotovines; salt crystals; calcareous; abrupt boundary.

FIGURE 63. Profile of a saline soil.

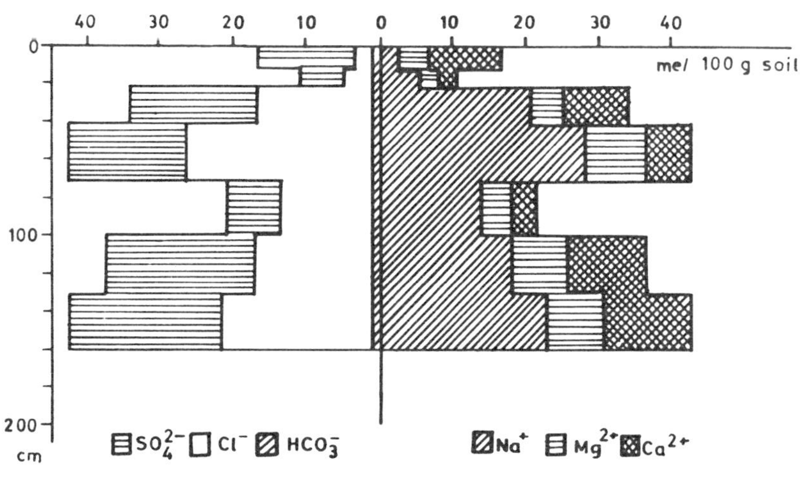

FIGURE 64. Salt profile of a saline soil in the Caspian Lowlands (a chloride-sulfate solonchak).[1]

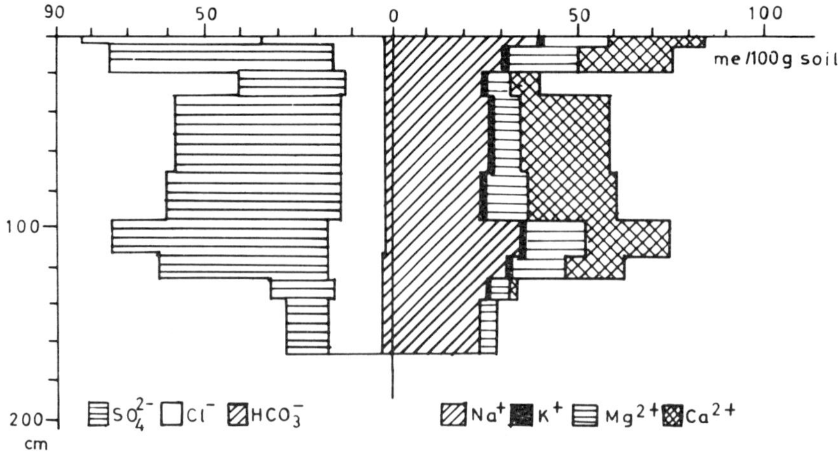

FIGURE 65. Salt profile of a saline soil (solonchak) in the Romanian Danube Plain, Iancs Valley (Romania).[1]

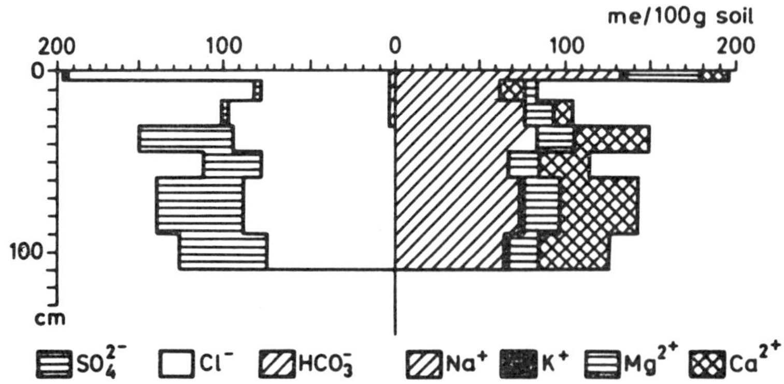

FIGURE 66. Salt profile of a saline soil (solonchak) in North Dobrogea, south of Lunca (Romania).[1]

IVAb′gsa (137 — 165 cm)	Another buried soil; dark gray (10YR 4/1) silt loam, gray (10YR 5/1) when dry; mottled; massive; thin roots, calcareous; abrupt boundary.
IVCb′g (165 — 195 cm)	Dark gray (10YR 4/1) silt loam, gray (10YR 5.5/1) when dry, dark reddish-brown (5YR 3/4) with gray (N5) mottles when moist, yellowish-red (5YR 4/6) mottles when dry; massive; calcareous.

The salt profile of this soil is shown in Figure 65. Figure 66 demonstrates the salt profile of a marine solonchak on the Black Sea shore in the Chituc Peninsula (Dobrogea) Romania.[1]

In general the profile of these soils is weakly developed, showing the following features:

A_{11}sa (2—8/5 cm)	Sand, loamy sand; black to dark grayish-brown (hue 1OYR with values of 2.5 to 3.5 and chroma less than 2) when moist, and grayish-brown when dry (1OYR 5 to 6/1); abundant fine iron spots; single grain; very friable, moist; slightly hard when dry; frequently small shell fragments occur; abundant thin roots; on surface white salt segregations; strongly calcareous, clear boundary.
A_{12}sa (6—24 cm)	Thick; sand with some small shell fragments; very dark grayish-brown, dark gray (hue 1OYR with values of 3 to 4 and chroma less than 2) when moist, and gray, grayish-brown (1OYR 5/1 to 1.5) when dry; sometimes the color may be olive-gray (5Y 3 to 4/2); abundant fine iron spots; other features similar to those above; clear boundary.

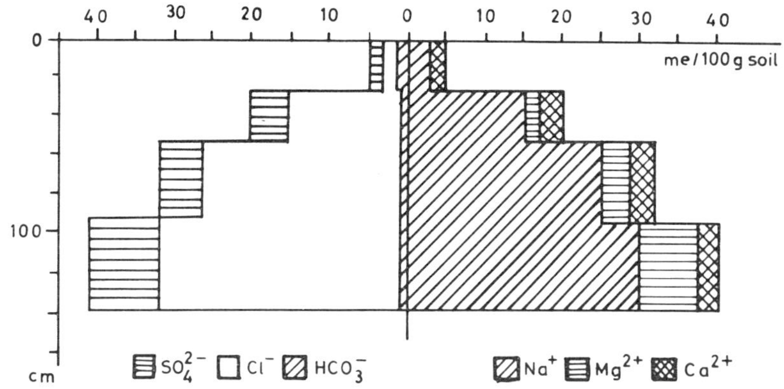

FIGURE 67. Salt profile of a saline soil (sulfate-chloride solonchak) in Marismas (Spain).[1]

A_s (8—20 cm)	Thick; sand; olive-gray, pale olive (5Y 4.5 to 5.5/3) or brown (1OYR 5/3) when moist, and light gray (1OYR to 2.5Y 7/2) when dry; abundant gray and brown spots (mainly along the root channels); small shell fragments; other characteristics similar to those above; gradual boundary.
G horizon	Occurs at 35 to 50 cm depth; loose sand; frequently very shelly; slightly cemented thin shell layers to be found; may be divided into two horizons:
G1	Light brownish-gray (2.5Y 6/2) or gray, light gray to pale olive (5Y 6/1-3) when moist, and light gray (2.5Y 7 to 7.5/2) when dry; abundant fine spots, dark reddish-brown (5YR 3/4) or brown; gradual boundary.
G2	Mottled; moist; greenish-gray (5GY 6/1), olive-gray or pale olive (5Y 5/2, 5Y 5/3) with abundant dark reddish-brown (5YR 2/2) and yellowish-red (5YR 4/6, 7.5YR 6/6) spots; sometimes when dry the color becomes light gray, white (2.5Y 7.5/2, 1OYR 8/1) with brownish-yellow (5YR 5/6) and pale brown (1OYR 7/4) spots; iron manganese segregations coating sand grains may be present; violent effervescence with diluted HCl.

The diversity of the shape of the salt profile in solonchak soils can be observed if, besides the above profiles, we take a look at the salt profile of a saline soil from the Guadalquivir River Delta in Spain (Figure 67). The description of this soil is as follows:

A_{11} (0—5 cm)	Dark grayish-brown (1OYR 4/2); calcareous fine clay; medium moderate laminar structure; abundant roots; slightly organic; moderately moist; very sticky and very plastic; abrupt, smooth boundary.
A_{12} (5—25 cm)	Dark grayish-brown (1OYR 4/2); calcareous fine clay; coarse moderate granular structure; abundant roots; very slightly organic; moist; very sticky and very plastic; gradual, smooth boundary.
AC (25—95 cm)	Dark brown (1OYR 4/3); calcareous fine clay; few roots; very coarse moderate prismatic structure; presence of slickensides; moist; very sticky and very plastic; common fine faint mottles about 5%, very dark gray (5YR 3/1) and dark reddish-gray (5YR 4/2), diffuse and smooth boundary.
C (95—135 cm)	Similar to the overlying layer but saturated with water; coarse yellowish-red patches, about 30% (5YR 5/8). Very few roots.
CG (135 + cm)	Gley, greenish-gray (5GY 5/1). Water appears. No roots.
Elevation:	5 m above sea level.
Water table:	140 cm.
Slope:	0%.

Another type of solonchak is that described by Servant, near Perpignan in the south of France (see also Tables 58 and 59).[1] Its profile is as follows:

Locality:	Bourdigou region 66, Station B_1, altitude 3 m.
Parent rock:	Alluviums of the Salanque. Water table at 2.20 m.
Vegetation:	"Sansoufre" bare soil.
Soil type:	"Sansoufre" soda solonchak.

Table 58
PHYSICAL AND CHEMICAL PROPERTIES OF A "SANSOUFRE" SOLONCHAK NEAR PERPIGNAN, FRANCE

Depth (cm)	Particle-size distribution CS	FS	CL	FL	Clay (%)	Organic matter (%)	$CaCO_3$ (%)	pH (1/2.5)	Gypsum, (meq/100 g)	EC (mmho)	ESP
0—3	0.4	3.3	29.7	45.4	21.2	1.4	3.4	8.8	4	91	80
3—5	0.1	3.0	31.6	51.7	14.6	0.8	1.2	8.7	10	48	75
5—10	0.1	3.0	32.0	51.1	14.8	0.6	1.9	8.7	0.5	57	70
10—30	0.2	3.8	31.5	50.9	13.6	0.6	2.0	8.7	0	31	65
30—60	0.0	1.0	23.1	60.9	15.0	0.6	1.0	8.7	0	25	65
60—90	0.0	1.2	27.2	54.9	16.7	0.6	0.6	8.5	u.	25	60
90—120	0.3	0.6	20.2	61.2	17.7	0.6	0.1	8.4	3	27	52
120—150	0.9	0.8	19.5	58.0	20.8	0.6	0.1	8.2	0	30	60
150—180	0.1	0.2	23.2	52.7	20.8	0.7	0.1	8.2	0	33	58
180—200	1.8	4.5	23.3	45.0	25.4	0.6	11.3	8.6	0	31	56

After Servant (cited in Szabolcs[1]).

Table 59
ANALYSIS OF SATURATION EXTRACT FROM STATION B_1, NEAR PERPIGNAN, FRANCE

Summer Ca	Mg	Na	Winter Ca	Mg	Na
36	484	2,174	45	281	1,260
45	82	691	45	113	696
27	32	328	32	51	385
24	26	270	28	49	325
25	28	262	28	42	265
24	29	264	24	38	225
27	27	250	32	43	225
35	29	238	31	40	200
38	33	240	31	44	230
39	38	236	32	38	235
39	40	240	37	40	252
39	47	256	21	35	250
43	108	440	37	86	408

Note: Values are meq/ℓ.

After Servant (cited in Szabolcs[1]).

A_1S (0 — 3 cm) Whitish surface with salt efflorescences; olive-gray (5Y 5/4) in the fracture; fresh; medium silt loam; organic matter not directly perceptible; porous; calcareous; medium cohesion; abrupt boundary.

A_1j (3 — 5 cm) Yellowish-brown (10YR 4/4); fresh, medium silt loam; laminar structure, plates from 2 to 4 mm thick; horizontal with undulations; gypsum deposits in round or almond shapes; organic matter not directly perceptible; very good porosity, calcareous; weak cohesion; abrupt boundary.

AC (5 — 10 cm) Yellowish-brown (10YR 4/4); fresh; medium silt loam; laminar structure identical to that of the horizon immediately above; but no gypsum crystallization; very good porosity; calcareous; weak cohesion; abrupt boundary.

C_1 (10 — 30 cm) Yellowish-brown (10YR 4/4); fresh; medium silt loam; structure tends to be massive but substructure angular blocky; fragile; good porosity; slightly calcareous; cohesion medium to weak; gradual boundary.

C_2 (30 — 60 cm)	Olive-brown (2.5Y 5; 5/4); fresh; medium silt loam; massive structure with blocky substructure; fragile; with gray-brown coating on the block surfaces; fairly good tubular porosity; slightly calcareous; medium cohesion; gradual boundary.
C_3 (60 — 90 cm)	Horizon identical to the preceding, from which it differs in the appearance of a few rust mottles made by redux processes.
C_4gj (90 — 120 cm)	Same characteristics as the immediately preceding, only with intensification of ochre spots and gypsum crystallizations of round or elongated shape.

III. ALKALI SOILS

Sodium salts capable of alkaline hydrolysis are dominant in the alkali soil-forming processes, and they determine the properties, water economy, and chemical charactersitcs of the soils.[4,21]

While the effect of neutral sodium salts on soils is mainly associated with arid or semiarid regions, alkali soils formed under the influence of sodium salts capable of alkaline hydrolysis may also be found in areas with a moderate or even humid climate.[21,300,301]

When sodium salts capable of alkaline hydrolysis are discussed — both in the literature and in practice — the effect of sodium carbonate or bicarbonate is usually emphasized. However, attention must not be focused solely on the effect of these compounds. It is fairly evident that in addition to Na_2CO_3 and $NaHCO_3$, other compounds capable of alkaline hydrolysis also play an important role in the formation of alkali soils, and our knowledge of the nature and dynamics of these other compounds and processes is still insufficient. Many authors[1,3,12,18,238,302,303] have pointed out the importance of the silicon compounds contained both in soils and in surface and groundwaters, in the formation processes of alkali soils. Although it has been generally accepted that silicon compounds, especially those formed with sodium and capable of alkaline hydrolysis, play an important role in alkali soils, no thorough study of these compounds has ever been undertaken. Most of the methods of chemical analysis merely express alkalinity as Na_2CO_3. The exact roles played by other compounds capable of alkaline hydrolysis in alkali soil formation have not been defined, though they may be of importance even where the dominance of Na_2CO_3 is unquestionable.

It is the presence of sodium salts capable of alkaline hydrolysis which determines the poor agricultural properties of alkali soils. The high alkalinity of the soil solution inhibits plant growth, or alkalinity causes physical soil properties which interfere with the water supply. Often both processes exert their harmful influences at the same time, but in alkali soils without a structural B horizon it is the former effect which usually dominates, whereas in alkali soils with a structural B horizon, it is the latter.[4,304]

As a rule, in alkali soils without a structural B horizon a notably higher concentration of water-soluble sodium salts capable of alkaline hydrolysis is found even in the top layers. In alkali soils with a structural B horizon, the concentration of water-soluble salts is often very low, and, particularly in the A horizon, the pH may be neutral or even slightly acidic. There are considerable differences between the properties of soils with and without structural B horizons, thus their characteristics and definitions will be treated separately.[305]

The surface of an alkali soil is shown in Figure 68. It can be clearly seen that during the drying of the surface, the organic matter accumulates in micro-depressions due to the leaching effect of the surface water, and after drying it forms a film-like layer which cracks and is later eroded.

Figure 69 also shows the surface of an alkali soil. The formation of cracks, particularly on patches lacking plant cover, is very characteristic of alkali soils, and these largely contribute to the formation of a surface layer with very poor water physical properties.[306]

The effect of sodium compounds capable of alkaline hydrolysis is indispensable for the development of alkali soils, as was mentioned above. However, a distinction must be made between the influence of comparatively high concentrations and that of comparatively low

FIGURE 68. Surface of an alkali soil with organic matter film in microdepressions.

FIGURE 69. Cracking of the surface of an alkali soil after drying.

FIGURE 70. Surface of an alkali soil without a structural B horizon in the Hungarian Plain.

concentrations of such compounds because they result in the formation of different types of alkali soils.

A. Formation of Alkali Soils under the Influence of Comparatively High Concentrations of Sodium Compounds Capable of Alkaline Hydrolysis (alkali soils without a structural B horizon)

Where the concentration of sodium carbonate in the upper horizon of the soil profile is high, soils without structural B horizons develop (Figure 70). These soils are very similar to the saline soils formed under the influence of neutral sodium salts in that they have no readily distinguishable horizons separated by abrupt boundaries. In several classification systems, the former are described as sodium carbonate solonchaks, the latter as sulfate or chloride solonchaks. It is quite evident that in spite of the similarities, these soils differ in many of their properties. The most important differences are that the reaction of sodium carbonate solonchaks is strongly alkaline, while that of sulfate or chloride solonchaks is nearly neutral, and that clay particles are more dispersed in the former than in the latter. 'Sigmond and Kelley described all of these soils as alkali soils,[12,18] but — according to the proposal of 'Sigmond — they are referred to in the literature as saline-alkaline or alkaline-saline soils.[12,10] Richards and Bonnet also used the definition.[2,180] Naturally, the differences must be taken into account in selecting methods for the amelioration and utilization of these soils.

These alkali soils resemble saline soils in having a saline (salic) horizon which can be found within 125 cm of the surface (125 cm where the texture is coarse, 90 cm for medium texture, and 75 cm where it is fine), or in having an electrical conductivity of more than 4 mmho in at least some part of the soil within 25 cm of the surface. But the pH (H_2O 1:1) should be more than 8.5 somewhere in the 0 to 25 cm layer. It should be noted that a phenolphthalein analysis is not adequate to determine the pH, because finely dispersed

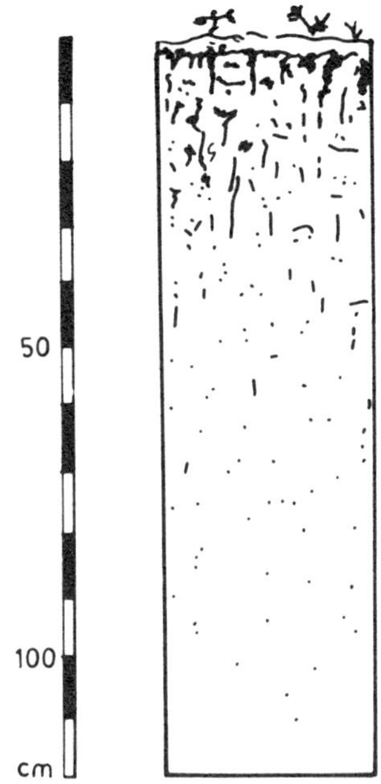

FIGURE 71. Schematic profile of an alkali soil without a structural B horizon.

calcium carbonate also produces a slightly pink color with phenolphthalein. A more precise pH determination or titrimetrical analysis must be made.

This determination may also require some modification according to local circumstances. In lowland areas alkalinity in the top layer may not fully correspond to the definition of alkali soils without structural B horizons, and part of the salic horizon may be lacking. The electrical conductivity, even when these soils have a pronounced profile, sometimes hardly reaches or exceeds 4 mmho; in these cases the pH values and morphological features of the soil decide whether or not they should be included in this class.

In numerous classification systems these soils are referred to as alkali-saline or saline-alkali soils. Neutral sodium salts may also be present among the water-soluble substances in the soils, but the dominant role is played by sodium carbonate because of its high alkalinity.[307]

Figure 71 shows a schematic profile of an alkali soil without a structural B horizon. Alkali soils without structural B horizons are found in various climatic zones. In western Siberia, central and southeastern Europe, Argentina, and China, as well as in many subtropical and tropical areas (Cuba, Puerto Rico, Kenya, Tanzania, Peru, Indo-China, etc.), these soils have developed under the controlling influence of mineralized groundwater under comparatively moderate climatic conditions. In many cases this influence is proved by the similarity of the chemical composition of the soil profile to that of the groundwater.[308]

The precentage of sodium carbonate and bicarbonate in these soils may vary from a few tenths of 1% to about 5%. Depending on local conditions, the maximum salt content may occur at the surface or in the lower layers.[309]

The high alkalinity of these soils results in top layers which are rather compact, structureless, and virtually impermeable — a very important factor from the point of view of drainage or chemical reclamation.

Most alkali soils without a structural B horizon are very poor in humus and plant nutrients, particularly in N and P, but humus top layers may be found in humid or semi-humid areas. When the soda-forming processes are associated with temporarily waterlogged conditions, the humus content of the top layer may reach 7% or more, as in the wadis of Lake Chad, while in other cases it may mount to only a few tenths of 1%.

Because high alkalinity is accompanied by high salinity, it governs not only the chemical but also the physico-chemical processes in these soils. Consequently, the determination of exchangeable cations is not always necessary, and it may often prove to be difficult during a routine analysis.

If the concentration of sodium salts capable of alkaline hydrolysis decreases, the transitional types which fall between alkali soils with and without structural B horizons may develop. These are identified in several classification systems as solonchak-solonetz or solonetz-solonchak. Similar processes occur during the desalinization of saline soils.

The following description of an alkali soil without a structural B horizon is from the Danube Valley on the Hungarian Plain.

Surroundings:	Pasture of extremely poor quality with bare spots.
Topography:	Flat.
Vegetation:	*Lepidium cartilagineum, Camphorosma ovata, Artemisia monogyna, Atropis (Puccinellia) limosa, Nostoc commune.*
Depth of the profile:	120 cm.
Effervescence (with dilute acid):	To the surface.
Alkalinity against phenolphthalein:	To the surface.
Thickness of the humus layer:	12 cm.
Water table:	64 cm.
Genetic horizons:	
A (0 — 12 cm)	Gray (pale, mouse-gray), whitish when dry, slightly moist, sandy silt; undeveloped weak structure; relatively plentiful roots; gradual boundary.
B_1 (12 — 30 cm)	Light gray, slightly moist, silty loam; slightly prismatic structure (especially when dry); gradual boundary.
B_2 (30 — 52 cm)	Whitish-gray, loamy silt; moist; prismatic structure when dry; horizon of lime accumulation; abrupt boundary.
C (52 — 120 cm)	Gray, slightly silty sand; loose; there are a few small bits of gravel; large, white, silty spots and gray, very loose sandy spots; the silty spots become rare with depth.
Soil type:	Sodic solonchak on calcareous Danube alluvial sand (alkali soil without a structural B horizon).

In the Hungarian system of profile description, the symbol "B" is used to indicate the horizon below the A horizon, even if it is structureless, and if it is not clearly distinguished from the A horizon by a sharp boundary. Figure 72 gives the salt profile of the foregoing alkali soil without a structural B horizon from the Hungarian Danube Valley (Apaj-4).

Figure 73 shows the chemical composition of the groundwater, which corresponds quite clearly to the salt composition of the soil described in the profile above.

The diffuse occurrence of alkali soils without a structural B horizon is demonstrated by the similarity between the above-described profile of a Hungarian Danube Valley soil to that of a soil in Karnal, Haryana (India) (see also Table 60):[300]

A1 (0 — 12 cm)	Pale yellow (5Y 7/3D) to olive brown (2.5Y 4.5/4 M); sandy loam; weak, coarse platy, friable, sticky and plastic; concretions and mottles absent; strong effervescence with dilute HCl; many roots; very low permeability; clear and smooth boundary; white soft salt incrustation present on the surface.
$A2_1$ (12 — 34 cm)	Light yellowish-brown (2.5Y 6/4 M); sandy clay loam; weak, medium subangular blocky; hard, friable, sticky and plastic; no concretions; few, fine olive yellow (2.5Y 6/8 M) mottles; strong effervescence with dilute HCl; plentiful roots; low permeability; gradual and smooth boundary.

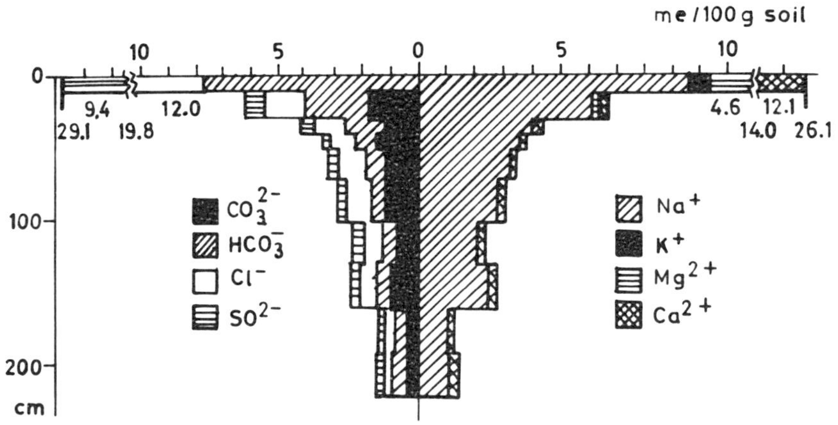

FIGURE 72. Salt profile of an alkali soil without a structural B horizon in the Hungarian Danube Valley (Profile Apaj-4).

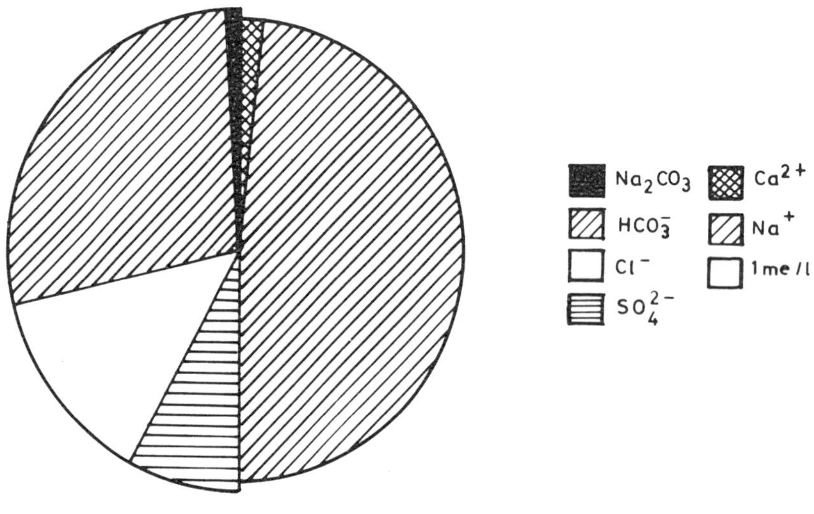

FIGURE 73. Chemical composition of the groundwater at Profile Apaj-4 (Hungary).

$B2_1$ (34 — 61 cm) Light olive brown (2.5Y 5/4 M); silty clay loam; moderate, medium subangular blocky; very hard, firm, very sticky and plastic; ferromanganese concretions up to 3 mm in diameter, about 2% by volume (field estimate); few coarse olive yellow (2.5Y 6/8 M) mottles; weak effervescence with dilute HCl; plentiful roots; low permeability; few prominent clay skins on ped faces and in the root channels; gradual and smooth boundary.

$B2_2$ (61 — 93 cm) Light olive brown (2.5Y 5/4 M) and olive brown (2.5Y 5/6 M) almost 50 to 50%; clay; strong, coarse, angular blocky; very hard, friable, very sticky and plastic; ferromanganese concretions up to 5 mm in diameter, about 5% by volume (field estimate); weak effervescence with dilute HCl; few roots; very low permeability; clay skins present on ped faces and in the root channels; clear and smooth boundary.

$B2_3$ (93 — 114 cm) Olive brown (2.5Y 4/4 M); sandy clay loam; weak, medium subangular blocky; friable, sticky and plastic; ferromanganese concretions up to 3 mm in diameter, about 2% by volume; many coarse, olive yellow (2.5Y 6/8) mottles; weak effervescence with dilute HCl; few fine roots; low permeability; abrupt and smooth boundary.

The above-described profile and the analytical data (see Table 60) clearly demonstrate

Table 60
SOME PHYSICO-CHEMICAL CHARACTERISTICS OF SOILS (KARNAL, HARYANA, INDIA)

Depth (cm)	Saturation	pH	EC (mmho/cm)	Analysis of saturation extract (meq/ℓ) Na	Ca	Mg	K	CO_3	HCO_3	Cl	SO_4
0—12	33.3	9.0	102.30	1,084.3	0.8	0.4	3.6	560.0	236.0	273.6	19.6
12—34	30.0	9.7	6.03	60.0	0.8	0.6	0.3	22.4	22.0	4.0	12.2
34—61	30.6	9.4	2.60	23.5	0.6	2.4	0.3	4.8	14.4	3.6	14.2
61—93	36.0	8.8	1.41	12.8	0.4	0.2	0.1	2.4	9.6	1.6	0.3
93—114	36.6	8.6	0.92	10.9	0.4	0.1	0.2	2.4	7.2	2.4	0.2
114—192	41.1	8.6	0.93	9.1	0.4	0.1	0.1	2.1	3.5	4.0	0.2

After Abrol and Bhumbla.[300]

that these soils show many similarities to the soils described previously in this chapter, and should be classed as alkali soils without a structural B horizon. These soils may be found under very different environmental conditions, wherever a comparatively high concentration of sodium salts capable of alkaline hydrolysis dominates the soil-forming processes.

Figure 74 (A and B) gives a clear demonstration of the change in the surface layer of an alkali soil during drying. In Figure 74A the micro-depressions are still waterlogged and a moldy, sticky surface layer exists, while in Figure 74B a very hard, cracked surface layer appears after drying.

B. Formation of Alkali Soils With a Structural B Horizon under the Influence of Comparatively Low Concentrations of Sodium Compounds Capable of Alkaline Hydrolysis

The definition of these soils is based on the prismatic or columnar structure of the B horizon, which is accompanied by a high proportion of exchangeable sodium ions (ESP > 15) (Figure 75).

The horizon always has well-developed structural elements, mainly columns. It can be easily distinguished from the horizon above it, which is less compact and has a less developed structure.[310] The presence of this B horizon denotes the genetic type of the soils, their main physical, chemical, physico-chemical and biological properties, and their fertility and potential for agricultural utilization. These are called solonetz soils in the Soviet literature, and this appellation has been generally accepted. As regards their development, solonetz-forming and solod-forming processes can be distinguished. Solod formation will be discussed later in this chapter.

In Figure 76 schematic profiles of alkali soils with structural B horizons (solonetz) are shown. The structural B horizon is situated at various depths, depending upon local circumstances. In some cases — often through the effects of erosion — it is at the surface, and the A horizon is completely lacking. These soils are called "Naga" and "Harde" soils in Chad and northern Cameroon.[71]

In the following description of a soil in Bol, near Lake Chad, reported by C. Cheverry,[72] the parent material has a decisive influence on pedogenesis and on the formation of the structural horizon (Tables 61 and 62):

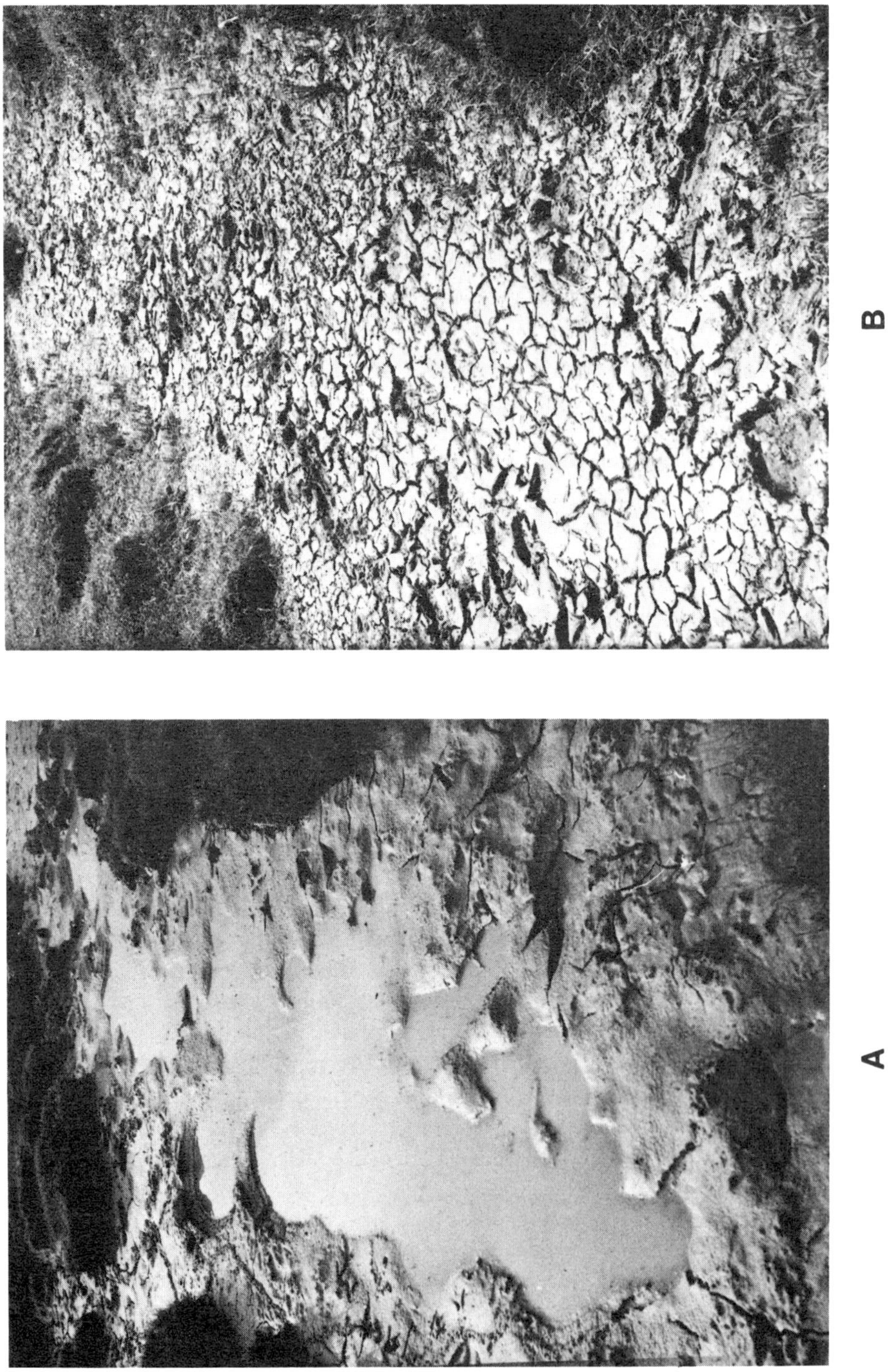

FIGURE 74. Surface of an alkali soil under waterlogged conditions (A) and after drying (B).

FIGURE 75. Columnar structure in the B horizon of an alkali soil.

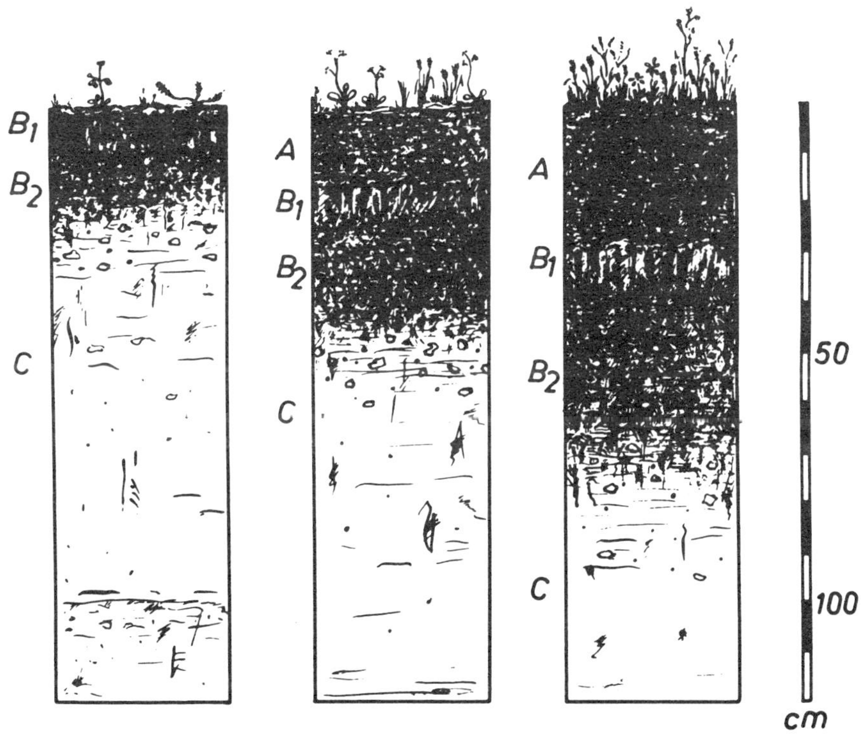

FIGURE 76. Schematic profiles of alkali soils with structural B horizons (solonetz).

Table 61
CHEMICAL CHARACTERISTICS AND SALINE COMPOSITION OF A PRESSED SOIL EXTRACT OF A SALT-AFFECTED SOIL NEAR LAKE CHAD

			Cations in meq/ℓ of solution					Anions in meq/ℓ solution							
Depth	pH	EC (mmho)	Ca	Mg	Na	K	Total	Cl	SO_4	CO_3	HCO_3	Total	SO_4^{2-} in % of anions	Ca/Mg	Na/K
Surface water (8/1970)	9.2	6.2	tr	tr	73.0	4.5	77.5	4.4	1.4	16.5	52.2	74.5	1.9	—	16.2
Soil solution															
10—15 cm	8.2	5.4	3.3	2.7	53.0	4.0	63.0	4.8	19.0	7.0	31.0	61.8	30.7	1.2	13.2
16—26 cm	8.3	3.1	2.0	2.7	30.0	1.9	36.6	2.9	2.0	4.0	27.0	35.9	5.6	0.7	15.7
26—36 cm	8.0	2.0	2.2	2.8	15.0	1.1	21.1	2.1	0.5	2.4	15.2	20.2	2.5	0.8	13.5
41—51 cm	7.7	1.2	1.0	1.2	9.6	1.0	12.8	1.4	3.0	0.0	7.9	12.3	24.4	0.8	10.0
Groundwater	7.0	1.7	4.5	3.7	8.4	1.0	16.6	0.7	0.2	0.0	16.2	17.1	1.2	1.2	8.4

Note: tr, trace.

After Cheverry.[71]

Table 62
SALINE COMPOSITION OF THE AQUEOUS EXTRACTS (1:2) OF A SALT-AFFECTED SOIL NEAR LAKE CHAD

			Cations in meq/ℓ of solution					Anions in meq/ℓ of solution							
Depth (cm)	pH	EC (mmho)	Ca	Mg	Na	K	Total	Cl	SO_4	CO_3	HCO_3	Total	SO_4^{2-} in % of anions	Ca/Mg	Na/K
0—3	10.0	17.5	tr.	tr.	48.0	2.6	50.6	2.7	2.6	6.0	35.0	46.3	5.6	—	19.4
3—6	9.7	12.5	0.05	0.05	37.0	2.0	39.1	2.4	3.3	5.2	26.0	36.9	8.9	1.0	18.5
10—15	8.8	5.9	0.8	0.5	13.0	0.9	15.2	1.0	4.8	0.6	8.0	14.4	33.3	1.6	14.4
16—26	8.3	5.3	0.7	0.4	10.0	0.7	11.8	1.5	4.0	0.05	6.2	11.75	34.0	1.8	14.3
26—36	7.6	2.9	0.5	0.3	5.2	0.3	6.3	1.1	1.6	0.0	3.1	5.8	28.0	1.7	17.0
42—51	7.4	2.6	0.9	0.6	3.4	0.4	5.3	0.6	2.4	0.0	2.3	5.3	45.0	1.5	8.5
55—66	7.6	3.6	3.3	1.3	3.2	0.5	8.3	0.2	5.6	0.0	1.8	7.6	74.0	2.5	6.4

After Cheverry.[71]

Macroscopic data:	Profile No. 1 is 330 m from the foot of the dune slope, in the central, nearly flat part of the polder. It is situated on the axis of the water table. At the time of these observations (November, 1970), the profile had not been flooded for seven months. A few thickets of reeds (*Typha australis*) about 20 — 50 m^2 each, inhabited the surface of the soil which was otherwise — about 95% — bare, dry, and run through with a mesh of shrinkage cracks in polygons. A few peeling scales about 1 mm thick and 1 — 2 cm long, with raised edges, broke off. The water table is maintained at a depth of 51 cm, the upper limit of the structured clays. Its piezometric level is established at a depth of 13 cm. Sodium and bicarbonate are the dominant ions in the waters. Sulfates are scarce (0.1 meq/1).
0 — 8 cm	Surface horizon. Formed from the juxtaposition of prisms (length and width, 10 to 15 cm, height 8 cm) separated by shrinkage cracks from 0.5 to 5 cm wide. These prisms are composed of two parts: The upper part, fragmented into dry angular blocks of sombre gray color (10YR 4/1) in their mass and gray-black on their upper faces; some blocks are run through with a secondary network of very fine shrinkage cracks. Outside the micro-cracks, the blocks are barely porous and hard. Very strong effervescence with HCl. The middle and lower part of the prism, which has kept its cohesiveness, is humid and gray-brown in color (10YR 5/2) when dry; on the walls of the prism, at its base, a fine white saline powder is seen; fairly porous, pores half-closed with shiny walls and darker color than at the bottom. Strong effervescence with HCl. This horizon corresponds to the drying and the structuration of surface muds.
8 — 15 cm	Humic horizon. Gray-brown; no structural formation except for the beginning of a net-work of fine shrinkage cracks in the prolongations of the large cracks of the preceding horizon; rare cylindrical pores with shiny walls; malleable consistency; plastic; sticky; strong effervescence with HCl; gradual and regular boundary. This horizon corresponds to gray muds which have lost part of their water, but whose sedimentary organization is little modified.
15 — 40 cm	Gray to gray-brown mud, saturated with water; round black sulfide stains 5 to 10 cm in diameter, clearly delineated, fairly contrasting, without visible relations to the other characteristics; no structural formation; accentuation of the plastic and sticky qualities; weak porosity; effervescence with HCl. This is a sedimentary level and no longer a soil horizon. This holds true for the entire lower part of the profile.
40 — 50 cm	Brown mud saturated with water; slightly reddish tint; no structure; elastic consistency; oily feel; very weak porosity; strong effervescence with HCl.
50 — 66 cm	Structured clay (former carbonation level); drowned under the groundwater; slightly brownish-gray (2.5Y 6/2 when wet), without stains, clear, generalized fragmental structures, grainy, coase, semirigid consistency; very strong effervescence with HCl.
66 — 210 cm	Structured clay, drowned by the groundwater; gleyish color (5GY 4/1 when wet); some black stains toward 120 cm; very coarse subangular structure, not very clear, whose development diminishes with depth; very low porosity; effervescence with HCl to 90 cm, diminishing with depth.
210 — 260 cm	In the same material, appearance of very numerous little black sulfide stains of about 1 cm^2. Towards 260 cm, material becomes progressively less consistent, oilier to the touch.

During the soundings, odors were perceived at various depths, particularly at about 80 to 120 cm. This was marsh gas (methane), which was recognizable by its inflammable character, and which died out after about 10 sec.

The structural B horizon always markedly differs from the A horizon not only in morphology, color, and structure, but also in its physical, chemical, physico-chemical, and biological properties. Figure 77 gives a schematic illustration of some of the chemical, physical and physico-chemical properties of a solonetz soil.

It can be seen that at a given depth below the surface (in this case about 30 cm) an illuvial or accumulation horizon occurs. This is called a B or B_1 horizon. It is in this horizon that the accumulation of clay particles and sesquioxides may be observed; the content of water-soluble organic matter and the ESP value appear at their maximum there, while the ratio of $SiO_2:R_2O_3$ is lowest, signifying that the silicon compounds are comparatively minimal. In Figure 77 this horizon is found between 20 and 30 cm, as is often the case in nature, though

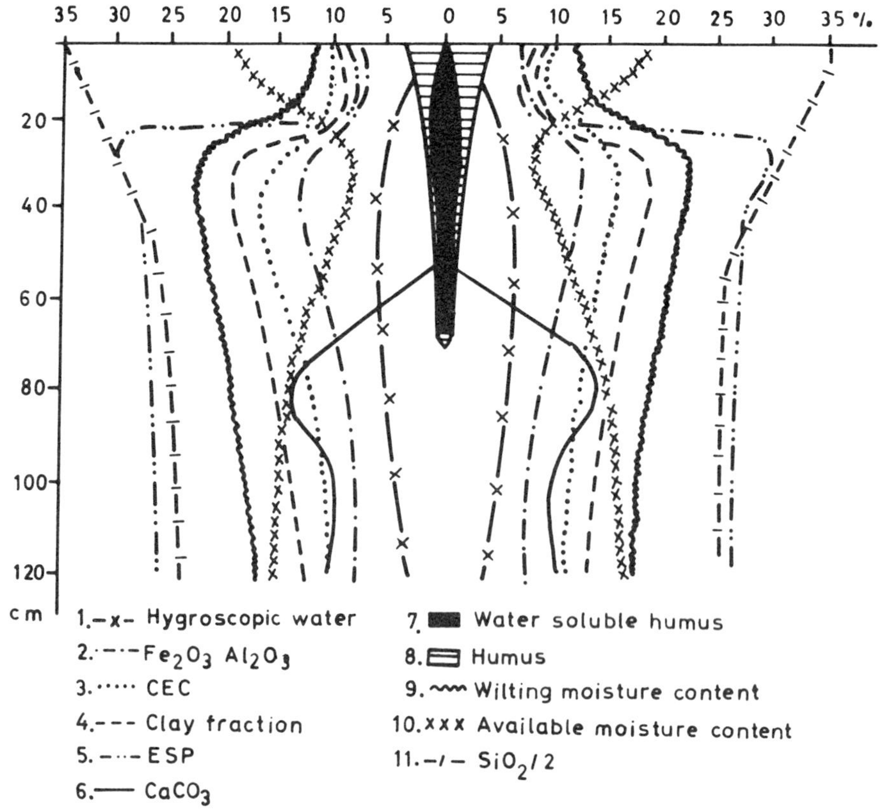

FIGURE 77. Schematic representation of some of the physical, chemical, and physico-chemical properties of an alkali soil with a structural B horizon (solonetz). (After Kovda, in modified form cited in Szabolcs.[1])

it also appears at other depths. Naturally, these maximum and minimum values are found at the depth where the B horizon develops.

The A horizon is called the eluvial horizon. It may be seen in Figure 77 that the movement and accumulation of materials here is exactly the opposite of that in the alluvial B horizon. The thickness of the A horizon is a very important feature of these soils. Their fertility is in direct proportion to it. For one thing, it determines the amount of water retained and made available to plants. The B horizon is much more compact than the A, and the penetration of plant roots is always impeded. Besides, as is clearly shown in Figure 77, the quantity of harmful substances — exchangeable sodium and water-soluble sodium salts — is lower in the A than in the B horizon. In Figure 78 the surface and profile of an alkali soil with a structural B horizon is shown.

Obviously the thickness of the A horizon has always been a consideration in the grouping and classification of solonetz soils. The limit values of its thickness (the basis on which the solonetz soil is classified) depend to a large extent on local conditions — climate, geochemistry, plant nutrition, and other factors. Accordingly, A horizon of various thicknesses correspond to a given water and nutrient economy in the soil profile. The following division of solonetz soils according to the thickness of the A horizon is intended only as a general indication:

FIGURE 78. Surface and profile of an alkali soil with a structural B horizon.

Type	Thickness of the A Horizon (cm)
Shallow solonetz soil	0 — 7
Medium solonetz soil	7 — 15
Deep solonetz soil	>15

In European countries where alkali soils with structural B horizons are widespread, this or similar classification systems are adopted for both surveying and reclamation purposes.

One of the important characteristics of the B horizons of solonetz soils is their high exchangeable sodium content. It is generally held that this factor is responsible for their poor physical and water regime properties, and for the compact structure of their B horizons. The exchangeable sodium content is usually expressed in meq/100 g of soil or, even more frequently, as a percentage of the cation-exchange capacity (ESP). As was mentioned earlier, the limit value for alkali soils with a structural B horizon is ESP > 15; when the ESP value is about 5 to 7, the first signs of the development of the compact B horizon may be observed in the profile. Naturally, the limit values are approximate and may vary according to the soil properties and local conditions.[1,311]

The data available are not sufficient to give a precise ESP limit value as a criterion for the term "solonetz", or even for "alkali soils" in all regions. In many countries, soils having an ESP value of 15 to 20, or even more, do not manifest solonetz morphological features, and may be fairly fertile, e.g., in the Sudan, India, etc. In other regions, such as some areas of the U.S.S.R., a solonetz profile may develop where the ESP value is less than 15. The basic criterion for the identification of these soils is always the above-described morphology of the B horizon.

Varying with the local circumstances, some alkali soils with a structural B horizon may contain considerable amounts of water-soluble sodium salts even in the upper layers (more than 4 mmho), while others are practically devoid of salt in the entire profile. Of these salts it is sometimes bicarbonates, sometimes sulfates, or sometimes even chlorides which are dominant. The maximum accumulation of water-soluble salts in the profile usually occurs in the lower part of the B_2 horizon, but depending on the conditions of soil formation, it may also appear in other layers, either above or below the B horizon. The pH value of these soils may also vary to a considerable degree.[251,312]

In some cases a strongly alkaline pH may be observed at the surface, while sometimes the pH of the top layer is neutral or even slightly acid. There are even solonetz soils in which a strongly alkaline pH does not occur at all, anywhere in the profile. In the B horizon, however, where the maximum exchangeable sodium percentage (ESP) is found, the pH is in most cases over 7. Exceptions are found in some of the Australian solonetz soils.

In the B horizon the sodium ions are mainly in exchangeable form, adsorbed onto the soil colloids which occur maximally there. These exchangeable sodium ions — depending upon the conditions of equilibrium between the solid and liquid phases — are capable of alkaline hydrolysis. This phenomenon is influenced especially by the CO_2 tension, by the CEC and ESP values, by the chemical composition and concentration of the soil solution, and by many other factors. The alkaline hydrolysis of the soil colloids, saturated with sodium to a higher or lower degree, brings about a more or less alkaline condition in this horizon. Alkaline hydrolysis usually takes place in the B horizon, because of the presence there of sodium compounds capable of alkaline hydrolysis, and because of the interaction of the solid and liquid phases of the soil — except where the soils are affected by strongly developed solod-forming processes.[4]

Environmental influences are very important in the formation of alkali soils with a structural B horizon, particularly that of mineralized groundwaters.

In Europe the solonetz soils in humid or semi-arid regions can be subdivided into more detailed grouping units as follows:[21,24]

- A subclass without a structural B horizon
- A subclass with a structural B horizon
    1. Solonchak-solonetz and calcareous solonetz
    2. Noncalcareous solonetz with an A horizon < 15 cm
    3. Solodized and/or deeply leached solonetz and solod
    4. Solonetz-like and slightly salt-affected soils with minor structure formation

While a further subdivision of these soils was deemed absolutely necessary by experts of the European Working Group of Subcommission A of the International Society of Soil Science, they agreed that such subdivisions for regions other than Europe should be elaborated according to the specific local conditions.

One of the indispensable conditions for solonetz formation is the action of comparatively diluted solutions of sodium salts. If the concentration of these salts is high, the processes described above take place. Regardless of how these concentrations arise (the solutions may gradually approach saturation, or highly concentrated solutions may be diluted), they constitute a condition for solonetz formation. It is difficult to argue with Gedroitz,[238] who

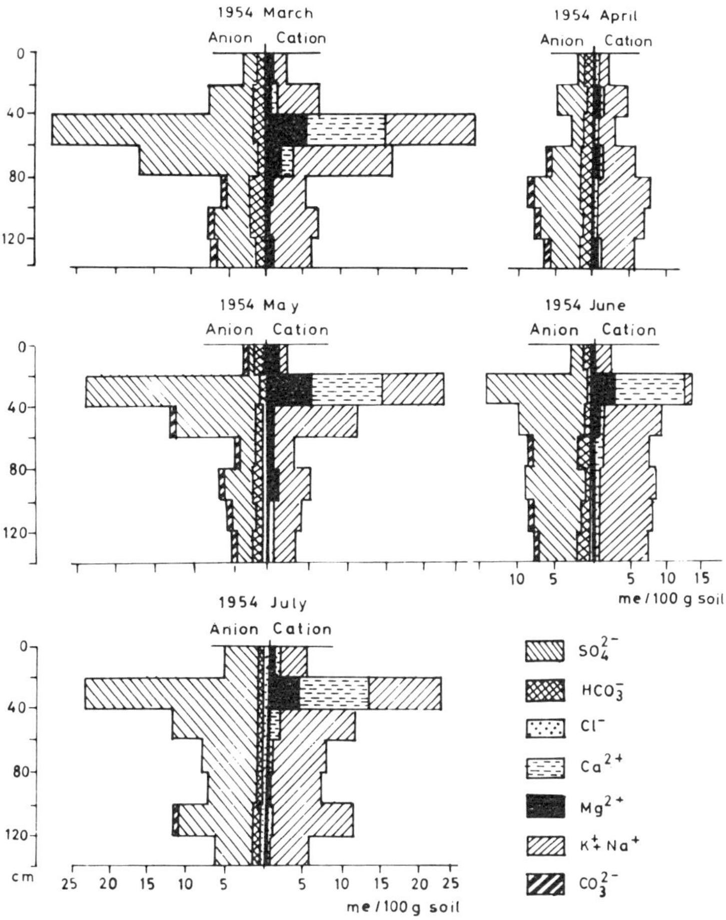

FIGURE 79. Seasonal dynamics of salt solutions in soils under irrigation in Szarvas (Hungary).

associated solonetz formation with the effect of sodium salt solutions diluted during the leaching of a solonchak, but the explanation given by Antipov-Karataev,[313] and Vilyams,[314] who consider it to be the result of increasing concentrations of sodium salt solutions in the course of soil formation, is equally valid.

The presence of sodium salts in a certain concentration in the liquid phase of the soil does not in itself necessarily result in the evolution of a solonetz profile. Another precondition is that the salt solutions migrate alternately upwards and downwards in the soil profile. This is how salt solutions of varying concentrations communicate with the various soil layers from one period to another. Moreover, the soil layers involved become alternately drier and more humid. All these factors induce substantial, reversible and irreversible changes in the soil profile. These cycles were observed by Glinka,[11] Gedroitz,[238] 'Sigmond,[12] Treitz,[14] and many other experts, all of whom concluded that the periodical alternation between dryness and humidity is necessary for the evolution of a solonetz soil. The soil changes due to the interaction of the solid and liquid phases. During this up-and-down migration, the soil solution alternately dissolves and accumulates certain compounds; as the migrations continue, the interaction between the solid and liquid phases of the soil induces greater and greater alterations. Thus, these movements cause not only the translocation of the various compounds within the profile, but also changes in the soil materials themselves.

On the basis of examinations conducted on the Hungarian Plain, Figure 79 demonstrates how intensive the seasonal dynamics of soluble salts is. It may be clearly seen in the figure that in consequence of the changing natural conditions, not only the concentration but also the chemical composition of the soil solution moving between the various soil layers is altered for several months. It may also be seen that, of the sodium salts capable of alkaline hydrolysis, the carbonates, for example, are present in changing quantities, and they migrate within the profile. It should be noted that the examinations shown in Figure 79 were conducted in soils under irrigation. Numerous observations have proved that the dynamism of salts in solonetz formation may also be experienced without irrigation, but of course it is far less intensive.[315]

Affected by the migration of relatively dilute sodium salt solutions, ion-exchange processes take place and soil colloids absorb a certain quantity of sodium ions, according to the rules of equilibrium for ion-exchange reactions. This phenomenon has been thoroughly studied by many experts, and various equations have been formulated to characterize these processes. Even a schematic review of this problem is far beyond the scope of this book. Although in theory the various ion-exchange processes are reversible, in fact they may induce irreversible processes in the soil.

The ion exchange not only affects the surface properties of colloids, but in all probability results in greater changes in the lattice layers of clay minerals and in the structure of the organo-mineral colloid compounds in the soil. The extent of these changes depends on the structure of the colloids, on the equilibrium between the solid and liquid phases, on the pH of the medium and on many other factors. The joint effect of these factors may cause irreversible changes which are decisive both in solonetz and solod formation.[316]

It is as a consequence of one of these irreversible changes — the transformation of certain sesquioxides and silicon compounds into a mobile form and their migration within the profile — that the differing properties of the A and B horizons characteristic of solonetz formation develop. As seen in Figure 77, the percentage of silicon compounds increases in the A horizon and decreases in the B horizon, while that of sesquioxides decreases in the A horizon and increases in the B horizon. The clay fraction and the water-soluble organic matter also decrease in the A horizon and increase in the B.

Naturally these processes are profoundly influenced by the quantity and quality of the clay fraction of the soil. In soils of coarse mechanical composition without clay fractions, solonetz does not usually develop. This also explains why the ESP values indicative of a preliminary or advanced stage of solonetz formation may differ markedly from the 5 to 7 to 20 to 25 limit values due to differences in the quantities and qualities of the clay fractions. In such cases it may be more expedient to give the ESP values in terms of the colloid fraction rather than 100 g of soil.

Over a period of time, which may not necessarily be long, these processes may induce profound changes in the clay minerals and in their distribution and mineralogical properties, which remain even if the equilibrium of the cation-exchange process in the soil is altered. For instance, even if the concentration of sodium salt solutions considerably exceeds the limit values for solonetz formation, once a soil has developed a structural B horizon (characteristic of solonetzes), it will only develop into a salt-affected soil without a structural B horizon over a very long period, and may not do so at all. Here the salt content may be very high in the upper soil layers, just as in the case of salt-affected soils without a structural B horizon, but the residual structural B horizon remains in the profile. In the Soviet literature these soils are often called solonchak-solonetz, and some soils referred to as saline-alkali in the English literature also belong to this group.

Similarly, the removal of sodium salt from solutions of relatively low concentration (which have caused the formation of the solonetz structure) will often leave the profile unchanged. In order to alter the compact B horizon, it is recommended to treat solonetz soils with amendments having a positive acid reaction, or capable of acid hydrolysis, such as HCl, H_2SO_4, $Fe_2(SO_4)_3$, $Al_2(SO_4)_3$, gypsum, or other acidic products. One of the most important effects of these compounds is that they induce ion exchange. They induce other changes, too, for example, in the colloid fraction of the soil.[317]

When excess water-soluble salts — particularly NaCl and/or Na_2SO_4 — are leached out of a saline soil with a high colloid content, solonetz or alkali soil formation takes place as described by Gedroitz.[238] Kovda[3] suggests the application of compounds capable of acid hydrolysis (e.g., gypsum) concurrently with the reduction of the sodium salt concentration, to ensure that the amelioration of a solonchak leads to the formation of a nonsaline and nonalkali soil and not to the formation of a solonetz.

Where sodium salts of relatively low concentration cause the formation of solonetz soils, two sets of circumstances may be essentially distinguished.

In the first, the soil profile and the top layers are capillarily linked with the mineralized groundwater. Kovda states that in this case, solonetz formation is intertwined with solonchak formation, because if the water table rises, capillary action causes salts from the groundwater to accumulate in the soil layers, and a solonchak soil develops.[3] It is the fluctuation of the water table that causes the upward and downward migration of sodium salt solutions, described above as one of the most important conditions of solonetz formation. In the Soviet literature this type of solonetz soil, which is permanently affected by the groundwater, is called meadow solonetz.

Naturally, apart from the groundwater, precipitation also influences the migration of solutions in the soil. Figure 46 (Chapter 3) shows the groundwater and wetting conditions in a solonetz area of the Hungarian Plain over a 6-year period. It may be seen that during most years, the precipitation penetrating the soil comes into contact with the capillary groundwater at certain times. The accumulation of soluble salts is usually found at the depth where the upward and downward currents of the solutions meet. Kovda,[3] Antipov-Karataev,[313] Darab,[312] and others point out that the distribution of the soluble-salt accumulation within the profile depends on whether the upward or downward solution movements are dominant.

The following is the description of a shallow meadow solonetz profile in the Hortobágy region in Hungary, exemplifying this type of soil.

Surroundings:	Pasture of good quality.
Topography:	Slightly uneven, flat.
Vegetation:	*Artemisia monogyna, Polygonum aviculaire, Festuca pseudovina.*
Depth of the profile:	120 cm.
Effervescence (with dilute acid):	32 cm.
Thickness of humus layer:	56 cm.
Alkalinity against phenolphthalein:	50 cm.
Groundwater table:	230 cm.
Genetic horizons:	
A (0 — 3 cm)	Pale gray, moist, weak, solodized sandy loam with an ash-like texture, many fine roots: abrupt boundary.
B_1 (3 — 15 cm)	Gray, dry, extremely hard; distinctly columnar structure; clay loam; roots relatively plentiful; tops of columns and some places on the sides are discolored, solodized; abrupt boundary.
B_2 (15 — 31 cm)	Gray, somewhat darker, moist, slightly hard; fine prismatic structure; clay loam; moderate roots; in $CaCO_3$ abrupt, otherwise gradual boundary.
B_3 (31 — 55 cm)	Brownish-gray, somewhat lighter in color; moist, slightly hard; fine prismatic structure; clay loam; few roots; iron mottles and iron concretions growing more frequent with depth; white lime spots; fine lime concretions; abrupt boundary in color.
C_1 (55 — 93 cm)	Grayish-yellow, moist, moderately hard; coarse prismatic structure; loess-like clay loam; dark, clayey humus streaks, white lime mottles, many lime concretions; stains of iron, soft iron concretions; gradual boundary.

FIGURE 80. Surface and profile of a shallow meadow solonetz soil in Hortobágy (Hungary).

C_2 (93 — 110 cm)	Grayish-yellow, moist, moderately hard; loess-like clay loam; lime and iron concretions, rusty stains of iron; greenish-gray mottles of gley.
(110 — 130 cm)	Yellow loamy clay, many lime and iron concretions.
(130 — 150 cm)	Yellow silty clay, lime and iron concretions.
(150 — 180 cm)	Yellowish-gray clay, segregated lime and many iron concretions.
(180 — 210 cm)	Gray clay; some segregated lime and iron; gley formation.
(210 — 220 cm)	Reddish-brown, sticky clay.
Soil type:	Shallow meadow solonetz on calcareous loess-like clay loam.

Figure 80 illustrates the profile of the shallow meadow solonetz soil described above. Figure 81 demonstrates the water-soluble salt content in the 1:5 aqueous extract of a profile developed under the influence of the groundwater. In Figure 82 the close similarity of the chemical composition of the groundwater to that of the soil profile is quite visible. The sodium carbonate content is considerable in both. It should be added that the depth of the sodium carbonate accumulation may change within the same profile, depending on the seasonal dynamics of the soluble salts.

In most of these soils the B horizon is situated near the surface; the A horizon is relatively

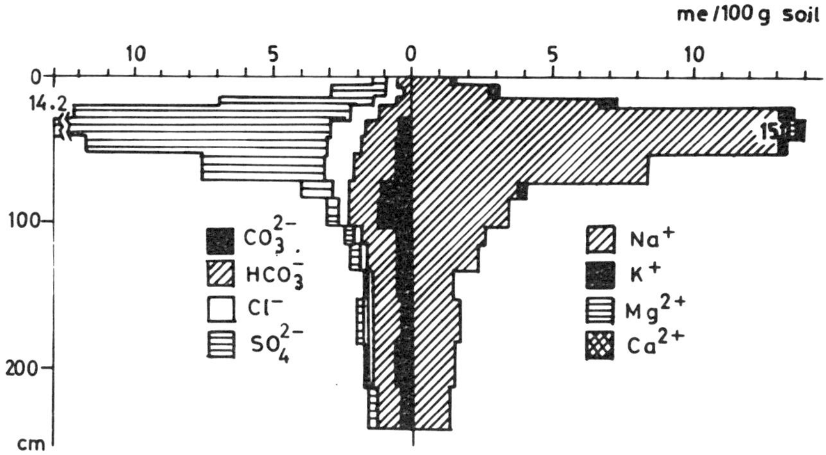

FIGURE 81. Water-soluble salt content of a 1:5 aqueous extract from a shallow meadow solonetz soil in Hortobágy (Hungary).

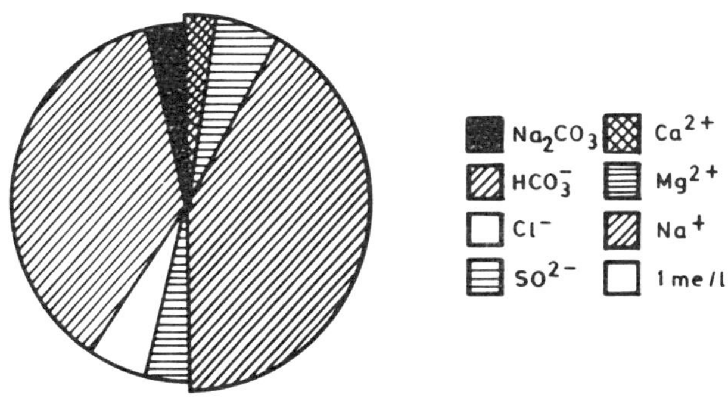

FIGURE 82. Chemical composition of the groundwater under a shallow meadow solonetz soil in Hortobágy (Hungary).

shallow as a consequence. Note that in certain cases, the B horizon is at the surface and there is no A horizon.

The great majority of sodium carbonate-containing solonetz soils have developed through the above-described processes, under the influence of mineralized groundwaters. They often occur in areas where the climatic conditions are far from arid, e.g., in tropical Africa, some states in the U.S., Argentina, Canada, Victoria (Australia), the Hungarian Plain, etc.

The sodium carbonate content of these soils ranges from considerable to nonexistent, according to the geochemical characteristics of the given areas, especially the chemical composition of the groundwaters. As a rule, the groundwaters contain sodium hydrocarbonate, but they may be rich in sodium sulfate, as in Alberta (Canada), the Hungarian Plain, etc.

In the Argentine solonetz described below, the morphological characteristics are very typical of humic solonetz (see also Tables 63 and 64).

Location:	8 km NE of Las Flores, Buenos Aires Province, Argentina.
Altitude:	28 m.
Physiography:	Depressed pampa.
Drainage:	Imperfect.
Vegetation:	Eastern pampa, grassland.

Table 63
PHYSICAL AND CHEMICAL PROPERTIES OF A SOLONETZ IN BUENOS AIRES PROVINCE, ARGENTINA

Horizon	Depth (cm)	pH (H_2O)	Cation exchange, meq/100 g					$CaCO_3$ (%)
			CEC	Ca	Mg	K	Na	
A_p	0—15	8.0	19.6	9.7	4.7	1.5	2.9	0
A_1	15—20							
A_2	20—40	8.9	7.6	2.6	1.4	1.1	2.0	0
B_{21t}	40—52	8.7	23.0	8.1	5.0	2.3	6.5	0
B_{22t}	52—85	9.1	21.6	7.5	6.5	2.2	5.4	0.8
B_{31x}	85—150	8.9	15.4	6.2	4.3	2.1	3.4	0.7

Table 64
PROPERTIES OF A SOLONETZ IN BUENOS AIRES PROVINCE[1]

Horizon	Soluble salts (E.C. — mmho)	Organic matter			Particle size distribution (%)			
		C (%)	N (%)	C/N	Coarse sand	Fine sand	Silt	Clay
A_p								
A_1	—	1.9	0.25	8	0	53	36	11
A_2	—	0.2	0.03	7	0	71	22	7
B_{21t}	—	0.3	0.06	—	0	48	25	27
B_{22t}	0.7	0.2	0.05	—	0	48	41	11
B_{31x}	1.4	0.1	—	—	0	58	38	4

Climate: 5,112, typical pampean.

Profile description:

A_p (0 — 15 cm) Very dark gray (10YR 3/1) moist, sandy loam; weak medium subangular blocky structure breaking to granular; very friable; abrupt smooth boundary.

A_1 (15 — 20 cm) Very dark gray (10YR 3/1) moist, sandy loam; weak medium subangular blocky structure breaking to granular; very friable, nonsticky, nonplastic, crotovines, gradual wavy boundary.

A_2 (20 — 40 cm) Dark brown (10YR 3/3) moist, sand; weak medium and coarse platy structure, breaking to single grain; loose; crotovines; abrupt smooth boundary.

B_{21t} (40 — 52 cm) Brown to dark brown (10YR 4/3) moist, heavy clay; strong, coarse prismatic and columnar structure; hard, firm, very sticky, very plastic; many iron-manganese concretions; abundant, prominent medium brown mottles (7.5YR 4.5/4); abundant very dark gray brown (10YR 3/2) clay skins; gradual wavy boundary.

B_{22t} (52 — 85 cm) Dark yellowish-brown (10YR 4/4) moist, clay loam; coarse prismatic structure breaking to subangular blocky; friable, very plastic, slightly sticky; moderate calcareous concretions; many prominent iron mottles; gradual wavy boundary.

B_{31z} (85 — 150 cm) Dark yellowish-brown (10YR 4/4) moist, sandy clay loam; weak, coarse subangular blocky structure breaking to single grain; friable, very hard when dry, plastic, slightly sticky; many iron-manganese and calcareous concretions; few clay skins; common mottles; weakly developed fragipan.

C (150 cm +) Water level.

The first step toward the amelioration of a solonetz soil developed under the influence of mineralized groundwaters must be the lowering of the water table, which cuts off the communication of the groundwater with the upper layers of the soil profile. But even if this is achieved, the tendencies of the subsequent soil formation processes may differ from case to case; they may, for example, lead to solod formation.[318]

The second set of circumstances which may be distinguished when sodium salts form

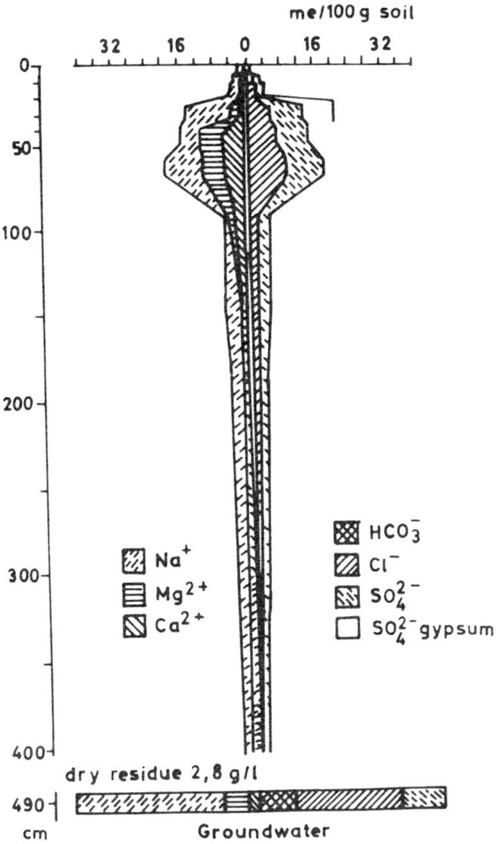

FIGURE 83. Salt profile of a solonetz soil turining into steppe formation in the U.S.S.R. (After Yulidov.[1])

solonetz soils is that in which the mineralized groundwater is at such a great depth (more than 6 m) that it exerts practically no influence on the soil formation processes. The sodium salt solutions affecting solonetz-forming processes penetrate the soil profile from other sources, in other ways. Under arid or semiarid conditions, the weathering of minerals, the decomposition of organic and inorganic compounds, or the transport of water-soluble salts by the surface waters may constitute such sources. Climatic conditions may stimulate the repeated downward and upward migrations of the sodium salt solutions which influence the profiles of these soils, and although they have all the features characteristic of solonetzes, they differ conspicuously from those developed under the influence of groundwaters.[319-321]

Figure 83 shows the salt profile of a steppe solonetz, as it is called in the Soviet literature. These soils are common in the Ukraine, the Trans-Volga region, Kazakhstan, China, Australia, and in many semi-arid countries.

In steppe solonetz soils, water-soluble salts often accumulate in deeper horizons than in meadow solonetz soils. This is frequently true of the B horizon as well, and consequently the A horizon is much thicker, if it has not been eroded. Naturally, the B horizons of these soils are characterized by an alkaline reaction, and may be considered as media capable of alkaline hydrolysis. Usually free sodium bicarbonate cannot be found in the upper layers, however, or even in the whole profile, and sometimes the content of sodium bicarbonate is low compared to that of sodium sulfate or perhaps sodium chloride. It must be noted that solonetz soils with high sodium carbonate contents may also develop under these climatic conditions if the parent material contains considerable amounts of soluble carbonates.

As regards the amelioration of these soils, the main tasks are to improve the properties of the soil horizons and to alter the dynamics of salts within the profile. If proper leaching is assured, the application of amendments capable of acid hydrolysis is very effective, and the loosening of the B horizon is often advantageous too.

C. Solod-Forming Processes

Solod soils appear differently in various soil grouping and classification systems. They are closely related to the planosols mentioned in the FAO/UNESCO classification system, and with the vertisols found in numerous other classification systems. However, both genetically and from a practical point of view they are most closely related to solonetz soils.

In spite of the numerous publications devoted to the subject, the genetic processes leading to the formation of the solod type of salt-affected soils have yet to be thoroughly studied and elucidated. In the literature, the scheme evolved by Gedroitz has been generally accepted for the formation of most of the solods: Solonchak → Solonetz → Solod.[238]

According to this scheme the process develops concurrently with the leaching of solonchaks, under the influence of more diluted, slightly alkaline solutions — or even water, in the last stages of the process.

Although the above scheme seems quite simple, the processes taking place in the course of solodization are by no means simple. The sodium concentration of the soil colloids decreases, and at the same time the H ions assume an important role.[108] The alternative prevalence of these processes — which are more intensive than in the case of solonetz formation — causes lattice layers of clay minerals to be loosened, altered, sometimes even destroyed, and complex organo-mineral compounds are decomposed. Consequently, the accumulation of silicon compounds may be observed on the soil surface and/or in the upper layers of the profile. This phenomenon was observed first — even before Gedroitz — by Popov in 1915 in the forest steppe zone in the European area of the U.S.S.R.[1]

Solods commonly form in the salt-affected soils of central and south-eastern Europe, throughout the U.S.S.R., in China, Canada, the U.S., South America, Australia, and Africa (Chad, Malawi, Zambia, etc.). The genetics of these soils have been explained in widely varying descriptions by different authors, in different places, observing very different solod soils. In some places, their formation does not appear to have started with the leaching of solonchak salts. In Chad, for example, it proceeded directly from the accumulated debris of granitic rocks rich in sodic minerals or from granitic alluviums of the Logone River in a catena or toposequence of soils.

On Figure 84 the profile of a solonetz soil with substantial solodization in horizon A_2 can be seen.

As indicated previously, large quantities of silicon compounds accumulate on the soil surface and/or in the upper soil horizons during solod-forming processes. Gedroitz[238] describes their chemical composition as "amorphous silicic acid". Several experts (Muraközy,[322] Tyurin,[323] Vilyams[314]) explain this accumulation mainly by biological factors. Kovda,[3] and Szabolcs[1] have detected relationships between the high silicon content of these soils and that of the ash of plants grown on them, and cite the particular biological cycle of silicon compounds as one of the possible causes of the formation of solod soils. Kovda notes that the silicon compounds in the groundwater often play a prominent role in solod formation.[3] Gerei and Szendrei[316] point out that the silicon compounds which accumulate in solod soils contain a considerable amount of finely dispersed quartz.

The quantity of accumulated silicon compounds is often determined by the method of Gedroitz, in 5% KOH. Extracted through this method they may amount to 1% or even more of the total weight of the soil in the upper layer. Gedroitz's 5% KOH extraction method can also be used to determine the ratio of silica to sesquioxides. A SiO_2:Al_2O_3 ratio higher than 2 indicates solod formation.

There is some controversy in the literature relating to the above statement; it should be noted that: (1) the silicon compounds accumulated in solod soils probably consist of several

FIGURE 84. Profile of a solodized solonetz soil (Hungary).

types of substances of both biological and nonbiological origin; and (2) the silicon compounds are not only formed in the course of solodization, but they are also mobile both on the soil surface and in the profile.

The mechanism of the solod-forming process has not yet been fully researched and described. As in a number of other hydromorphic processes of soil genetics, temporary waterlogging on the surface or in certain soil horizons plays an important role in the transformation and migration of compounds in the soil. The complex process described by R. Brinkman and given the name ferrolysis is in some respects similar to certain part processes occurring during the formation of solod soils, but it is also closely related to the solod-forming process under hydromorphic conditions.[324] In Figure 85 the scheme of ferrolysis is demonstrated.

As a rule, the most characteristic solod spots usually develop in micro-depressions, where the increased accumulation of silicon compounds on the surface is fairly evident, having been transported there from the surrounding area.

This also explains whey it can happen that the amount of SiO_2 extracted with 5% KOH and the SiO_2:Al_2O_3 ratio (both determined by Gedroitz's method[238]) do not always correspond to morphological observations of solod-forming processes in a given place.

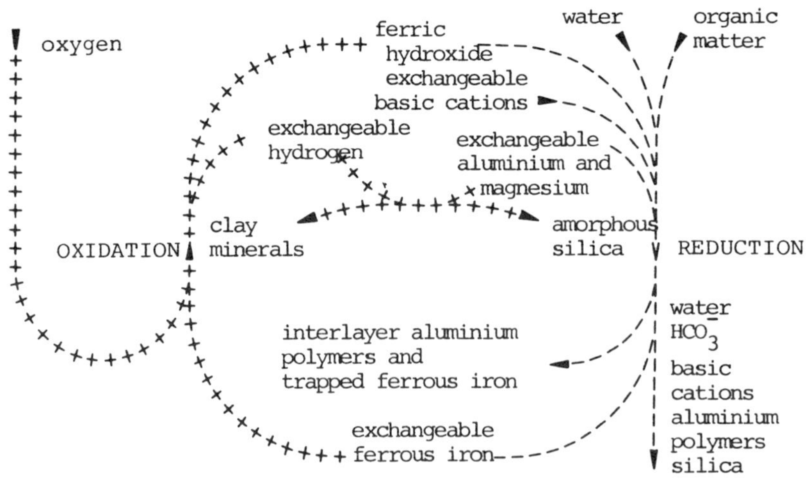

FIGURE 85. Scheme of ferrolysis. (After Brinkman.[324])

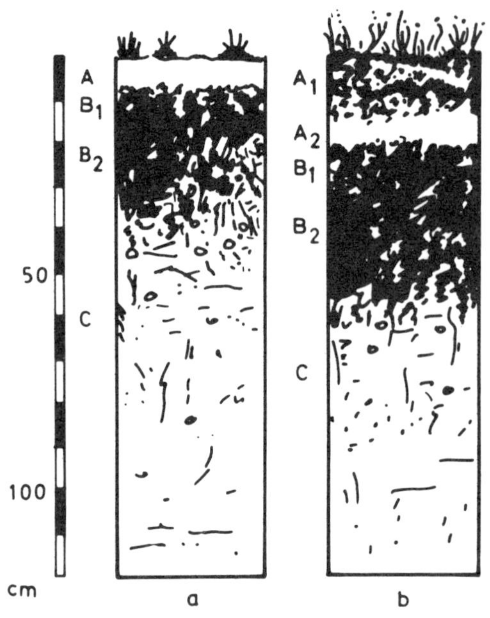

FIGURE 86. Schematic profiles of solod soils.

Two types of morphology can be distinguished among solod profiles, as seen in Figure 86 (A and B). Figure 86A represents a schematic solod profile where the accumulation of SiO_2 is on the surface. In this case it is the A horizon which is solodized. This occurs, as a rule, if the groundwater is capillarily linked with the upper layers of the soil profile, or if a shallow solonetz has been solodized. Thus, parallel with solod formation, a considerably high ESP value can also be measured in the A horizon, the pH of which is often alkaline. The statement that the solodized horizon of a solod soil shows an acid reaction cannot always be verified.

Microdepressions are often the sites of the alternate drying and waterlogging which provoke solod-forming processes. The solod formation represented in Figure 86A usually occurs in heavy-textured soils where, as a result of the slowness of solution movements, water-saturated

FIGURE 87. The upper part of a solod profile, affected by salt accumulation from below.

conditions are prolonged. The upper part of the B horizon is often solodized as well, and — unlike the B horizons of well-developed solonetz soils — its columns or crusts contain a considerable quantity of silicon compounds, and are lighter in color. Accumulations of sesquioxides can always be found under the solodized horizon and, as in solonetz soils, the clay content is always higher than in the A horizon.

Under the influence of mineralized groundwater, if it is capillarily linked with the upper soil layers, the solod profile may become salinized. It happens sometimes that solonchak-, solonetz-, and solod-forming processes, which are separated in the scheme presented by Gedroitz, are intertwined in the same profile (Szabolcs[315]) (see Figure 87).

This type of solod soil is fairly common in Australia, Idaho (U.S.), western Siberia, the Hungarian Plain, and many other places where an alkaline groundwater is linked with the soil profile.[325,326] The following is a description of a solod profile from the Hortobágy region, Hungary.

Effervescence (with dilute acid):	36 cm.

Table 65
ANALYTICAL DATA OF 5% KOH EXTRACT OF A SOLOD SOIL IN HORTOBÁGY, HUNGARY

Horizon	Depth (cm)	SiO_2 soluble in 5% KOH	Al_2O_3 soluble in 5% KOH	SiO_2:Al_2O_3 ratio in eq.	pH (H_2O)	pH (KCl)
A	2—10	3.96	0.260	13.0	6.9	6.1
B	16—24	2.40	0.272	7.2	8.0	7.1

Genetic horizons:

A (0 — 13 cm)	Light gray, heavy clayey loam; finely granulated structure, laminated in some places; a large quantity of dry, decayed roots; clear boundary.
B_1 (13 — 40 cm)	More compact than the A horizon, heavy, clayey loam; dark colored, columnar prismatic structure; the upper parts of the columns are covered with gray dust, sometimes observable in the prisms as well; the upper boundary of the columns is blurred; further down, the columns are intensely black; slightly damp residues of roots; gradual boundary.
B_2 (40 — 69 cm)	Brownish-black, downwards the color becomes lighter; slightly moist, heavy, loamy clay; on drying, it assumes a nut-like structure; few decayed roots; iron concretions from pea- to walnut-size, alternating with gley patches; gradual boundary.
C (69 cm +)	Yellow, loess-like, structureless clay with gley and rusty layers.

In Table 65 the analytical data of the 5% KOH extract of the above profile are presented together with the pH values.

In Figure 86B the schematic profile represents a soil in which solod formation manifests itself in the A_2 horizon. In cases like this, the A_1 horizon is humous, while beneath the solodized A_2 horizon displaying the maximum SiO_2:Al_2O_3 ratio, the B horizon is also affected by solod-forming processes to a greater or lesser degree.

As a rule, this type develops where the groundwater is rarely linked with the upper soil layers, or not at all. Consequently the downward movements of the soil solution can be more intense than in the former type. This is why the A_1 and A_2 horizons of solod soils developed in this way have acid pH and the profile does not have a high water-soluble salt content. The ESP values are comparatively low or even negligible in the A_1 and A_2 horizons, and often also in the B horizon. This type is characteristic of the forest steppe and steppe areas of the Russian Plain, Australia, the U.S., etc.

The solod-forming process is closely related to solonetz-forming processes, and the occurrence of purely solod soils is far rarer than that of slightly or strongly solodized solonetzes. These terms vary depending on which processes are dominant and more developed.[1]

Sometimes a large quantity of exchangeable Mg and a small quantity of exchangeable Na are found in the B horizon of solod or solonetz soils, especially when the profile is linked with the groundwater.[315]

El-Gabaly,[318] Kovda,[3] and Gerei[316] point out that the accumulation of magnesium ions is connected with the transformation of soil colloids — especially clay minerals — in the course of solod-forming and some solonetz-forming processes, e.g., with montmorillonite formation. When discussing solonetz and solod-formation, El-Gabaly calls attention to the significance of the simultaneous occurrence of sodium and magnesium ions in the course of the evolution of salt-affected soils.

Szendrei characterized the clay mineral stability and transformation related to solonetz-forming and particularly to solod-forming processes in soils with respect to the role of Mg.[253]

In the MgO-Al_2O_3-SiO_2-H_2O system, the solute activity functions for the secondary minerals were found in the stability field of Mg-montmorillonite. The majority of solute activity points were found in the stability field of chlorite. At lower magnesium activity (solodized

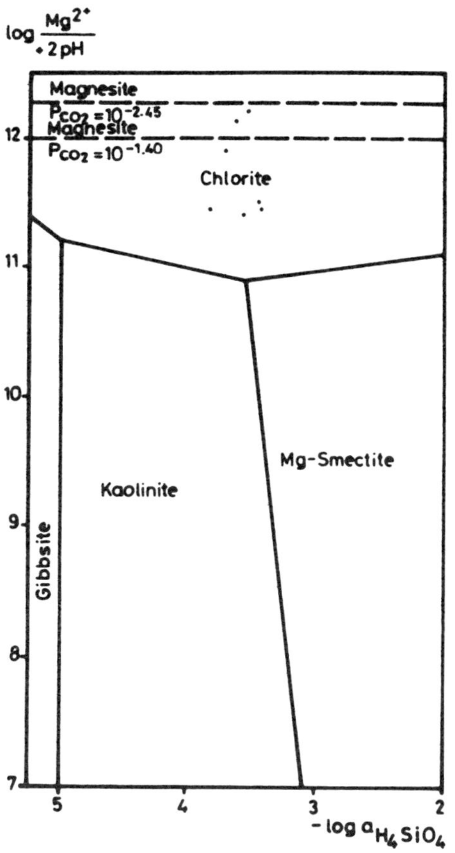

FIGURE 88. Stability diagram of minerals in the $MgO-Al_2O_3-SiO_2-H_2O-CO_2/g$ system. Points indicate the activities in saturation extracts of meadow solonetz soils.

meadow solonetz soil) chlorite also seems to be unstable (Figure 88). Some of the solute activity points fell into the stability field of Mg-smectite, indicating the stability of these minerals at lower Mg^{2+} and H^+ ion activities.

Szendrei also took into consideration the possibility of the occurrence of metastable phases. Evaluating the stability of secondary minerals in this system (Figure 89) the chlorite minerals are stable at higher Mg^{2+} and H^+ ion activities, while at lower values pyrophyllite is the thermodynamically stable mineral. Montmorillonite can also be expected as a metastable phase.

Solod-forming processes often occur also in irrigated soils, especially where the migration of slightly alkaline solutions within the profile of heavy-textured soils is intense and temporary waterlogging occurs, e.g., in the cultivation of rice. Thus, the conclusion may be drawn that secondary salinization (secondary formation of solonchak soils), secondary alkalization (secondary formation of solonetz soils), and secondary solodization (degradation of irrigated soils) frequently coincide in irrigated soils.[315,327]

IV. GYPSIFEROUS SOILS

Although gypsiferous soils (which are extensive mainly in arid and semi-arid regions of North Africa, the Near and Middle East, the Central Asian Republics of the U.S.S.R., China, India, Iraq, Iran, Afghanistan, the U.S., and Mexico, in arid areas of South America, and in dry regions of Australia) occupy vast territories, no reliable reports are available on

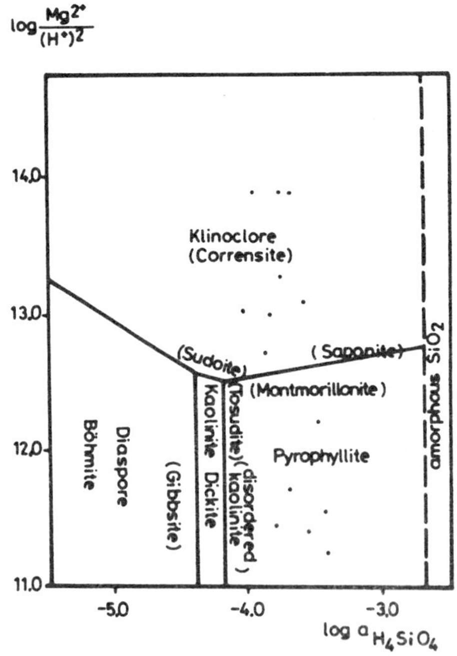

FIGURE 89. Stability diagram in the MgO-Al_2O_3-SiO_2-H_2O system. Points indicate the activities in saturation extracts of meadow solonetz soils.

the extension of this group of salt-affected soils except for some information on the UNESCO/FAO World Soil Map.[1,4,22]

The diagnostics and definition of gypsiferous soils are also not sufficiently elaborated. It is generally accepted that if the gypsum content reaches 1% in the top layers of a soil or exceeds 2% somewhere in the deeper layers (up to 60 to 80 cm) the soil must be classified as gypsiferous.

As was mentioned earlier in connection with saline soils, the accumulation of calcium sulfate is quite frequent in saline soils, and it occurs together with other salts. This is particularly true of sodium chloride and sodium sulfate, but even calcium carbonate may accompany gypsum in salt-affected soils. Accordingly, some properties of gypsiferous soils develop under the joint influence of existing salts. This is why, in spite of the acid reaction of gypsum, neutral or even alkali pH may occur in different soil horizons. The gypsum content of gypsiferous soils may be very high, sometimes exceeding 25% in the horizon of maximum occurrence. Gypsum concretions are also frequent in various horizons of gypsiferous soils.

In Table 66 some chemical characteristics are given for a gypsiferous soil from Iraq. In the soil described in the table, the gypsum content of the upper layer and of the concretions is signficantly high. The joint occurrence of gypsum and carbonate results in a slightly alkali pH in the soil suspension.

Besides calcium ions, sodium ions also play a significant role in gypsiferous soils. As is shown in Table 66, the ESP value is quite high and in spite of the dominance of calcium sulfate, the occurrence and role of other ions cannot be neglected. In Table 67 the chemical composition of a 1:5 aqueous extract is demonstrated for the same gypsiferous soil (Iraq). This shows that chloride ions are also found in the soil and that among the cations, Mg and K play a rather minor role. However, in other gypsiferous soils the magnesium content may be quite considerable.

It can be assumed that this phenomenon depends not only on local conditions but also on

Table 66
CHEMICAL CHARACTERISTICS OF GYPSIFEROUS SOILS IN IRAQ[328]

	pH in saturation extract	$CaCO_3$ (%)	SP (%)	CEC (meq/100 g)	ESP	Gypsum (%)
Soil	7.15	13.4	52.8	10.43	12.56	27.76
Concretion	7.5	3.74	—	—	—	43.20

the seasonal dynamics of salt migration in the soils. The reserves of magnesium in the gypsiferous soils studied are very high, as is shown in Table 68. Owing to local circumstances the bulk of magnesium compounds is currently in immobile form in the soils.

As has already been mentioned, gypsum dominates in all the soil properties. This can also be seen in Figure 90, which shows the X-ray diffractograms of the sample. Besides gypsum, calcite also appears on the X-ray diffractogram, as well as chloride feldspars, attapulgite, and a small amount of dolomite. Quartz also occurs in the sample. The mineralogical analysis also found small amounts of illite and illite-smectite minerals.

In Figure 91A the scanning electron microscope picture of a suspended sample demonstrates the shape of gypsum in the soil sample, while the elemental analysis (Figure 91B) proves that the formulae tested (points indicated in Figure 91A) consist of gypsum.

Transmission electron microscope pictures of drop preparations from untreated soils clearly show the needle-like crystals of gypsum (Figure 92). The occurrence of attapulgite in the sample was traced using the transmission electron microscope, as shown in Figure 93.

The location of attapulgite in the soil matrix was studied by means of electron microscope and elemental analysis (Figure 94). In Figure 94A the arrow shows attapulgite thorns protuding into gaps in the soil matrix, Figure 94B demonstrates the elemental analysis of the point indicated.

Such analyses are important because one of the dominant clay minerals in areas where gypsiferous soils frequently occur is attapulgite.

The grouping of gypsiferous soils must always be interpreted conditionally. Whether gypsiferous soils should be included in other soil classes such as saline soils, arid soils, etc. depends on the purpose of the survey. From the pedological point of view they could be included in other classes, but when the practical utilization, especially the irrigation of gypsiferous soils is to be considered, more diverse grouping and treatment are necessary. Gypsiferous soils are particularly vulnerable when irrigation and/or drainage canals are planned and constructed, because, although the water solubility of gypsum is commonly low, in the case of large quantities of water a large amount of gypsum may go into solution. In such cases considerable voids may occur, and the canals may even collapse. Therefore, the properties of gypsiferous soils must be thoroughly studied before designing irrigation and drainage systems in such areas.

Another practical issue related to gypsiferous soils is the chemical consequence of leaching.[329] Due to the high gypsum content, practically total removal of the salts from the soil is possible during leaching wihtout the adsorption of sodium ions on the surface of soil colloids, a phenomenon which is frequent and very hazardous during the leaching of some nongypsiferous saline soils.[330]

V. MAGNESIUM SOILS (By K. Darab)

The definition and diagnostics of magnesium soils have been given even less attention than gypsiferous soils.

Table 67
COMPOSITION OF THE 1:5 AQUEOUS EXTRACT OF A POWDERED GYPSIFEROUS SOIL SAMPLE

pH	Electric conductivity (mmho)	Dry residue (%)	Heating residue (%)	Alkalinity		Cl^-	SO_4^{2-}	Ca^{2+}	Mg^{2+}	Na^+	K^+
				CO_3^{2-}	Total						
7.1	4.29	1.937	1.341	—	0.010 (0.159)	0.138 (3.900)	0.634 (13.214)	0.385 (19.210)	0.008 (0.660)	0.025 (1.090)	0.006 (0.147)

Note: Values (except where noted otherwise) are percent, with meq given in parentheses.

Table 68
ELEMENTAL COMPOSITION OF THE EXAMINED SOIL SAMPLE, AS %

Sample	Air-dry soil loss	Heating loss	SiO_2	R_2O_3	Al_2O_3	Fe_2O_3	MgO	CaO	K_2O	Na_2O	P_2O_5	SO_3	MnO
A. Original powdered sample	6.58	15.52	34.77	9.40	6.54	2.86	4.21	20.75	0.95	0.78	0.12	11.49	0.06
B. After dissolution in hydrochloric acid													
In soil % after dissolution	2.35	7.90	62.25	18.68	13.37	5.31	5.72	0.79	1.75	1.07	0.11	0.37	0.05
In original soil %	—	—	34.85	10.45	7.48	2.97	3.20	0.44	0.98	0.60	0.06	0.21	0.03
C. Hydrochloric acid loss	—	44.0	—	—	—	—	—	—	—	—	—	—	—

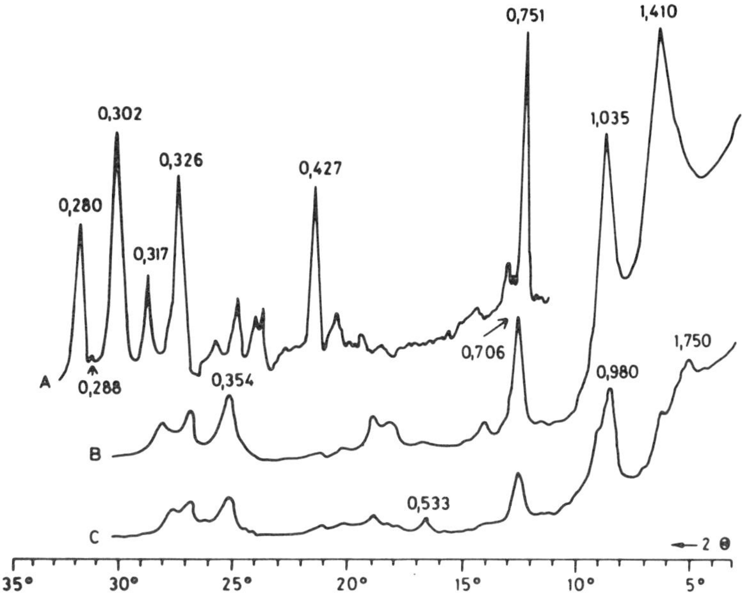

FIGURE 90. X-ray diffractograms. (A) Dust; (B) oriented; (C) oriented, treated with glycerine. The values indicated are given in nm.

A

FIGURE 91. Scanning electron microscope picture of a suspended sample. (A) Points 1 and 2 are the sites of elemental analysis. (B) Elemental analysis of the formula shown in (A). The results of the elemental analysis made at the points indicated demonstrate that the tested formulas are of gypsum.

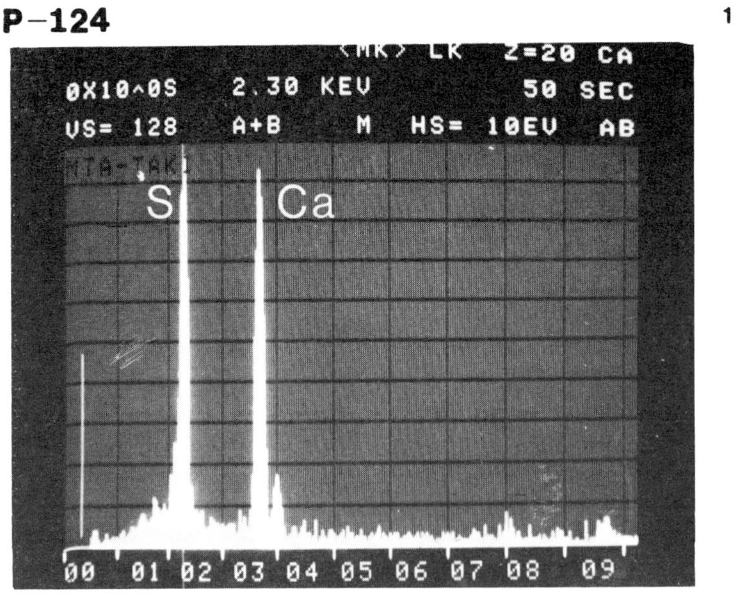

FIGURE 91B

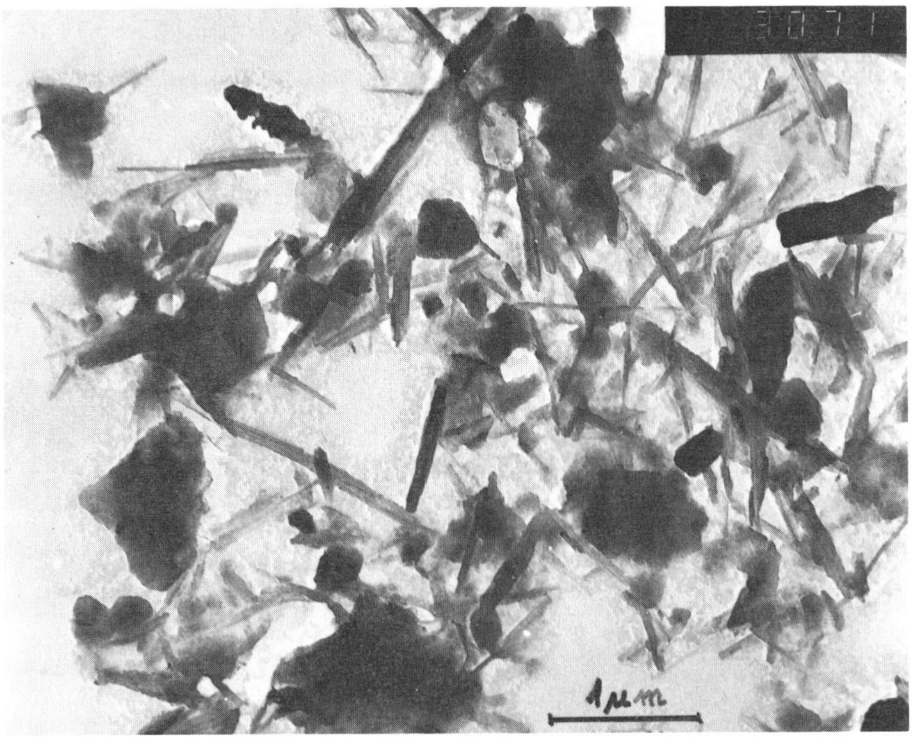

FIGURE 92. Transmission electron microscope picture of a drop preparation made from untreated soil. Among the big structural elements are needle-like crystals of gypsum showing thicker, dissolution traces, being long-shaped; and the unbroken contoured needle-like crystals are attapulgite.

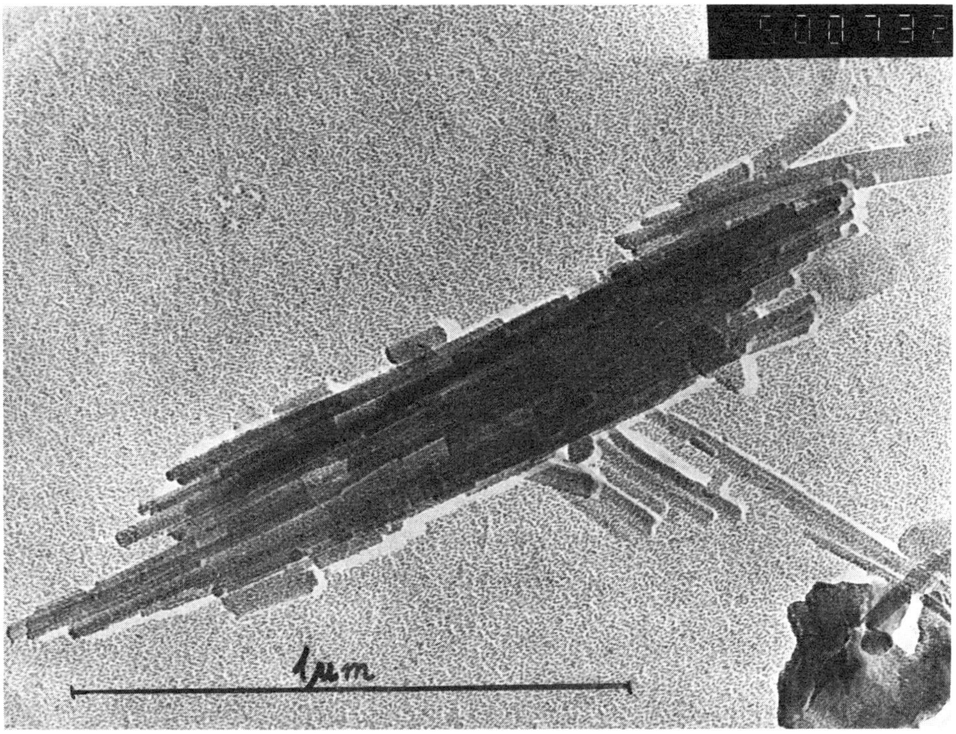

FIGURE 93. Transmission electron microscope pictures of attapulgite.

P – 147

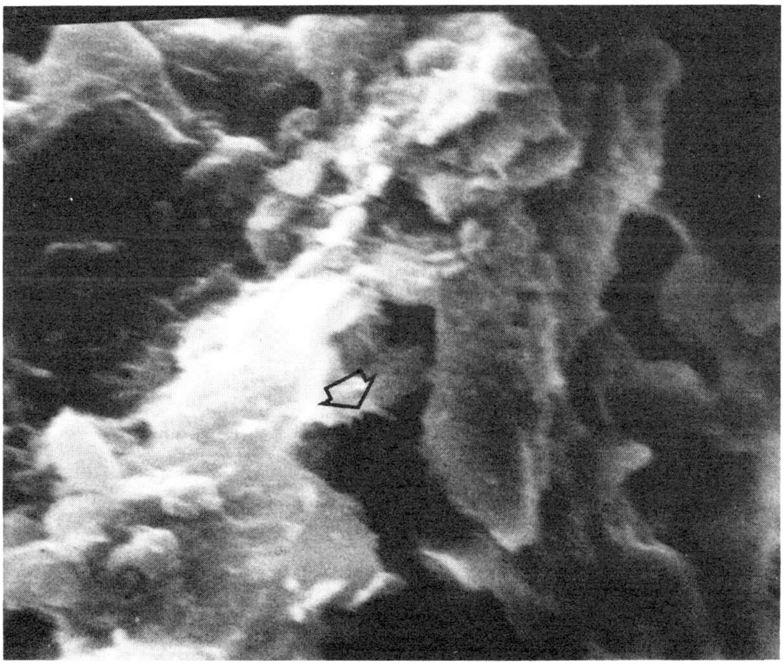

2

5000 X

A

FIGURE 94. (A) Attapulgite "thorns" protruding into the gap, obtained from an undisturbed soil sample; (B) elemental analysis of the formula shown in (A).

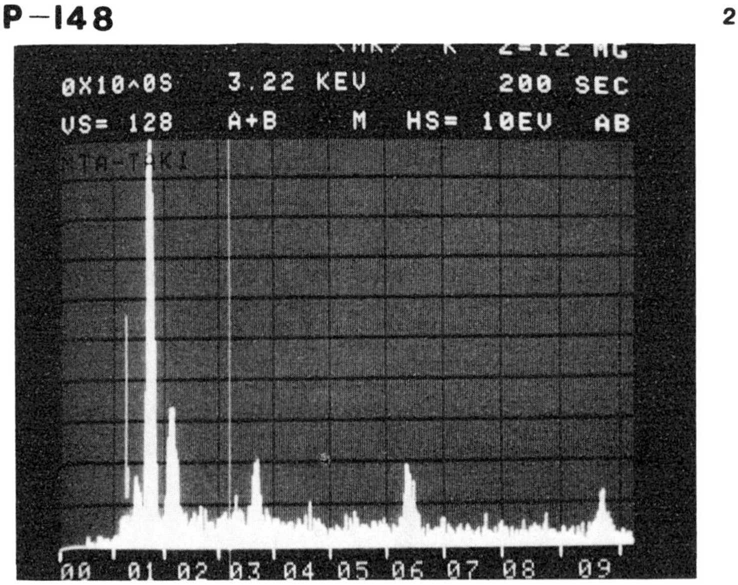

FIGURE 94B

Diverse opinions on the effects of magnesium on soil fertility are reported in the literature (Antipov-Karataev and Filipova,[331] Arany,[332] Gedroitz,[238] Kelley,[18] Mondal,[333] and Polynov[334]).

The diversity of opinions on this subject may be due to the fact that magnesium soils with different genesis and properties are included in a single group, regardless of the soil-forming processes, the sources and manner of magnesium enrichment, and the type and mobility of magnesium compounds accumulated.

The term magnesium soils includes:

1. Solonchak soils where magnesium accumulates in the form of inorganic salts (mainly magnesium sulfate and magnesium chloride). The high concentration of magnesium salt in the soil solution of magnesium solonchaks is toxic to plant growth, thereby having a direct physiological effect.[335]
2. Hydromorphic soils, with dark color and heavy texture, which contain increased quantities of total and exchangeable magnesium.[336] The fertility of these soils is influenced by the poor water conductivity and by the limited mobility of the soil moisture. Some authors relate the poor water-physical properties with an increase in the degree of magnesium saturation.[335,337] Other papers refer to the formation of magnesium-soil-organic matter complexes with high stabilities. These magnesium complexes, together with the layer silicates of the clay fraction, form compounds of great hydrability.[338] There is evidence in the literature indicating a relationship between the increase in the amount of swelling layer silicates and the inhibited water movement in these soils.
3. Some soils have a high degree of magnesium saturation associated with a large variation in the sodium saturation. These soils have every morphological feature of the solonetz profile. Some authors suggest that similarities may exist between the behavior of magnesium and sodium ions in solonetz formation.[339] Others regard the increase in the degree of magnesium saturation as the first stage in the solonetz-forming process.[332] Polynov[334] assumes that exchangeable magnesium accumulates during the leaching of salt-affected soils. Antipov-Karataev and Mamaeva[340] proved that the morphology of solonetz soils have a high degree of magnesium and a low degree of sodium saturation

is the relic of a former stage of solonetz soil formation connected with the enrichment of exchangeable sodium.

The only common feature present in all these soils is that they contain increased amounts of mobile magnesium compounds,[341] which may influence their fertility.[342]

The formation and accumulation of magnesium compounds are due to a large number of geochemical and soil-forming processes taking part in the upper strata of the earth's crust.

When discussing the influence of magnesium compounds on soil properties studies must be made on:

1. The types and sources of magnesium compounds
2. The stability of magnesium-bearing constituents and the mobility of the weathering products
3. The effect of mobile magnesium compounds on the properties of soils and plant growth

A. The Types and Sources of Magnesium Compounds in Soils and Soil-Forming Processes

Magnesium is a very common element, and its relatively high abundance in the crust and mantle of the earth leads to its occurrence in numerous silicate minerals (see also Chapter 2). Among the magnesium minerals the following are the most abundant in weathered sediments and soils:

1. Silicates. Magnesium is an important constituent of a large number of common rock-forming silicates. These include the orthosilicates with the olivine and garnet structures; chainsilicates with pyroxene and amphibole structure; and layer silicates as biotite, chlorites, and smectites.
2. Carbonates. Dolomite ($CaCO_3 \cdot MgCO_3$) and dolomitized calcite occur in some sediments, limestones, and soils.
3. Sulfates and chlorides. The relatively high concentration of magnesium in mineralized surface and groundwater on the Earth and in the seawaters leads to the accumulation of magnesium sulfate and magnesium chloride in cases where evaporation is prevailing over leaching.

B. The Stability of Magnesium-Bearing Constituents in Rocks and the Mobility of Weathering Products Containing Magnesium

Magnesium occurs in a high proportion of the minerals in mafic igneous rocks. In the case of magnesium silicates and ferromagnesian minerals, the degree of basicity is the factor which determines the relative stability of the crystal network. The stability of the silicate minerals increases as their basicity decreases. This decrease in the basicity and increase in the stability of the minerals in weathered mafic rocks are usually connected with the hydration and decomposition of primary minerals and with the formation of sedimentary rocks.

Magnesium is a constituent of the crystal network of trioctahedral layer silicates formed by the substitution of divalent magnesium for trivalent cations in the octahedral sheet. Among the magnesium-rich trioctahedral minerals, biotite is one of the representative minerals in sediments and soils. In biotites 20 to 60% of the occupied sites contain magnesium.

Trioctahedral micas are common constituents of igneous and metamorphic rocks. They are readily weathered, so trioctahedral micas and hydromicas are seldom found in sediments and soils.

Other magnesium-rich layer silicates in sediments and soils are vermiculites, smectites, and chlorites (Table 69).

Clay-sized vermiculites are frequently found in soils. They can be either dioctahedral or trioctahedral, but the dioctahedral form is more frequent than the trioctahedral variety.

Table 69
ELEMENTAL COMPOSITION OF SOME CLAY MINERALS

Elements	Allophane	Kaolinite	Halloysite	Montmorillonite	Vermiculite	Illite	Chlorite
SiO_2	33.96	45.44	44.08	51.14	35.92	49.26	26.68
Al_2O_3	31.12	38.52	39.20	19.76	10.68	28.97	25.20
Fe_2O_3	Trace	0.8	0.1	0.83	10.94	2.27	—
FeO	—	—	—	—	0.82	0.57	8.7
MgO	—	0.08	0.05	3.22	22.00	1.32	26.96
CaO	2.26	0.08	—	1.62	0.44	0.67	0.28
Na_2O	—	0.66	0.2	0.04	—	0.13	—
K_2O	—	0.14		0.11	—	7.47	—
H_2O	33.12	14.2	16.18	22.80	19.84	9.25	11.7

Note: Values are percent.

After Degens, E. T., (cited in Darab[239]).

One part of the magnesium can be found in clay vermiculite as an interlayer cation which is easily exchangeable by other cations from the intermicellar solution in soil-solution systems.

Smectites occur only as silt and clay-sized particles. The basal spacing of smectites depends on the composition of the interlayer cations and the amount of water associated with the exchangeable ions. Magnesium ions appear in smectites in exchangeable form and in the octahedral sheets of the crystal network, replacing aluminum ions. A typical example of magnesium-rich smectites is montmorillonite. The magnesium content in montmorillonite varies from 5 to 32%. The dioctahedral layer composition, involving 0.4 to 0.6 units of octahedral magnesium, can often be found among soil montmorillonites.

The abundance of exchangeable Ca^{2+} and Mg^{2+}, silica-rich media, and a hydromorphic environment promote the development of expansive forms of montmorillonites.

Among the magnesium-rich minerals, chlorites occur extensively in soils, partly inherited from the parent rocks and partly formed during the soil-forming processes. The MgO content of chlorites can be as high as 27%. Different types of chlorite minerals can be distinguished by their structure and chemistry. The octahedral sheets may have a dioctahedral or trioctahedral structure. Trioctahedral chlorites are more frequent in rocks than the dioctahedral form of crystal network, whereas the situation is opposite in the case of soils.

Most soil clay fractions contain mixtures of minerals including some transitional forms of layer silicates. By swelling chlorite, the product of the supposed process has a 2:1 structure with interlayer brucite islands and hydrated exchangeable magnesium ions. The ratio of octahedral magnesium hydroxide and the exchangeable magnesium ions depend on the pH value and on the magnesium ion concentration of the intermicellar solution. Alkaline reaction and high magnesium concentration favors the formation of chlorite, while a decrease in pH value and in the magnesium concentration of the solution promotes the destruction of the brucite layer and the desorption of magnesium ions.

In some cases the distribution of carbonate minerals among fractions of different sizes is taken as a measure of the distribution of carbonates of pedogenetic origin or inherited from the parent rock (Table 70). Due to the disintegration of dolomite rocks, disintegrated particles of primary minerals may occur in the fine fraction of sediments and soils. A better result can be obtained by comparing the dolomite contents in the coarse fractions of the parent material and in the different horizons of the soils (Figure 95).

Magnesium carbonate has higher solubility than dolomite and calcium carbonate.[344] The accumulation of magnesium carbonate in a free form is very rare in soils. This is due to the adsorption of magnesium by clay and also the formation of dolomitized calcites and dolomite

Table 70
THE DISTRIBUTION OF CARBONATE MINERALS IN THE SAND AND SILT FRACTION OF SALT-AFFECTED SOILS

Soil type and horizon		Depth of sampling (cm)	Sand fraction: Dolomite (%)	Sand fraction: Calcite (%)	Sand fraction: Total carbonate as $CaCO_3$ (%)	Silt fraction: Dolomite (%)	Silt fraction: Calcite (%)	Silt fraction: Total carbonate as $CaCO_3$ (%)
Solonchak-Solonetz	A	0—5	1.6	0.7	2.4	4.5	0.3	5.2
	B_1	5—15	1.5	1.2	2.8	5.0	0.9	6.3
	B_2	15—31	7.2	4.1	11.9	11.2	4.2	16.4
	BC	31—44	9.1	5.5	15.4	20.1	6.4	29.3
	C	44—55	7.9	5.7	14.3	18.5	6.0	26.1
Solonchak	A	0—3	12.6	0.6	14.4	9.5	1.3	11.6
		3—11	6.7	2.4	10.3	12.3	2.8	16.2
	B_1	11—17	7.6	3.2	11.4	—	—	—
		17—31	8.4	3.9	13.0	19.0	6.4	27.0
	B_2	31—55	12.8	5.6	19.5	26.1	8.8	36.1
	C	55—70	6.7	6.1	14.0	23.9	5.7	31.7

After Szendrei.[253]

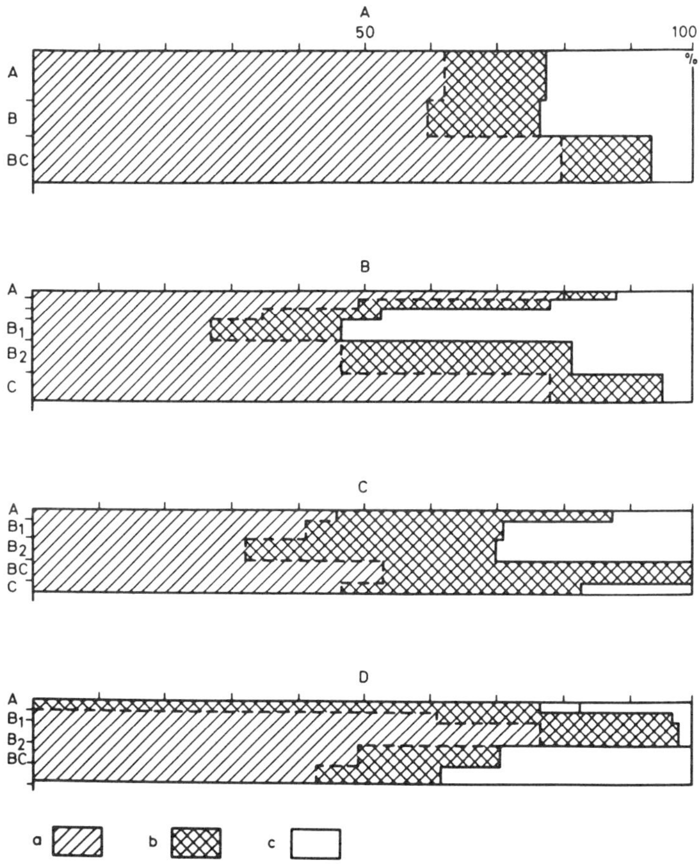

FIGURE 95. Primary carbonates as (a) $CaCO_3$, (b) dolomite content, and (c) calcite content of pedogenetic origin in the samples of hydromorphic soils with a diameter of >0.01 mm. (A) Meadow alluvial soil; (B) solonchak; (C) solonchak solonetz; (D) solonetz with shallow A horizon.

Table 71
MAGNESIUM CONTENT IN A SOLONETZ SOIL AND IN ITS CLAY FRACTION

Depth of sampling (cm)	Particles $\phi < 1\ \mu$ (%)	Total Mg in soil meq/100 g	Mg in clay		Exchangeable Mg (meq/100 g soil)	Soil Mg (meq/100 g soil)
			meq/100 g clay	meq/100 g soil		
8—16	53.52	79.37	69.44	31.17	17.12	0.30
26—37	60.44	81.85	71.93	43.47	30.30	0.24
48—55	53.72	40.18	65.48	35.17	—	0.31
75—85	39.22	85.32	79.37	31.13	—	2.84
94—105	40.62	76.89	77.38	31.43	—	2.52
117—126	39.17	147.33	71.93	28.17	—	0.25
135—145	42.29	102.70	81.35	34.40	—	—

during the accumulation of carbonates. The solubility of calcites dolomitized to different degrees is a complex function of the chemistry and structure of the carbonates.

Inorganic salts of magnesium are found mainly in seawater, and in the groundwater, sediments, and soils of deserts and semi-desert regions. The mobility and role of magnesium salts in soil-forming processes depend on the kind of anion to which the magnesium ion is bound. Saturated solutions of magnesium carbonate, magnesium sulfate, and magnesium chloride have concentrations of 1.2 g/ℓ, 262 g/ℓ, and 354 g/ℓ, respectively. Magnesium carbonate hydrolyzes in water to give an alkali reaction. Solutions of magnesium sulfate and magnesium chloride have neutral or slightly acidic reactions. During the decomposition of primary minerals, the magnesium readily goes into solution. Due to the good solubility of inorganic magnesium salts, the weathering products of magnesium-containing minerals have high mobility and accumulate in seawater, in surface and groundwaters, in soils, and in sediments if the climate and the geomorphological conditions favor the accumulation of salts. Magnesium sulfate and magnesium chloride usually precipitate together with other readily soluble salts, either as separate minerals or as complex crystals.

C. The Effect of Mobile Magnesium Compounds on the Properties of Soils and Plant Growth

The sources of magnesium compounds in soils are primary minerals or their weathering products in soil-forming rocks. The mobility of magnesium and of other elements is determined by the chemical bonds within the minerals. The magnesium reserves of soils can be distributed in three groups according to their mobility:

1. Mobile magnesium compounds. This includes the readily soluble inorganic magnesium salts and the exchangeable magnesium in soils.
2. Short-term reserves of mobile magnesium. The magnesium in the layer silicates of clay fractions, and the magnesium in the fine-grained, dolomitized calcite belong to this group.
3. The potential reserves of mobile magnesium. The magnesium of minerals in the nonclay fractions of the soil belongs to this kind of magnesium compound.

The quantities and ratios of soil magnesium with different mobilities vary over a wide range (Table 71). In a meadow solonetz formed under the influence of mineralized groundwater, the total magnesium has a minimum value in the B horizon, while the highest values were measured in the C horizon. The magnesium concentration in clay fractions is less than that in nonclay fractions.[345] The ratio of these two values decreases from the B_2 to the C horizon. The B_1 horizon of this soil is an exception, where the bulk of the soil magnesium

Table 72
MINERALOGICAL COMPOSITION OF THE CLAY FRACTION OF A SOLONETZ SOIL, AS %

Depth of sampling (cm)	Illite	Montmorillonite	Illite-montmorillonite	Chlorite	Illite-chlorite	Illite-chlorite-montmorillonite	Quartz	Feldspars
8—16	70	—	5	5	5	—	10	5
26—37	45	20	10	10	5	—	5	5
48—55	45	20	10	10	5	—	5	5
75—85	40	30	5	10	5	—	5	5
97—105	40	30	5	10	5	—	5	5
117—126	40	30	5	10	5	—	5	5

Table 73
MINERAL COMPOSITION AND TOTAL MAGNESIUM CONTENT IN SAMPLES WITH DIFFERENT SIZE RANGES OF THE "B" HORIZON OF A MEADOW SOIL

Size range φ μ	Particles (%)	Total Mg (meq/100 g sample)	Total Mg (meq/100 g soil)	Mobile Mg (meq/100 g soil)	Minerals[a]
<1	44.1	118.6	52.3	30.37	Mo > M > Chl = Fsp = Q
1—5	12.3	61.6	7.5	—	Q > M = Fsp > Chl
5—10	9.4	31.7	3.0	—	Q > Fsp > M = Chl
10—50	14.3	38.7	5.5	—	Q > Fsp > M = Chl = C
50—100	1.5	34.2	0.53	—	Q > Fsp > M = Chl = C
100—200	0.1	36.2	0.04	—	Q > Fsp > M > = Chl = C
Soil	—	—	99.2		—

[a] Q, quartz; Fsp, feldspars; M, mica and hydromica; Mo, montmorillonite; Chl, chlorite; C, carbonates.

is in the clay fraction and in the form of mobile magnesium. The same rule can be deduced from the mineralogical composition of the clay fractions in different soil horizons (Table 72).

The enrichment of magnesium in clay is not limited to the accumulation horizon of solonetz soils. In the B horizon of a meadow soil, mobile magnesium made up 30.6% of the total magnesium, and the magnesium in the clay fraction amounted to 52.7% of the total magnesium content of the soil (Table 73).

Heavy-textured magnesium soils may differ in their morphological features as well as in their fertility. One group of these heavy-textured magnesium soils has every morphological feature of a solonetz soil and is classified as solonetz. The other group of these soils is formed under hydromorphic conditions. They are dark and heavy-textured, but do not possess the morphology of solonetz soils. They can be grouped with the so called meadow soils and may be regarded as a kind of vertisol.[347]

A typical example of the distribution of these soils is given in Figure 96. The soil profiles are meadow soil (M-I), and medium and shallow solonetz soils. They form a cross-section at a distance of 100 to 500 m from the riverbed. The difference between the elevations is about 3 m. The depth of the groundwater table varies between 2.4 and 3.5 m from the surface.

In a Mg-solonetz soil with poor drainage the weighted average of mobile (exchangeable and soluble) sodium is the highest among the profiles in the cross-section and this soil has the lowest quantity of mobile magnesium (Table 74).

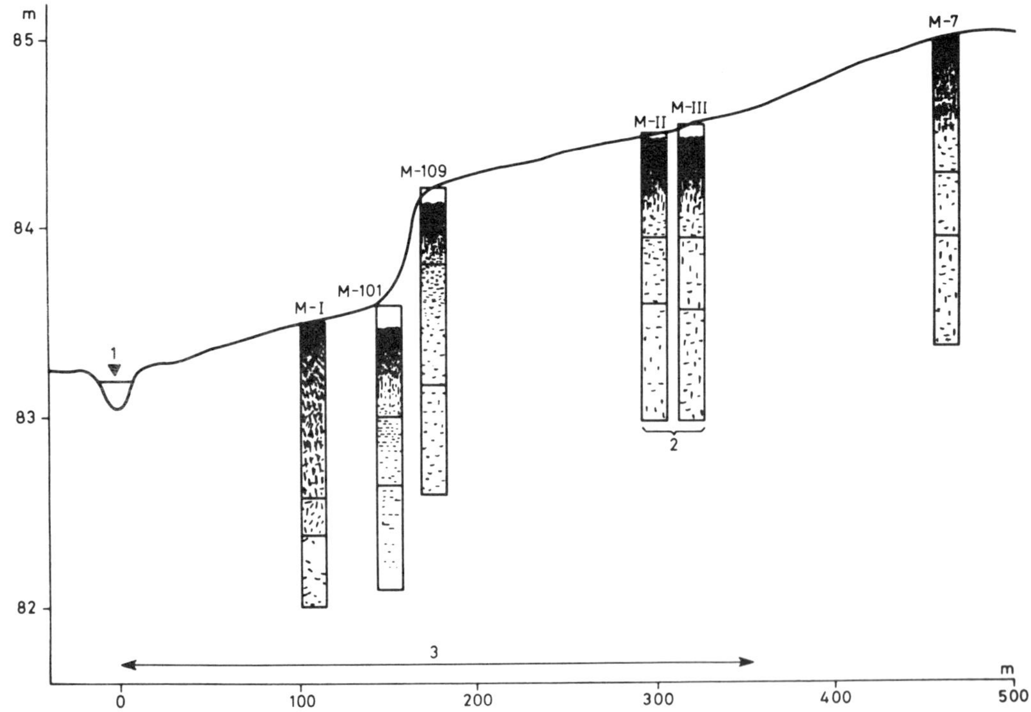

FIGURE 96. Soil profiles in vertical section "M". Vertical axis: height above sea level; horizontal axis: distance from the river bed.

Table 74
SOME CHARACTERISTICS OF SOIL PROFILES OF THE VERTICAL SECTION "M"

Profile no.	Soil type	Altitude above sea-level (m)	Distance from the rivet (m)	Max. total salt conc. (meq/ℓ)	Depth of salt accumulation (cm)	Weighted average of mobile Na⁺ (meq/100 g)	Weighted average of mobile Mg⁺⁺ (meq/100 g)
M-I	Meadow soil	83.5	100	50	80—120	5.12	8.05
M-101	Middle solonetz	83.6	150	150	20—60	6.20	15.25
M-109	Middle solonetz	84.2	170	40	40—80	3.79	17.94
M-II	Shallow solonetz	84.5	290	65	10—25	6.51	14.07
M-III	Middle solonetz	84.5	310	90	60—85	6.76	15.11
M-7	Shallow solonetz	85.0	450	20	20—80	20.94	6.73

In solonetz soils with poor drainage (Profiles M-II and M-7) the morphology and chemical properties of the soil are in harmony. In their accumulation horizons, which have a developed columnar structure, these soils contain a high amount of exchangeable sodium. In solonetz soils with better drainage (Profiles M-109, M-III) the weighted averages of mobile sodium are low, and the soils contain high quantities of mobile magnesium compounds. In these soils, except in the solodized A horizon and the columnar structured B horizon, the degree of sodium saturation is very low (Figure 97).

The exchangeable magnesium percentage (EMgP value) is more than 30% in the B_1 horizon and exceeds 50% in the B_2, BC and C horizons.

The solonetz morphology of soils with a low degree of sodium saturation may be regarded as a relic of an earlier stage in the soil formation process, connected with the enrichment of exchangeable sodium.

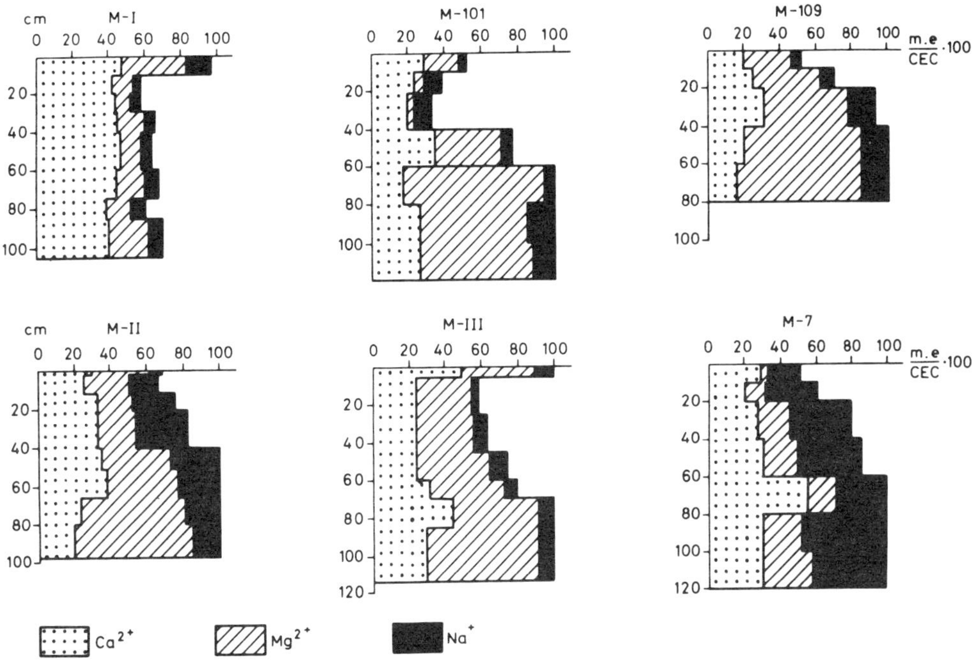

FIGURE 97. Exchangeable cations in the soil profiles of a cross-section.

The composition of clay minerals in Table 75 shows that a dynamic equilibrium exists between the soil solution, the surface of the clay particles, and the crystalline lattice of the layer silicates in the soils. This interaction leads to the enrichment of soil minerals with magnesium during the accumulation period in the formation of salt-affected soils. During the leaching process, the magnesium from the octahedron of chlorite and the trioctahedral position of montmorillonite goes into exchangeable form to replace the leached exchangeable sodium and magnesium. The resulting dioctahedral montomorillonite, with its high ability to swell, leads to the morphological features and poor water-physical properties of solonetz soil in spite of a decrease in the degree of sodium saturation.[348]

Magnesium salts usually accumulate in saline soils together with other readily soluble salts such as sodium sulfate and sodium chloride.

The chemistry of the groundwater and the chemistry of the salt crust at the top of a meadow solonchak are often similar to each other, as is shown by the chemistry of 1:5 aqueous extracts of a solonchak soil from the Kyzyl-Kum desert (Table 76). The ratio of equivalent concentrations of sodium to magnesium is practically the same in the groundwater and in the soil aqueous extract, but the soil is more chloridic and less sulfatic than the groundwater.

It seldom happens that the chemistry of soil salinization is characterized by the prevalence of magnesium salts. One example of this is given in Figure 98, which illustrates the chemistry of the soil solution in a crusty solonchak.

Magnesium sulfate and magnesium chloride are very toxic to plants. The toxic effects of magnesium chloride and magnesium sulfate are partly due to their ready solubility, and partly to changes in the ion uptake of the plant under saline conditions (Table 77).

* * *

This short review summarizes recent knowledge on magnesium soils. However, besides the soil formations mentioned, other soils are also frequently designated as magnesium soils.

Table 75
QUANTITIES OF MINERALS IN THE CLAY FRACTION OF THE INVESTIGATED SOILS

Profile no.	Depth (cm)	Illite	Montmorillonite	Illite montm.	Chlorite	Illite-chlorite	Illite-montm.-chlorite	Quartz	Feldspars
M-I	0—10	40	10	10	5	5	—	20	10
	10—20	30	35	10	5	5	—	5	5
	20—30	25	40	10	10	5	—	5	5
	30—40	15	40	15	10	5	—	5	5
	40—60	15	30	15	15	10	—	5	5
	60—76	20	30	15	10	10	—	5	5
	76—85	20	25	15	15	15	—	5	5
	85—106	20	25	15	15	15	—	5	5
	106—127	20	25	15	15	15	—	5	5
M-II	0—2	35	15	20	10	10	—	5	5
	2—12	20	15	20	10	15	—	10	5
	12—22	20	20	20	10	15	—	10	5
	22—42	20	20	20	15	10	—	5	5
	42—52	20	25	20	15	10	—	5	5
	60—81	20	20	20	15	10	—	5	5
	81—98	20	20	20	15	10	—	5	5
M-III	0—7	25	35	10	5	10	—	10	5
	7—27	25	35	10	5	—	—	5	5
	27—47	30	45	15	5	—	—	5	—
	47—60	15	50	15	5	—	—	5	5
	60—70	15	50	15	5	—	—	5	5
	70—85	10	60	15	5	—	—	5	5
	85—115	5	65	5	10	—	—	10	5
	115—150	5	65	5	10	—	—	10	5

Note: Values are g/100 g clay.

Table 76
CHEMICAL COMPOSITION OF 1:5 AQUEOUS EXTRACT OF A SOLONCHAK

Profile no.	Depth of sampling (cm)	Soluble salts	CO_3^{2+}	HCO_3^-	Cl^-	SO_4^{2-}	Ca^{2+}	Mg^{2+}	$Ca^{2+} + Mg^{2+}$
7255	0—0.5	20.297	0.056	0.093	8.135	3.650	0.427	0.842	4.97
	0.5—5	10.740	Trace	0.025	4.186	2.040	0.485	0.535	2.12
	5—10	11.379	—	0.034	4.030	2.210	0.482	0.560	2.06
	20—25	6.200	—	0.022	2.535	1.140	0.207	0.267	1.45
	35—40	4.730	—	0.023	1.850	0.902	0.148	0.230	1.03
	50—55	4.497	—	0.028	1.650	1.100	0.175	0.190	1.04
	85—90	2.475	—	0.028	0.955	0.625	0.052	0.112	0.65
	100—105	2.525	—	0.025	0.820	0.645	0.085	0.100	0.56
	125—130	2.292	—	0.028	0.821	0.560	0.040	0.091	0.59
Groundwater	545	31.140	—	0.850	8.230	11.430	0.610	1.520	7.54

Note: Values of chemical composition are % in the 1:5 water extract, g/ℓ in groundwater.

After Borovsky.[186]

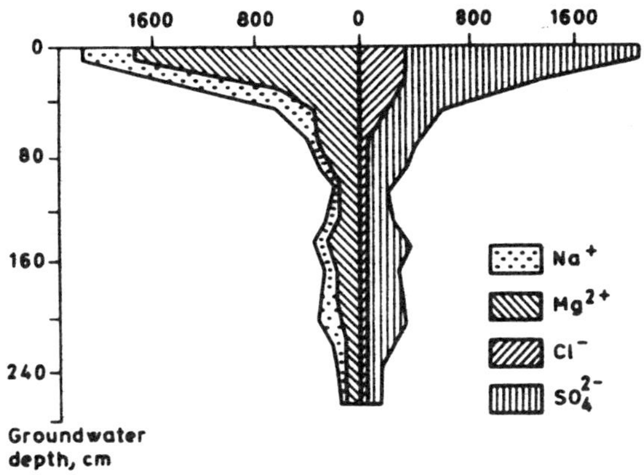

FIGURE 98. Salt content in the soil solutions of a crust solonchak (Kara-Kalpak Experimental Station).

Table 77
MINERAL COMPOSITION OF COTTON PLANTS DURING DIFFERENT PHASES OF VEGETATION

Condition of plant	Vegetation phase	Mineral substances	Cl^-	SO_4^{2-}	Ca^{2+}	Mg^{2+}	Na^+	Cl^-/SO_4^{2-}	$Na^+/(Ca^{2+}+Mg^{2+})$	Na^+/Mg^{2+}	Mg^{2+}/Ca^{2+}
Normal	Sprouting	16.10	1.54	3.83	3.35	0.81	0.41	0.54	0.076	0.27	0.40
	Budding	11.50	0.78	2.83	3.01	0.68	0.13	0.37	0.028	0.10	0.40
	Ripening	10.60	0.40	3.23	3.95	0.35	0.04	0.17	0.008	0.06	0.15
Depressed	Sprouting	17.00	1.56	2.03	1.83	0.87	0.54	1.04	0.144	0.33	0.78
	Budding	16.40	1.97	3.47	3.89	0.94	0.40	0.77	0.064	0.23	0.40
	Ripening	8.20	0.81	1.72	2.28	0.57	0.19	0.64	0.052	0.18	0.41
Strongly depressed	Sprouting	20.73	2.84	3.05	1.26	1.10	1.22	1.30	0.32	0.51	1.65
	Budding	17.14	2.04	5.94	2.10	1.47	0.63	0.50	0.12	0.23	1.15
	Ripening	11.78	0.81	3.06	1.97	1.15	0.43	0.36	0.09	0.19	0.96

Note: Values of mineral composition are percent of absolute dry weight.

After Kovda.[3]

One of these types is "smolnitza", which occurs frequently in the Balkan peninsula in Europe and is closely related to vertisols or grumosols. Smolnitza soils are always dark in color and have a heavy texture. Most of them contain a high amount of magnesium and are often solonetzic.

VI. ACID SULFATE SOILS

Acid sulfate soils are extensive in the seashore marshes of all continents.[1,29] However, they occur most frequently in the tidal marshes of northwest Europe, on the western coast of Africa — particularly in the Gulf of Guinea, in Madagascar, southern India, and Indo-China. The latter region has been studied thoroughly and the information available on acid sulfate soils refers mainly to these territories.

The preconditions for the formation of acid sulfate soils are as follows:

1. Presence or formation of sedimentary iron sulfide under permanently or temporarily waterlogged conditions. The sulfur accumulates from the seawater, where it occurs

FIGURE 99. Surface of an Alaun soil (Finland).

mainly as sulfate, by two processes: evaporite formation, and fixation in sediments as iron sulfide minerals.

2. Reduction in sediments followed by oxidation of sulfides into sulfates, partly through the activities of biological mediators.

The majority of the sulfur in tidal marsh sediments consists of ferrous sulfide minerals. Thus, iron and sulfur play the dominant roles in the formation of acid sulfate soils.[349]

In the literature a great number of papers are available on the transformation of sulfur-containing minerals under different environmental conditions.

The dominating sulfur-containing mineral in the marine sediments in areas where acid sulfate soils are formed is pyrite, the content of which may reach 5% in sediments, but which very frequently occurs at a rate of 1 to 4%. Elemental S can also be produced in considerable quantities by the chemical and microbiological oxidation of H_2S.[350] In marine masses affected by reduction, elemental S appears in small quantities not exceeding 2%. While in most soil types sulfur occurs mainly as a constituent of organic materials, in tidal sediments the quantity of organic sulfur is insignificant. It can be concluded that pyrite is the most important sulfide mineral both in current and in potentially acid sulfate soils. This conclusion is valid for all regions, not only in the tropical and subtropical areas of acid sulfate soil formation, but also on the coast of Sweden, where the local name for acid sulfate soils is Gyttya, and in Finland where they are called Alaun soils.[351] Besides pyrite, several other minerals, mainly of the alunite-jarosite group play a considerable role during the formation of acid sulfate soils.

In Figure 99 the surface of an Alaun soil from the western coast of Finland is demonstrated, and in Figure 100 the profile of the same soil is given.

Due to the effect of oxidation, which appears as soon as the soil material dries up, considerable changes take place. The most significant of these is the drastic alteration of the pH value from neutral or slightly alkali to strongly acid. A model experiment was carried out on Alaun soils from western Finland using a three-month treatment involving temporary and permanent waterlogging in order to investigate the transformation in iron and sulfur compounds.[350,351]

FIGURE 100. Profile of an Alaun soil (Finland).

Table 78
CHANGES IN CHEMICAL COMPOSITION OF THE TOP LAYER OF AN ALAUN SOIL FROM THE WESTERN COAST OF FINLAND IN A MODEL EXPERIMENT

Treatment	Fe^{+++} (mg%)	Fe^{++} (mg%)	Fe^{+++} (in % of total Fe)	H_2S (mg%)	pH	SO_4 (mg%)
Temporary waterlogging	17.77	19.81	46.1	Trace	3.28	18.43
Permanent waterlogging	2.86	100.80	2.7	5.4	6.68	5.69

The results of the study are demonstrated in Table 78. Data of the table show that permanent waterlogging did not affect the chemical properties of the soil, while temporary waterlogging drastically changed practically all the properties measured. The pH value became much more acidic, while the ferric sulfate content sharply increased parallel with a decrease in sulfides and ferrous ions. In nature similar processes result in large areas of acid sulfate soils.

The influence of iron and sulfur compounds play the dominant role, as described above. The main environmental factors influencing this process are the pH and oxidation reduction potential of the media.[349,350] These relationships are demonstrated in Figures 101 and 102 for iron and sulfur compounds, respectively.

In acid sulfate soils the depth of the groundwater determines both the physical and chemical properties of the soil horizons. The strongly acid reaction and relatively high amount of water-soluble salts (ferric and aluminum sulfates) dominate in the upper layers and in layers

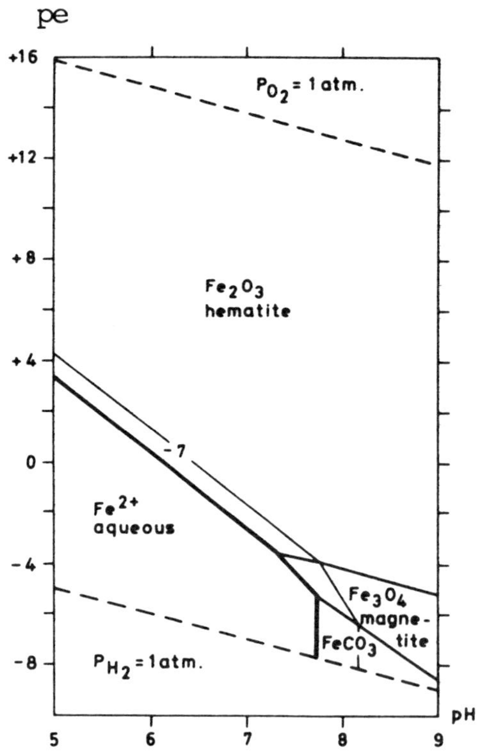

FIGURE 101. Relationships in the $Fe\text{-}H_2O\text{-}O_2\text{-}CO_2$ system as a function of pH and pe at a total dissolved iron activity of 10^{-6} and a total dissolved carbonate activity of 10^{-2}, 25°C and 1 atm total pressure.[349] The boundaries at log $\Sigma Fe = -7$ are shown.

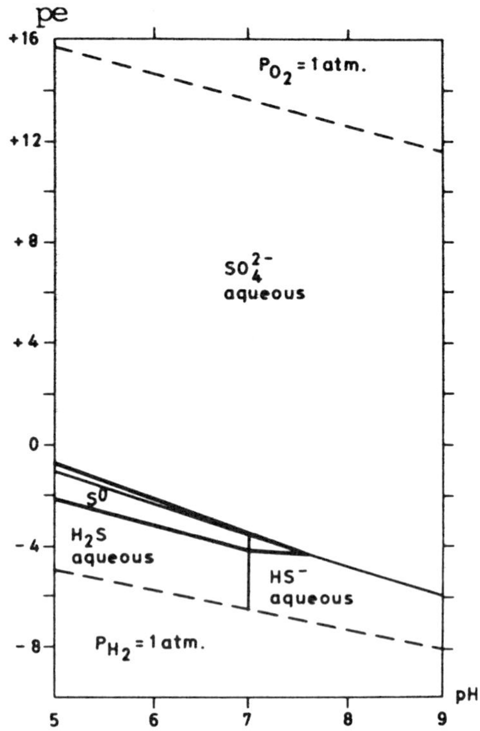

FIGURE 102. Relationships of sulfur species in water at 25°C and 1 atm total pressure, as a function of pH and pe.[350] The stability field of elemental sulfur is limited by a total dissolved sulfur activity of 10^{-1}.

which are not linked with the water table. It is frequently observed that acidic reaction can be measured in the upper layers simultaneously with neutral or alkali reaction in deeper layers (which are under the influence of water). The results of such observations are illustrated in Figure 103 which shows the pH profile of soils on the western coastline of Finland at Ylistaro. While on the top layer the pH value is nearly 3, at a depth of 1.5 m it changes to a pronounced alkaline pH.

The morphology of the profiles of acid sulfate soils is determined by the environmental conditions; this means that they can be sandy (Katerzand), loamy and clayey (cat clay) and of very different color, depending on local circumstances. As a rule their color is darker before drying and quite light after the acid sulfate horizon has formed.

As was mentioned above, acid sulfate soils develop in different climatic zones from the Arctic circle to the Equator. In the hierarchy of various soil classification systems, these soils are placed at various taxonomic levels. In Table 79 a correlation system is given for the placement of acid sulfate soils in different international classification systems, namely the U.S. Soil Taxonomy and the FAO/UNESCO World Soil Map.[352]

From Table 79 it is clear that in spite of the diversity in their placing in the taxonomical hierarchy, due to their specific properties acid sulfate soils are grouped into separate units in the various soil classification systems. It is also clear that all the adverse properties of acid sulfate soils are caused by the high amounts of iron sulfate and partly of aluminum sulfate, which provoke a strongly acidic reaction and toxicity for plants. It is well-known that acidity stronger than pH 4.0 causes direct damage to plants. This statement is valid

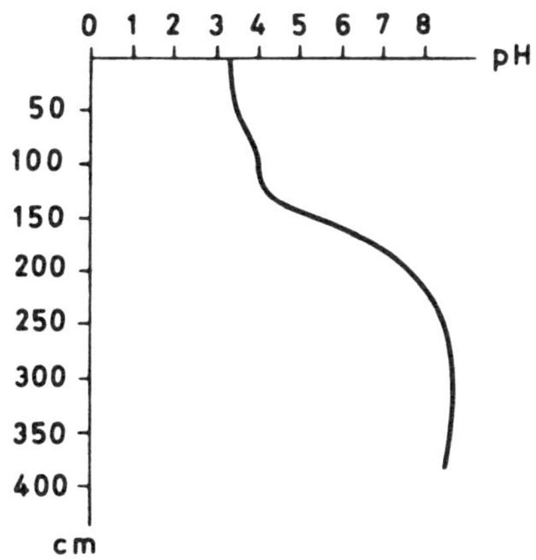

FIGURE 103. pH of a soil profile from Ylistaro (Finland).

even for rice, which is a major crop in the tropical and subtropical areas where acid sulfate soils are formed. Rice is quite a tolerant crop, but it suffers from an excess of water-soluble iron and particularly of aluminum. The toxic effect of aluminum ions is a consequence of the strongly acidic pH. The application of lime materials mitigates or offsets the adverse properties and harmful effect of acid sulfate soils on plants.

Table 79
PLACEMENT OF ACID SULFATE SOILS IN THE AMERICAN CLASSIFICATION CORRELATED WITH UNITS OF THE SOIL MAP OF THE WORLD (FAO/UNESCO)

Climatic zone	USDA	Main characteristics	World soil map	Main characteristics
World-wide	Typic sulfohemists	pH (1:1 water) <3.5 within 50 cm	Thionic fluvisols	pH (KCl) of dried soils <3.5 within 100 cm
World-wide	Typic sulfaquepts	pH (1:1 water) <3.5 within 50 cm	Thionic fluvisols	See above
Wet tropics and monsoon	Sulfic tropaquapts	pH (1:1 water) <4.0 between 50 and 150 cm and/or 3.5 to 4.0 within 50 cm	Thionic fluvisols	See above
			Humic gleysols	pH (KCl) >3.5; umbric or O horizon
			Dystric gleysols	pH (KCl) >3.5; ochric A horizon
Wet tropics	Histic sulfic tropaquepts	Acidity like sulfic tropaquepts histic epipedon	Thionic fluvisols	See above
			Humic gleysols	pH (KCl) >3.5; O horizon
Monsoon and semi-arid	Vertic sulfic tropaquepts	Acidity like sulfic tropaquepts; cracks (1 cm) at 50 cm when dry	Thionic fluvisols	See above
			Humic gleysols	pH (KCl) >3.5; umbric or O horizon
			Dystric gleysols	pH (KCl) >3.5; ochric A horizon
Humid temperate	Sulfic[a] humaquepts	Acidity like sulfic tropaquepts; umbric or histic epipedon	Thionic fluvisols	See above
			Humic gleysols	pH (KCl) >3.5; umbric or O horizon
Temperate	Sulfic haplaquepts	Acidity like sulfic tropaquepts; ochric epidedon	Thionic fluvisols	See above
			Dystric gleysols	pH (KCl) >3.5; ochric A horizon
Monsoon and semi-arid	Sulfic[a] haplaquepts	Acidicity like sulfic tropaquepts; Sodium saturation >15 in half or more of upper 50 cm; mostly saline in some part of the year	Thionic fluvisols	See above
			Gleyic solonchaks	pH (KCl) >3.5; salic horizon and/or saline at some time of year

[a] Subgroups not recognized in American classification.

After van der Kevie.[352]

Chapter 5

SECONDARY SALINIZATION AND/OR ALKALIZATION OF IRRIGATED SOILS AND POTENTIAL SALT-AFFECTED SOILS

I. IRRIGATION AND SALINITY

Irrigation, which is one of the oldest methods used in agriculture, has a long history of both favorable and unfavorable results. Among the latter the problem of salinity has always been, and remains, one of the most important.[353,354]

Secondary salinization is the most harmful and widespread phenomenon among the effects of irrigation on soils and the environment. This process (secondary because it is caused by human action) has been known since ancient times, and its effects are still present today. More than half of all the irrigated soils in the world are affected by secondary salinization and/or alkalization.[355,356]

Many authors have described the consequences of the extension of salinity, which has developed in the irrigated areas of the world due to primitive or incorrect technologies.[353,356-358]

Very often, in the course of many thousands of years, irrigation caused the accumulation of harmful salts, both in soils and waters, and had a deteriorating effect which led to the decline not only of agricultural production but also of ancient culture and civilizations.[359]

Particularly in arid and semi-arid regions, irrigation is as old as agriculture itself.[360] The lack of rainfall made the application of irrigation imperative in many ancient agricultural systems. The problem stemmed from the fact that such systems developed mainly in arid regions where, due to the landscape geochemistry of deserts and semideserts, salts tend to accumulate.[361] This problem was closely related to nearly all the ancient cultures which employed irrigation.[3,353,362]

It is well-known that in ancient Mesopotamia, an area of fertile soils between the Rivers Tigris and Euphrates, quantities of grain and other crops sufficient to feed large populations were produced for a long time.

In modern times, this area has been known only as a bare desert. It is also well-known that in ancient China, in the Indus Valley,[362] and in South America vast territories were turned into deserts because of salinity problems resulting from irrigation by ancient societies. The problem of secondary salinization runs through the whole history of mankind.[363] Evidently, there was neither sufficient knowledge nor technical means to predict, explain, and combat salinization for many thousands of years. Consequently, the degradation of the fertility of these soils and other adverse effects were recognized too late to prevent them from ruin.[353] The process forced people to leave the land as it became saline, and to shift their irrigation techniques to another place which, in many cases, also became salinized. As long as new territories were available, the shifting of irrigated agriculture offered a temporary solution. However, as the population density increased or new lands were exhausted, tragic consequences became apparent. More than one such example is known from history.

Figure 104 shows secondary salinization along a drainage canal in a European irrigation system.[364-366]

In order to estimate the global development and extension of secondary salinization and alkalization, as well as to predict the possible consequences of an increase in irrigation, the causes and dynamics of the process should be characterized.[367]

The main reservoir of saline water in deserts and semideserts is the groundwater. The soluble salts accumulated in the groundwater, as a result of the geochemical processes

FIGURE 104. Secondary salinization along a drainage canal in a European irrigation system (Hungary).

described in previous chapters of this book, can cause salinity problems when the groundwater is used for irrigation. As long as the groundwater table is deep and the moisture cannot come up through capillary flow to the soil profile, even saline groundwaters do not induce immediate salinization. As the effect of irrigation, however, the groundwater may rise so high that it can reach the surface layers and cause salinization even when good quality water is used for irrigation.[368] Unfortunately, this simple rule is often ignored during the planning and exploitation of many recent irrigation systems. The hazard of raising the groundwater table by irrigation is underestimated because even when the water table is 10 to 20 m below the surface before the construction of the irrigation system, it can easily be elevated to within 1 to 2 m of the surface, particularly in cases where drainage is poor.

The natural or artificial drainage of irrigated land, or land to be irrigated, is also a major factor in the processes of secondary salinization.[369] Referring again to the experiences of ancient agriculture, irrigation was practiced for thousands of years in the Nile Valley under a dry climate without secondary salinization. This phenomenon may be explained by the good natural drainage of the land strips along the Nile because of the suction effect of the river. Similar examples have been found recently in many places, e.g., in Syria, Central Asia, etc.[370-372] As early as the 19th century perennial irrigation was introduced in Egypt. Due to the change in the salt and water regime, the Nile Delta became salinized, and an artificial drainage system became necessary to get rid of excess salts.[318]

It can be concluded that, with the exception of places with good natural drainage, which is rather rare,[373] in deserts and semideserts the lack of artificial drainage leads, sooner or

Table 80
DEVELOPMENT OF IRRIGATION IN THE WORLD

Year	Irrigated land (million ha)
1800	8
1900	48
1949	92
1959	149
1980	200

Table 81
IRRIGATED SOILS IN SOME DRY COUNTRIES IN 1900

Country	Irrigated territory (million ha)
Indian subcontinent	15.49
U.S.S.R.	3.80
U.S.	3.01
Japan	2.72
Egypt	2.00
Italy	1.30
Spain	1.00
Chile	0.30
Total:	**30.12**

later, to secondary salinization. In this book the term secondary salinization is used in its conventional meaning, as differing from primary salinization which develops without human interference. Secondary salinization develops as a result of human activities, mainly irrigation.

In non-arid areas, where the rainfall and consequently the leaching of substances through the soils is substantial, secondary salinization and alkalization can be avoided, even without artificial drainage.[374]

II. GLOBAL EXTENSION OF IRRIGATION AND ITS INFLUENCE ON THE SECONDARY SALINIZATION OF SOILS AND WATERS

Thorne and Peterson,[356] Kovda,[3] Szabolcs,[1] Kovda and Szabolcs,[39] described the regular features mainly observed in relation to irrigated soils. The expansion of irrigation not only resulted in more areas being irrigated, but also affected the properties of soils, both positively and negatively. Although irrigation dates back to prehistoric times, its rapid development only started about 200 years ago (Table 80).

From Table 80 it can clearly be seen that the area of irrigated land grew from 8 million ha in 1800 to 48 million ha in 1900, and has more than doubled in the last 50 years. This trend is very remarkable and has resulted in large increases in the production of agricultural goods, but has also resulted in a number of technical and environmental problems.[1,375] Only a few of these will be discussed in this review.

1. At the turn of the century, when the irrigated lands of the world did not exceed 50 million hectares, they were mainly distributed among a few dry countries,[376,377] as shown in Table 81. At the beginning of the century the countries listed in Table 81 accounted for 2/3 of all the irrigated soils of the world. At that time irrigation was only practiced in dry regions. However, in recent times, irrigation has been extended to many semihumid regions in addition to the arid and semi-arid regions.[378,379] The problems of salt accumulation and secondary salinization differ from one climatic region to the other, according to the chemistry of salt accumulation.[380,381]
2. In many countries irrigation has been introduced as a new practice within the last few decades. The experience gained in countries where irrigation has long traditions were not always known and applied to combat secondary salinization.[382]
3. The extension of irrigation affected not only the land actually irrigated, but also the neighboring nonirrigated territories. As long as irrigation was concentrated in small areas, its environmental effect was evidently much less than when large-scale irrigation systems affected large surrounding areas.[383]

4. The effect of irrigation on the biosphere (besides the irrigated crop),[384] increased parallel with the sharp development of this method. These adverse effects included the salinization and contamination of drinking water, waterlogged and saline areas as a breeding ground for parasites and diseases, toxic effects on soil microorganisms, etc.

In many places these phenomena, and others not listed here constitute a barrier not only to the further development of agriculture and human civilization, but also to the maintenance of the present level of production.[385]

There has been a remarkable increase in irrigated territories in different parts of the world and in different countries and regions. In the U.S., for example, the area under irrigation was doubled between 1949 and 1973 to 21 million hectares. Another example: in the U.S.S.R. 1 million hectares of newly irrigated land are brought under cultivation every year. In Kenya the area of irrigated land was doubled between 1959 and 1969, and further development is envisaged. In Hungary irrigated areas have shown a more than ten-fold increase since the Second World War ended in 1945.

In many countries where irrigation was introduced mainly under non-arid conditions in densely populated areas, its side-effects were different from those appearing in the majority of arid countries. In dry countries, for example, the area surrounding massive irrigated lands is a vast desert, which means that the adverse consequence of irrigation, such as the secondary salinization of adjoining areas can be tolerated. In countries where the utilization of land is over 70%, as in Hungary, side-effects of this type would be catastrophic.[386]

In some countries, like Egypt, nearly 100% of the agricultural land is irrigated. The corresponding figures are 70% for the Malagasy Republic, 26% for Thailand, and 50% for Pakistan. Similar ratios exist in many arid and semi-arid countries. In less arid or semihumid countries the irrigated land often accounts for only a few percent, but this ratio is sharply increasing even in those countries. For example, in 1980, nearly 13% of the total agricultural land was irrigated in France, more than 10% in Spain, and nearly 15% in Greece.

In spite of the availability of many sources of information, accurate data concerning countries where irrigation has recently been introduced are very scarce. Widely different accounts and estimates can be found in various papers and records. Estimates of the total irrigated lands in the world range from 150 to 250 million hectares. The explanation for such a diversity of information is probably the fact that it is one thing to register the existing irrigation systems in the world and quite another to keep a record of those which are in permanent operation. This, in all probability, is why the data published by FAO (Food and Agricultural Organization of the United Nations) and ICID (International Council for Irrigation and Drainage) are always different regarding the area of irrigated land.

It is evident that neglected or abandoned irrigation systems are quite frequent and account for a very high percentage of all existing systems. According to the estimates of FAO and UNESCO (United Nations Educational, Scientific and Cultural Organization), as many as half of all the existing irrigation systems of the world are more or less influenced by secondary salinization, alkalization, and waterlogging. This phenomenon is very common not only in old irrigation systems, but also in areas where irrigation has only recently been introduced.

According to estimates made by all the above-mentioned agencies, 10 million hectares of irrigated land are abandoned yearly as a consequence of the adverse effects of irrigation, mainly secondary salinization and alkalization.

In Figures 105 and 106, salt accumulation is demonstrated in the vicinity of irrigation systems in the Hungarian Plain. In such cases not only the irrigated land but a considerable part of the surrounding territories must be withdrawn from agricultural utilization due to the harmful effect of salts.

The losses and damage mentioned are not evenly distributed among the irrigating countries.

FIGURE 105. Salt accumulation in the vicinity of an irrigation system in the Hungarian Plain.

FIGURE 106. Salt accumulation along the canals of an irrigation system in the Hungarian Plain.

In some of them the damage is low, but in others it may be high enough to constitute a major problem for agriculture or even for the national economy of the country. Unfortunately, the world abounds with such sad examples. In Pakistan Nazir Ahmad carried out statistical analyses with respect to secondarily salinized land.[387] According to his data, out of a total of 13.8 million hectares of irrigated territory, salinized areas accounted for 2.1 million hectares after a few years of irrigation. Among the causes of secondary salinization in Pakistan he indicated the joint effect of irrigation and groundwater. It is known from FAO reports,[388] and from the papers of Kovda,[3] that more than 50% of the irrigated soils in Iraq and Iran is affected by secondary salinization.

According to FAO estimates for Iraq, 50% of the river plain is affected by salinity. The same authors estimate the extension of salinity and alkalinity in Iran at more than 15% of the total land surface, exceeding 25 million hectares. As a result of irrigation, the salinity problems are increasing year by year.

In Turkey more than 25% of the alluvial soils are already affected by salinity, and practically all of them may become saline due to the envisaged increase in irrigation.

FAO reports on salinity in the Euphrates Valley in Syria estimated the adverse effects as follows:[20]

1. On more than 20,000 ha, salinity had reached a level where the soils had to be taken out of cultivation, with an estimated loss of 30,000 tons of cotton (Gossypium hirsutum) per year.
2. On about 30,000 hectares, the yield decreased by 50%, with the total loss estimated at 20,000 tons of cotton per year.
3. On about 60,000 hectares, the yield decreased by 20%, and the total loss was estimated at about 18,000 tons of cotton per year.

The Near and Middle East region depends largely on irrigated agriculture. Although only 36% of the arable land of these regions is irrigated, 70% of the value of all the crops grown is produced on this area. It is envisaged that by the next decade irrigation will increase to 51% of the arable area. However, the problem of salinity is also expected to increase on these soils, because they have a high potential for becoming saline.

At present no continent is free from very serious occurrences of salinization and alkalization. According to Zavaleta,[64] practically all irrigated alluvial soils in Peru show the features of salinity and alkalinity. In Argentina 50% of the 40,000 hectares of land irrigated in the 19th century are not salinized. In Australia secondary salinization and alkalization take place in the valley of the River Murray and in Northern Victoria, where some 80,000 hectares are affected.[389] The same phenomena can be observed in Alberta, Canada. Similar processes have been recorded in the northern states of the U.S., where irrigation was introduced much later than in the dry west. It should be noted that the last four examples and many other irrigated regions are far from being arid, and the majority of salts accumulating are associated with sodium salts capable of alkaline hydrolysis and not with the neutral sodium salts generally found in desert and semidesert areas.

Figure 107 demonstrates the accumulation of sodium salts capable of alkaline hydrolysis in the top soil layer and alongside irrigation canals in the Hungarian Plain. The damage caused by alkalization on the surface layer of irrigated soils on a rice plantation is shown on Figure 108.

The alkalization of irrigate soils may damage quite considerable areas in densely populated and agriculturally densely utilized areas. This phenomenon can be seen in Figure 109, showing an alkalized and consequently degraded area previously irrigated.

On the basis of the available data, it can be concluded that the processes of secondary salinization and alkalization accelerate parallel with the accelerating increase in irrigation in many parts of the world.[390]

FIGURE 107. Accumulation of sodium salts capable of alkaline hydrolysis on the top soil layer alongside irrigation canals (Hungary).

FIGURE 108. Damage caused by alkalization on the surface layer of irrigated soil on a rice plantation (Hungary).

The secondary salinization and alkalization of irrigated areas are often combined with waterlogging and bog-forming processes.[391] In such cases saline bogs and swamps develop on and around irrigated soils. Figure 110 shows the development of this process in irrigation systems, and Figure 111 shows the situation when, due to nearly permanent waterlogging, a saline swamp has already developed.

FIGURE 109. An alkalized and degraded area (previously irrigated).

FIGURE 110. Development of waterlogging and bog-forming processes in an irrigation system.

FIGURE 111. Saline swamp developed due to permanent waterlogging.

The greater part of irrigated areas consists of soils which are currently nonsaline and nonalkali. Part of this land may be potentially saline or potentially alkaline.[392] Potential salinity or alkalinity means that a soil which is neither saline or alkaline at the moment can be turned into one or the other due to the application of incorrect methods of irrigation.

While reliable records are available on the world-wide extension of salt-affected soils, proper data on the global extension of secondary salinized or alkalized soils are, unfortunately, still lacking; we have to make our own estimates. The UNESCO/FAO data cited above, suggesting that salinization and alkalization exist on 50% of irrigated areas, provide some guidance for these calculations. Based on these and other figures, it can be concluded that the area of secondary salinized and alkalized soils which were once fertile but have since turned into bare deserts as a side-effect of incorrect irrigation, is of the same order of magnitude as those currently existing salt-affected soils which have never been irrigated. Such estimates are essential if predictions are to be made with regard to the future.

III. PROSPECTS OF THE FUTURE EXTENSION OF IRRIGATION AND OF POTENTIALLY SALT-AFFECTED SOILS

The soils considered potentially salt-affected are those which are neither saline nor alkaline at present, or only to a very low degree, but where human intervention — especially irrigation — may result in severe salinization and/or alkalization. It is well known that the majority of the irrigated territories in the world are exposed to the hazard of secondary salinization, alkalization, and waterlogging.

Theoretically, many factors contribute to the salinization of soils; but it is primarily irrigation which has led to the formation of the millions of hectares of saline and/or alkali soils throughout the world. It is paradoxical that many irrigation systems which have been thoroughly planned and constructed at great expense have failed to increase the productivity of the land but have rather transformed it into barren, saline fallow. This transformation is known as secondary salinization and/or alkalization, and it is as old as irrigated agriculture itself.

While existing salt-affected soils can be recognized on the basis of a few morphological, chemical, and physico-chemical observations and determinations, the recognition of potentially salt-affected soils, as well as the determination of the long-term hazard of salinity or alkalinity on any given territory, necessitates special surveys and methods.

Secondary salinization and alkalization are likely to occur where there is a risk of salt accumulation due to the following causes:

1. Irrigation water of poor quality.
2. A rise in the level of the groundwater: (a) the salt content of the groundwater accumulates in the affected layers; (b) the rising groundwater transports the salts from the deeper soil layers to the surface or surface layers; or (c) the rising water table hinders natural drainage and the leaching of salts.[393]
3. The appearance of deep saline material at the surface or around the irrigated sector through erosion or agricultural activity.

The degree of risk is determined by the following:

1. Climatic factors such as temperature, wind, rainfall, humidity, vapor pressure and evaporation, and their fluctuation and dynamics.
2. Geological, geomorphological, geochemical, hydrological, hydrogeological, and hydrochemical factors such as natural drainage, the depth and fluctuation of the water table, the direction and velocity of the horizontal groundwater flow, the salt content and composition of the groundwater, salt content and composition of the groundwater, etc. and whether or not saline materials, deposits, or fossil soils are present, even at deeper levels.
3. Soil factors such as profile, texture, structure, degree of saturation with water, conductivity, soluble salt content, salt composition and salt profiles, exchangeable cations, pH, etc.
4. Agrotechnical factors such as land use, crops, cultivation methods, subsoiling, etc.
5. Factors such as the amount of irrigation water, the method, frequency and intensity of irrigation, the salt content and composition of irrigation water, natural and artificial drainage, etc.

According to these factors the aims and methods of the preliminary survey are selected to determine the existence or the degree of potential salinity and/or alkalinity of a soil. Obviously, the environmental conditions cited in 1, 2 and 3, and the methods of utilization cited in 4 and 5 above should be taken into consideration, and appropriate limit values and methods — based on uniform principles — should be selected in the course of this procedure.

In desert and semidesert regions, the above-listed factors combine to render virtually all irrigated areas potentially salt-affected. This is not the case in moderate and humid climates, where the basic aims of surveys and studies are to signal the areas which are in danger of salinization or alkalization, to warn against harmful practices which might lead to them, and wherever possible, to elaborate methods for preventing them.

The wider use of irrigation remains one major prospect for increasing yields and for nourishing the world population. If irrigation systems are carefully exploited, and if, during the planning and construction phases, the necessary preliminary surveys are made and precautions taken, the production of food and raw materials can be multiplied in the future on a world-wide scale.[354,356]

Many prognoses are available with respect to the development of irrigation at the end of this century and the beginning of the next. Some of them are local or national reports, but some of them are on a global scale, like the well-known Report to the President, ICID prognoses, etc. (Aleksevsky [359]). It is evident that different figures are to be found in different sources, but on average about 400 million hectares of irrigated land are predicted for the first part of the 21st century. Unfortunately, no reliable predictions are available on the risk that secondary salinization will result from such a sharp increase in the territory of irrigated land in the near future. Based on past experience, there can be little doubt that, in general, the increase in the risk of secondary salinization and alkalization is not in linear proportion with the increase in the acreage of irrigated land; the correlation is closer to logarithmic. An exact analysis is yet to be made of the level of risk involved in current systems and in the predicted extension of irrigation in different countries.[394]

Evidently, this danger will be very diverse in different areas, regions, and districts, but if secondary salinization increases in space and extent as a world-wide process, it is more than probable that its global importance will sharply increase in the future.

Wherever possible, studies should be made and their results applied in the local, national, and even international planning of new irrigation systems as soon as possible. It is also necessary to extend such studies to the predictable joint effect on the environment of existing and future irrigation systems.

During the future development of irrigation, not only the pedological effect but also the general environmental influence of irrigation will assume new dimensions. For instance, the concentration of CO_2 in the atmosphere, and the rapid change in this value as the result of human activities, is a widely discussed problem today. A great number of books, reports, and prognoses are available on this subject, some of them threatening consequences which would be tragic for mankind.[395] If the territory of irrigated land doubles or trebles, the irrigated plants will be able to produce higher amounts of biomass and yields, and to consume as much as 30 to 40 billion tons of CO_2 annually in the course of photosynthesis instead of the current 15 to 20 billion tons.[213]

The literature contains many examples clearly showing that the success of irrigation is also interrelated with numerous vital problems. As is clear from the example mentioned above, the aim of irrigation development is to improve the food situation as well as the environment. It is also evident that the hazard of secondary salinization and alkalization will be one of the major obstacles in the path of this development if the study of this risk is not intensified and if methods for its prediction and prevention are not applied.

Up to now, no proper data and records are available on the extension of potentially salt-affected soils. First of all this term must be clarified. As mentioned above, potentially salt-affected soils are those which are nonsaline or alkaline on the top layers at the moment, but may be salinized due to irrigation. Evidently, such a definition is relative, because any soil can be salinized, e.g., when irrigated with saline water or when lacking drainage.

Kovda stressed that widely varying soil and subsoil conditions exist in different regions of the world.[213] The relationships between irrigation, crops, and soil must still be developed fully under such conditions.[397] In soils to be irrigated, a salt and water balance exists between soil, crop, and irrigation; and this is influenced by rainfall, stream flow, groundwater, evaporation, and transpiration. This balance is altered during irrigation.

According to Kovda and other authors (Anayan,[398] Borovsky,[307] Boumans and van der Molen,[399] Szabolcs et al.,[381] Zavaleta[64], the groundwater plays a significant role in salinity due to seepage from channels during irrigation. Large amounts of irrigation and drainage water seep into the groundwater from unlined canals, raising the water table to near the surface both in the irrigated area and in adjoining soils. In some situations this contribution

Table 82
EXISTING AND POTENTIAL SALT-AFFECTED SOILS IN SOME EUROPEAN COUNTRIES[24]

Country	Salt-affected soils (ha) Existing	Salt-affected soils (ha) Potential
Austria	500	2,500
Czechoslovakia	25,000	80,000
Hungary	740,000	885,000
Italy	500	400,000
U.S.S.R.	28,000,000	18,000,000

to the groundwater may equal or sometimes even exceed the quantity actually utilized by the irrigated crops.

This new accretion to the groundwater may be withdrawn from the subsoil, wholly or in part, by wells or tube-wells. Under good subsoil drainage conditions, it may seep into rivers. Under other conditions, it may raise the subsoil water level, fill pockets, or form perched water tables and cause local waterlogging, necessitating drainage in order to restore healthy conditions in the soil.

Consequently, irrigation also changes the salt content and salt distribution in the soil profile.[1,394] This is why the definition of secondary salinization and alkalization should always be examined against the background of the methods of irrigation, the soil and water properties, the farming pattern, the existence of natural or artificial drainage, etc. For this reason, it is always difficult to diagnose secondary salt-affected soils within a simple soil survey and this step is often omitted during the planning stage of irrigation. This omission has caused a great deal of unexpected damage during the first years or later periods after the introduction of irrigation systems in many countries of the world. Evidently, under different climatic conditions secondary salinization should be interpreted differently. Its close relationship to the biogeochemical processes of salt accumulation means that the hazard of secondary salinization is greater in desert areas than in humid regions, where natural leaching processes remove soluble salts.

Surveys and estimations on potentially salt-affected soils have been carried out in a number of European countries,[24] as shown in Table 82. The data in Table 82 clearly demonstrate that even in European countries where the hazard of salt accumulation is far less than in some of the countries discussed earlier, the area of potentially salt-affected soils is similar to, or even greater than, the present area of salinization. Under arid conditions this ratio will obviously be even higher.

In many arid and semi-arid areas practically all the soils, or at least a high percentage of them, can be termed as potentially saline. As a consequence of the features described above, the determination, grouping, characterization and mapping of secondary salt-affected soils must be performed within the context of the local environmental and economic conditions.

The importance attached to economic aspects,[400] in addition to environmental conditions, can be explained by the fact that, although the prevention of secondary salinization is theoretically possible in many developing countries, in most cases the need to construct expensive drainage systems makes the economic expediency of introducing irrigation questionable.[236]

Although irrigation is the principle cause of secondary salinization on a global scale, it is worthwhile listing other causes that have been studied and documented.[3,213]

When Nonirrigated Land is Utilized:

1. The formation of secondary alkaline or saline soils resulting from overgrazing and the compactness of sodic meadow land — along the contact zone of mountainous foothills and plains, and on low terraces in valleys where flooding by rivers has been excluded by barrage construction.
2. The formation of secondary saline soils as a result of the disposal of brackish water pumped from petroleum wells, coal mines, and industrial plants.
3. Salinization of soils after seawater invasion under the influence of land subsidence, or after heavy tsunami or storms and earthquakes.
4. Accelerated formation of saline alluvial soils on deltaic and tidal wave territories after periodic floodings have stopped as the result of dam construction in the middle and upper reaches of the river valley.

When Irrigated Land is Utilized:

1. The formation of waterlogged and saline soils along nonlined canals as the result of water seepage, groundwater elevation and evaporation.
2. The formation of first spotty saline fields, then totally saline fields after 5 to 15 years of irrigation without appropriate drainage installations for the evacuation of saline subsoil water.
3. Application of incorrect methods when irrigating soils with brackish (or alkaline) irrigation water taken from saline rivers, tube wells installed into saline groundwater or after overpumping good quality subsoil water, or sea or gulf sources.
4. Appearance of saline soils on valley terraces above and after the construction of a dam (as the result of the submergence of subsoil water following reservoir formation).

* * *

It can thus be seen that man-made salinization caused the destruction of the global biospheric mechanism not only through its direct influence on the soil, but also indirectly, through several processes ranging from photosynthesis to the cycling of bioelements (C, O, N), etc. Such influences must also be taken into account with respect to soil organic matter, energy resources, soil bioprocesses, etc. The negative consequences of soil salinization are not only social and economic problems but lead to the global destruction of the biosphere of our planet.[213]

Some recent studies describe the adverse effects of salinity not only on plants, but on animals, too. The increased electrolyte contents of the environment, particularly of waters and soils, result in excessive salt intake by animals, disturbing the metabolism of their life functions. In Australia, for example, sheep and other livestock die from time to time after drinking salty water in hot regions. The kidneys and other organs suffer from saline and sodic waters; the total body water increases, and there is an expansion of the extracellular volume. Other disorders also occur in the wake of the salinization of soil and water, particularly in dry areas.[298]

Besides the accumulation of electrolytes in soils due to irrigation, other secondary processes also occur in irrigated soils. One of these processes is the degradation of irrigated soils, caused by temporary waterlogging, particularly on heavy-textured soils. This phenomenon has been observed mostly in southwest Asia by different authors, but it has also been described by Szabolcs and Darab on the Hungarian Plain,[302] and by other authors in the Ukraine,[382] as well as in several countries where rice has been produced for years on heavy soils with a slightly alkaline medium. The process of degradation itself is characterized

FIGURE 112. Degradation in a waterlogged alkali soil on the Hungarian Plain.

by a diminishing of the humus content in the upper layers of the soil, combined with the accumulation of silicic acid in the top layers and the migration of sesquioxide into the B horizon, which is very similar to the solod-forming process described in Chapter 4. This is why the term secondary solod-forming process is also found in the literature.

In irrigation systems where such degradation processes develop, the white dust on the surface consists not of soluble salts but of compounds of silicic acid, which under such conditions are termed amorphous silicic acids and denoted in pedologic literature by the symbol SiO_2.

Since such soils occur mainly in the flood areas of rivers and in old moorlands, it is worth investigating the way they develop. The top layer of these soils exhibits extremely unfavorable properties, such as a very low humus content, pulverulent laminated structure, and restricted water permeability.

Figure 112 illustrates the degradation in the waterlogged alkali soil on the Hungarian Plain, with visible signs of cracking and the loss of the humous top layer.

Due to the degradation process described above, the fertility of rice fields sometimes decreases in the course of a few years, even when no substantial salt accumulation occurs. In such cases the soil loses its color, and the physical, chemical, and biological properties deteriorate to a considerable extent. The land must thus be left fallow because the soil is "worn out". As was mentioned previously, this process is similar in many respects to the process of solodization, but, owing to the permanent activity of the factors provoking it, it is more intensive and affects larger areas.

The characteristic chemical properties of soils degraded due to permanent rice production are also similar to those of solod soils. For instance, the increase in the 5% KOH-soluble silicon content is high in the upper layers of the soils, and the SiO_2:Al_2O_3 ratio is also very high. Four years of rice monoculture on the alkali soils of the Hungarian Plain caused pronounced soil degradation compared with the utilization of similar soils for grass. The results of these investigations aer demonstrated in Table 83. The data show that the 5%

Table 83
5% KOH SOLUBLE SiO_2, AND Al_2O_3 IN ALKALI SOILS UNDER 4 YEARS OF RICE AND UNDER GRASS

	Depth (cm)	SiO_2 (%)	Al_2O_3 (%)	SiO_2 (meq/ℓ) / Al_2O_3 (meq/ℓ)
Rice	3—8	3.26	1.84	14.95
	18—22	2.42	0.07	29.40
Grass	4—14	2.04	0.21	6.36
	22—33	1.46	0.73	1.71

KOH-soluble SiO_2 content sharply increased in soils under rice and the SiO_2:Al_2O_3 ratio also increased intensively; however, both characteristics had much lower values under grass.

The most degraded horizons are located in the top layer of irrigated soils, another proof that irrigation is the cause of degradation.

The process of degradation is shown not only by the analysis of the 5% KOH-extract but also by the investigation of the exchange complex of the soils. In such cases the S value, i.e., the proportion of all exchangeable cations, decreases in the degraded horizon, whereas the T-S value, i.e., the proportion of adsorbed H ions, increases.

The intensive development of these processes can be observed in areas where natural alkaline conditions are, in any case, inducive of soil degradation. Irrigation develops and intensifies this process rather than initiating it.

The process is unfavorable partly because the organic matter is rendered soluble and part of it is lost, and partly because, owing to the degradation of the colloidal complex of the soil, both the chemical and physical properties of the soil, including the possibility of structure formation, etc. deteriorate. Excessive anaerobic conditions also have an unfavorable effect on the microbiology of the soil; and this, in all probability, makes itself felt in the availability and uptake of nutrients and in other phytobiological processes. As shown by experimental data, dressing with organic manure and the inclusion of grass and clover, in the crop rotation counteract degradation to a certain extent.

Pedogenic chemical processes are closely related with the vegetative cover. Although they are not classed as important plant nutrients, silicon compounds play an important part in the constitution of plants.

Rice, like hydrophilic plants in general, accumulates large amounts of silicic acid in its organs. There is an obvious relationship between the substantial amount of silicic acid in rice and the release of notable proportions of silicic acid in soils under rice.

In some cases the processes of salt accumulation and soil degradation overlap in irrigated soils. In such cases the adverse effects of excess salts combine with soil degradation to reduce soil fertility.

Apart from the most frequent processes described above, the secondary formation of different types of salt-affected soils is also possible in other cases due to various side-effects of irrigation or drainage. For instances, irrigation water with a high magnesium content can produce magnesium soils, and secondary processes also occur in areas where acid sulfate soils are formed. In tidal marsh areas, the secondary formation of acid sulfate soils combined with signs of soil degradation is often observed after soils are dessicated and introduced into production due to the effect of irrigation and drainage, as was described in the Korean Peninsula, southeast China, Southern India, etc.

Potentially acid sulfate soils are also fairly widespread in Indo-China. Their placement in the U.S. Soil Taxonomy System is demonstrated in Table 84, which shows that the potentially acid sulfate soils must be placed into various groups of the U.S. soil classification system.

Table 84
PLACEMENT OF POTENTIALLY ACID SULFATE SOILS IN THE AMERICAN CLASSIFICATION, CORRELATED WITH UNITS OF THE SOIL MAP OF THE WORLD (FAO/UNESCO)[29]

Climatic zone	USDA	Main characteristics	World soil map	Main characteristics
World-wide	Typic sulfaquents	pH (1:1 water) of dried soil <3.5 within 50 cm if n >1.0; within 30 cm if n <0.7	Thionic Fluvisols	pH (KCl) of dried soil <3.5 within 100 cm
World-wide	Sulfic Hydraquents	pH (1:1 water) of dried soil <4.5 in upper 25 cm, or more acid between 50 and 100 cm if n >0.7 between 20 and 50 cm	Thionic Fluvisols	Some that are less acid than sulfaquents but sufficiently acid to be included in sulfic hydraquents; non-saline
			Gleyic Solonchaks	Salic horizon and saline at some time of the year: part that meets acidity requirements for sulfic hydraquents but not for thionic fluvisols
			Dystric Fluvisols	Not sufficiently acid for thionic fluvisols but meets requirements for sulfic hydraquents; nonsaline
Wet tropics and monsoon	Sulfic[a] Tropaquents	Acidity requirements of sulfic hydraquents but n <0.7 between 20 and 50 cm	Thionic Fluvisols Gleyic Solonchaks Dystric Fluvisols	(See above)
Temperate	Sulfic Fluvaquents	Same as sulfic tropaquents	—	—
World-wide	Typic Sulfihemists	Sulfidic materials within 100 cm; pH (1:1 water) of dried soil <3.5	Thionic Fluvisols	(See above)

[a] Subgroups not recognized in American classification.

IV. COMBAT AGAINST SECONDARY SALINIZATION AND ALKALIZATION, INCLUDING THE CONTROL OF IRRIGATED SOILS AND IRRIGATION WATER

It is essential to intensify both theoretical and technical activities related to the hazard of secondary salinization and alkalization. Up to now, the influence of geochemical and hydrogeochemical processes has been underestimated in many places, resulting, sooner or later, in salt accumulation due to the irrigation of the given territory. While an abundance of studies have been made on irrigation water, including the determination of quality requirements,[392] the effect of the groundwater on soil salinity is often left out of consideration. For this reason, a comprehensive preliminary study not only on soils and surface waters, but also on underground waters and layers is necessary before new irrigation systems are planned or existing systems extended.

When studying the influence of the groundwater on salt accumulation, the critical depth of the groundwater table should first be determined. This critical depth is the level below which leaching prevails, owing to natural or irrigated conditions, and above which salt accumulation takes place in the soil profile. In other words, the salt regime of the given territory is in equilibrium at the critical depth.

A. Salt Regime

The salt regime of the soil denotes the accumulation and leaching of soluble salts, and the periodical changes in these processes in the soil. An examination of the salt regime, i.e., of the periodical changes in the quantity and quality of salts, reveals the prevailing direction of salt movements and whether the accumulation or leaching of salts takes place under the given conditions.

B. Salt Balance

Although a study of the salt regime supplies important data on the dynamics of a salt-affected soil, it does not in itself present any new information. Therefore, the next step is to compare the salt content of the soil at given times and to express these values in terms of salt balances.[401,402] Salt balances reveal the effects, or expected effects, of amelioration, of agrotechnical measures on changes in the degree of salinity and akalinity of the soil, and the effect of irrigation on the soluble salt contents. The establishment of salt balances requires an exact knowledge of the cause, rate, and degree of accumulation and leaching.[392]

The following data are needed for the establishment of the salt balance, regardless of the extent of the area concerned:

1. Total amount of soluble salts at the beginning and end of the observation period
2. The increase in soluble salt contents during the observation period
3. The decrease in soluble salt contents during the observation period

Three types of salt balances may be distinguished:

1. Stable salt balance
2. Balance of salt accumulation
3. Balance of leaching

The salt balance of soils may be established for both larger (e.g., a drainage basin) and smaller territorial units.[402]

A salt balance may be established either for the whole soil profile or for given genetic horizons.

The salt balance of a soil depends on the joint effect of numerous factors, partly treated below.

C. The Depth and Chemical Composition of the Groundwater

When considering the water table, the fact that it fluctuates periodically must be taken into account. The depth of the water table is influenced by local factors such as the amount of precipitation in the given location, and the micro- and meso-relief of the area. It is also affected by the amount of water entering the catchment area from various sources and thus contributing to the groundwater, and by the time it takes for the water to manifest itself at the site examined. The depth of the groundwater also changes when the natural conditions are artificially altered due to the effect of drainage or irrigation.

An analysis of the chemical composition of the groundwater must be carried out simultaneously with the determination of the depth of the water table.

The following qualitative data must be given for the groundwater:

1. The total salt contents of the groundwater, in mg/ℓ
2. The relative amount of sodium salts in the groundwater
3. The relative proportion of magnesium salts within the alkali earth metal salts
4. The distribution of soluble salts in the groundwater with respect to the anions

The physical and water regime properties of a soil also affect its salt regime to a considerable extent. Of these, two should be specifically mentioned: the water conductivity of the soil profile and, if present, the depth of the accumulation horizon with a heavier mechanical composition or disadvantageous water regime properties.[392]

D. The Influence of Relief

Factors that should be determined:

1. The infiltration of rainfall or irrigation water into soils with different relative depths
2. The wetting of the soil profile
3. When the downward movement of rainfall or irrigation water is followed by the upward movement of the groundwater, the moisture content of deeper layers

E. Irrigation Techniques

Under irrigated conditions, the salt regime of the soils also depends on the method and technical standard of irrigation management.[403]

In the case of surface irrigation, as a function of the amount of irrigation water applied, excessive leaching may result in a rise in the water table, a reduction in the possibility for leaching soluble salts, and even in an increase in the salt balance of the soils. This is particularly true when the drainage conditions of the irrigated area are unsatisfactory.

F. Calculation of Salt Balance

Given a knowledge of the factors influencing increases and decreases in the salt contents of soils, the salt balance can be calculated on the basis of the following equation:

$$b = a + \left(d + \frac{cv}{M\,t_{fs}} \cdot 10^{-5}\right) \tag{35}$$

where:

b = soluble salt content of the soil at the end of the observation period, in mg/100 g soil,

a = soluble salt content of the soil at the beginning of the observation period, in mg/100 g soil,
c = the salt concentration of the irrigation water, in g/ℓ,
v = the quantity of irrigation water applied during the observation period, in m^3/ha,
M = the thickness of the soil layer for which the salt balance was established, in m,
t_{fs} = the bulk density of the soil, and
d = the salt regime coefficient of the soil, in g/100 g soil.

The change occurring in the salt contents of a soil during the observation period is expressed by the salt regime coefficient, which gives the difference between the amount of salts leached from the soil and that of salts entering the soil from sources other than the irrigation water.

As it is rather difficult to calculate the value of d, it is advisable to determine the change in the salt content of the soil and calculate the salt regime coefficient on the basis of the following correlation:

$$d = b - \left(a + \frac{cv}{M\,t_{fs}} \cdot 10^{-5}\right) \qquad (36)$$

Having completed calculations for several irrigated soils, the following characteristic cases can be distinguished:

1. The salt regime constant is negative.
2. The salt regime constant is negative, but the salt content of the soil increased during the observation period. This means that although leaching took place to a certain extent, more salt was introduced with the irrigation water than could be leached out.
3. The salt regime constant is positive and the soluble salt content of the soil increased during the observation period. This indicates that more salt entered the soil not only from the irrigation water but also from the groundwater and perhaps from other sources, than could be leached out.
4. The salt regime coefficient of the soil may be zero. In this case the salt balance of the soil is stable, and the current depth of the groundwater table is practically tantamount to the critical depth.

To calculate the salt regime coefficient with the above simple equation requires the regular determinations of salt movements in the given territory.

Where even such determinations are difficult, Kovda recommends a simple empirical estimate of the critical depth of the groundwater:[404]

$$L = 170 + 8t \pm 15 \qquad (37)$$

where: L = the critical depth, in cm, and t = the mean annual temperature, in degrees centigrade.

G. Salinity Hazard

In order to control the possible hazard of salinization and/or alkalization in irrigated areas, or areas to be irrigated, the following factors should be studied and determined:[402]

1. Climatic factors, such as temperature, rainfall, humidity, vapor pressure, evaporation, and their fluctuations and dynamics;
2. Geological, geomorphological, geochemical, hydrological, hydrogeological, and hydrochemical factors, such as natural drainage, the depth and fluctuation of the water table, the direction and velocity of horizontal groundwater flow, the salt contents and composition of the groundwater, etc.

Table 85
SCHEME OF METHODS RECOMMENDED FOR THE CONTROL OF SALINITY AND ALKALINITY IN IRRIGATED AREAS

Before Construction of Irrigation System —
Preliminary Survey

Landscape	Planned irrigation
Climate	Available irrigation water quality and quantity
Hydrology	Groundwater depth and quality
Hydrogeology	Technology of irrigation
Geomorphology	Cropping pattern tolerance

During Irrigation — Monitoring

Salinity and alkalinity of soil and groundwater table
Chemical composition of groundwater
Chemical composition of irrigation water, filtration
Physical soil properties
Toxic elements, if any, in soil and water

3. Soil factors, such as soil profile, texture, structure, saturated and unsaturated water conductivity, soluble salt contents, salt composition and salt profiles, exchangeable cations, pH, etc.
4. Agrotechnical factors, such as land use, crops, cultivation methods, etc.
5. Irrigation practices, such as the amount of irrigation water, the method, frequency, and intensity of irrigation, salt contents and composition of irrigation water, natural and artificial drainage, etc.

H. Preliminary Survey and Control of Irrigated Soils

The factors mentioned above determine the aims and methods of the preliminary soil survey made in order to define the degree or existence of potential salinity and/or alkalinity.

The soil properties, groundwater depth and chemical composition, as well as the salt balance and the potential risk can be demonstrated on maps with recommendations for the techniques of irrigation and water use.[24,381,392]

Evidently, both the environmental conditions and the methods by which the territory in question is to be utilized should be taken into consideration when an area is evaluated in this respect. Consequently, different limit values and different methods, based on uniform principles, should be selected in the course of this procedure.[381]

Table 85 presents a scheme of methods recommended for the control of salinity and alkalinity on irrigated areas. The table shows that the prediction of secondary salinization and alkalization of soils to be irrigated should be based on a preliminary survey of the landscape and soils before the construction of the irrigation system. In this way, it is possible to take the necessary steps for the prevention of adverse processes.[381]

During irrigation, the soil and water properties should be carefully monitored in order to record any changes and to indicate whether precautions are necessary. Monitoring methods and the timing and location of sampling depend upon local conditions.

In the course of surveying and monitoring, if a reliable method is to be developed for the prediction of salinization and alkalinzation, the following problems must be solved: (1) the main sources of water-soluble salts (irrigation water, groundwater, surface waters, salty deep soil layers, etc.) must be identified; and (2) the main features of the salt regime must be characterized (salt balance), and the whole range of natural factors influencing the salt regime must be analyzed.

Consequently, an exact salinity and/or alkalinity prognosis must be based on the evaluation of many natural and human factors and a thorough knowledge of the existing soil processes.[368]

An example of the prediction of salinization and alkalization processes on irrigated areas and on areas to be irrigated is briefly described below based on studies made by the author and his colleagues, on the Hungarian Plain. The study consisted of the following steps.[1,21,392]

1. The determination of possibilities of effective irrigation from the viewpoint of general existing soil conditions on the Hungarian Plain — shown on a map with a scale of 1:100,000 (Figure 113).
2. The determination of possibilities and preconditions of effective irrigation from the viewpoint of more specific existing soil conditions on smaller areas within the territories in question — shown on maps with a scale of 1:25,000 (Figures 114, 115, 116, 117, 118, 119).

1. General Possibilities of Effective Irrigation From the Viewpoint of Soil Conditions on the Hungarian Plain

A series of maps with a scale of 1:100,000 were prepared or adapted as follows:

1. Genetic soil map (soil type and subtype)
2. Map of the average depth of the groundwater level
3. Map of the highest groundwater level
4. Map of the average salt content of the groundwater
5. Map of the chemical composition of the groundwater (dominating cations and anions)

On the basis of the above-mentioned materials, a map was prepared (with a scale of 1:100,000) indicating the general possibilities of effective irrigation from the viewpoint of soil conditions.

On this map (shown in Figure 113) three categories were distinguished:

1. **Areas to be irrigated.** In these areas irrigation will not result in hydrological and soil processes that significantly decrease soil fertility or adversely affect soil properties.
2. **Areas that may be irrigated conditionally.** In these areas harmful hydrological and soil processes — reducing soil fertility and unfavorably influencing the soil properties — may be prevented only by fulfilling certain requirements.
3. **Areas not to be irrigated.** In these areas irrigation is not advisable for the time being, because it would result in the development or intensification of undesirable hydrological and soil processes that reduce soil fertility. To prevent these phenomena, expensive technical and ameliorative measures are required.

2. Possibilities and Preconditions of Effective Irrigation From the Viewpoint of Soil Conditions in Smaller Areas

Having a general knowledge of the natural conditions, groundwater, soils, and irrigation waters of a large irrigation project (including present and future irrigation projects) a more detailed survey was necessary for the planning of irrigation districts, technical establishments, drainage systems, and other main irrigation practices. For this purpose, a more detailed survey was elaborated including the preparation of a series of maps with a scale of 1:25,000. On these maps the possibilities of irrigation were indicated more thoroughly and exactly, and more detailed information was given about the preconditions and measures necessary to guarantee the prevention of undesirable soil processes.

As a first step, the following maps were prepared with a scale of 1:25,000 (See Figures 114, 115, 116, 117 illustrating maps prepared — as an -example — for part of the Southern Transtisza region):

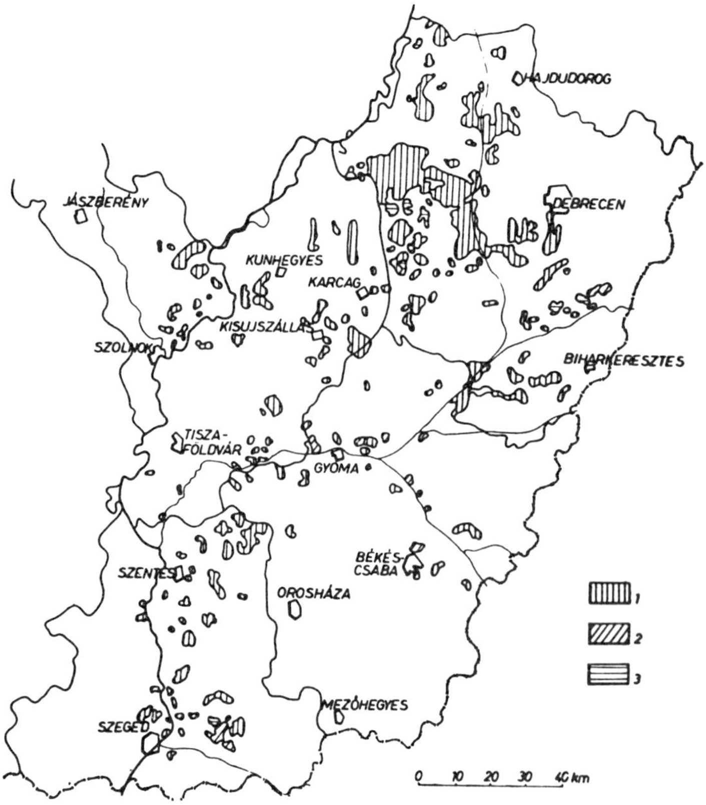

FIGURE 113. General possibilities for effective irrigation from the viewpoint of soil conditions in Szolnok, Hajdu-Bihar, Békés, and Csongrád counties (Eastern part of the Hungarian Plain).[1] (1) Areas to be irrigated; (2) areas that may be irrigated conditionally; (3) areas not to be irrigated.

Soil map (Figure 114) — This map indicates the soil type and subtype and gives a general picture of existing soil processes. The map is necessary for the interpretation and evaluation of soil properties and for forecasting prospective changes in soil conditions due to irrigation.

Map of soil texture and water properties (Figure 115) — This map indicates the soil texture (sand, sandy loam, loam, clay loam, clay, and organic soils) and the categories of soil water properties on the Hungarian Plain.

Map of salinity status (Figure 116) — This map indicates:

1. The average total salt content in the soil between the surface and the water table (as a percentage)
2. The maximum total salt content of the soil profile (as a percentage)
3. Depth to the salt maximum in the soil profile (in centimeters)
4. Soil reaction of the B_1 horizon

Map of groundwater conditions (Figure 117) — This map indicates:

1. The average depth to the water table (in meters)
2. The average salt content of the groundwater (in g/l)
3. Na content of the groundwater, as a percentage (Na%):

$$Na\% = \frac{Na^+}{Ca^{2+} + Mg^{2+} + Na^+ + K^+} \cdot 100 \text{ (expressed in meq/l)}$$

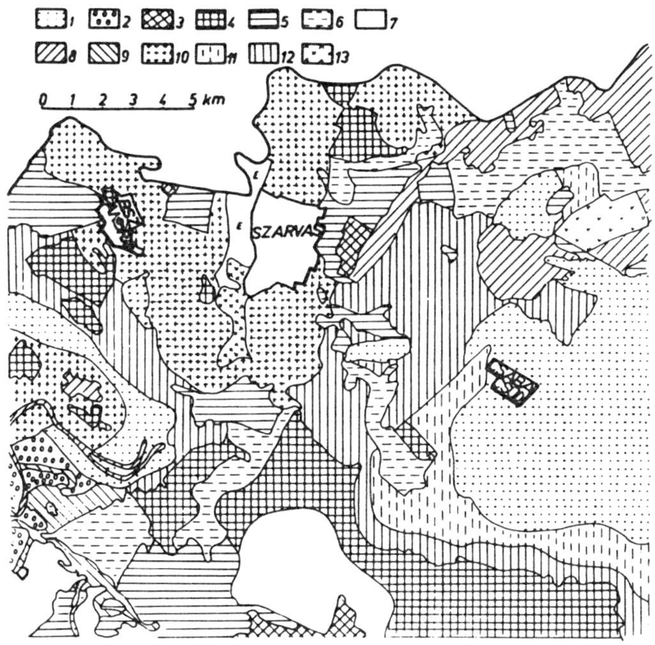

FIGURE 114. Soil map.[21] (1) Calcareous meadow chernozem soil; (2) meadow chernozem soil, solonetzic in deeper layers; (3) shallow meadow solonetz soil; (4) medium meadow solonetz soil; (5) deep meadow solonetz soil turning into steppe formation; (6) solonetzic meadow soil; (7) strongly solonetzic meadow soil; (8) meadow soil; (9) meadow soil, salty in deeper layers; (10) alluvial meadow soil; (11) calcareous chernozem meadow soil; (12) chernozem meadow soil, salty in deeper layers; (13) humous alluvial soil.

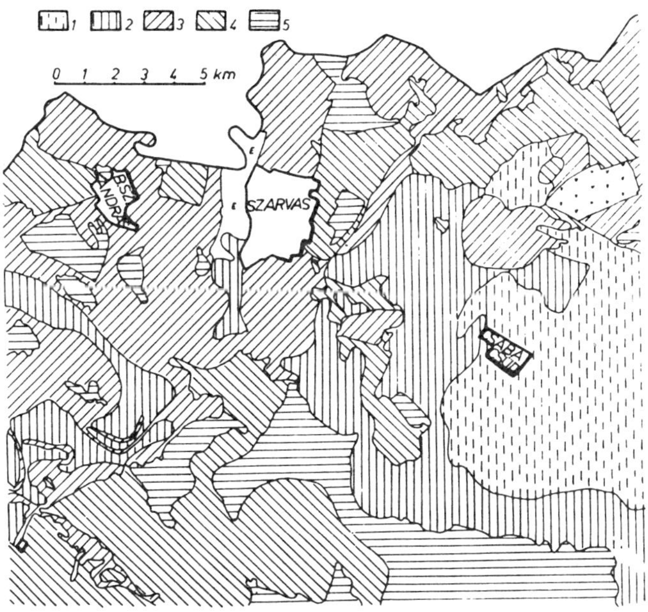

FIGURE 115. Map of soil texture and water properties.[21] (1) Soils with medium water holding capacity and high permeability; (2) soils with high water holding capacity, a high available moisture content and medium permeability; (3) soils with high water holding capacity, a low available moisture content and medium permeability; (4) soils with very high water holding capacity and low permeability; (5) soils with very high water holding capacity and very low permeability.

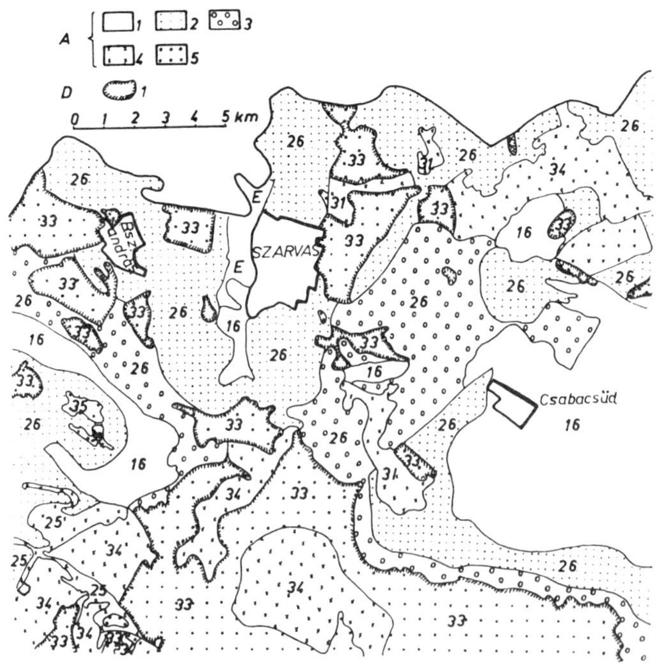

FIGURE 116. Map of salinity status.[21] (A) Average total salt content in the soil between the surface and the water table, in percentage: (1) 0.050 to 0.075%; (2) 0.075 to 0.10%; (3) 0.10 to 0.15%; (4) 0.15 to 0.20%; (5) more than 0.20%. (B) Maximum total salt content of the soil profile, in percentage (the first cipher of the two-figure numbers on the map): (1) less than 0.10%; (2) 0.10 to 0.20%; (3) 0.20 to 0.40%; (4) more than 0.40%. (C) Depth to the salt maximum in the soil profile, in cm (the second cipher of the two-figure numbers on the map): (1) 0 to 5 cm; (2) 5 to 25 cm; (3) 25 to 50 cm; (4) 50 to 100 cm; (5) 100 to 150 cm; (6) more than 150 cm. (D) Soil reaction of the B_1-horizon: (1) pH above 9; (2) (no border) pH 8-9.

On the basis of these maps two more maps were prepared on the same scale (1:25,000) (See Figures 118 and 119).

Map of the "critical depth" to the water table (Figure 118) — The critical depth of the water table is influenced by many factors:

1. Precipitation and potential evaporation
2. Soil texture, structure, and water properties — especially the water-holding capacity, wilting point, available moisture content, permeability, hydraulic and capillary conductivity, etc.
3. Present salt content, salt profile, and salt composition of soils and the harmful salinity limit
4. Salt content and composition of the groundwater
5. Stagnancy, supply, and horizontal flow of the groundwater
6. Human factors (irrigation, drainage, amelioration, agrotechnics, etc.) influencing the above-mentioned factors and the water movement in the soil

Map of the possibilities and preconditions of irrigation (Figure 119) — First, this map shows the possibilities of effective irrigation from the veiwpoint of detailed soil conditions. Using the same evaluation basis as used for the general map (1:100,000), the following were distinguished on the 1:25,000 map: (1) areas to be irrigated; (2) areas that may be irrigated conditionally; and (3) areas not to be irrigated. Second, Figure 119 indicates — on the basis of a comparative evaluation of the groundwater map (Figure 117) and the map of the critical depth to the water table (Figure 118) — the measures that must be taken for

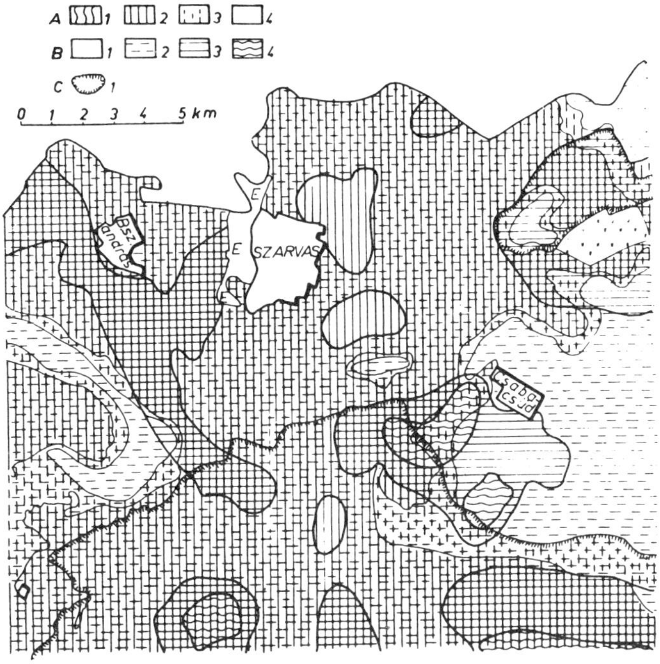

FIGURE 117. Map of groundwater conditions.[21] (A) Average depth to the water table, in m: (1) 1.5 to 2.5 m; (2) 2.5 to 3.5 m; (3) 3.5 to 4.5 m; (4) 4.5 to 60 m. (B) Average salt content of the groundwater, in g/ℓ: (1) 0 to 1; (2) 1 to 2; (3) 2 to 4; (4) 4 to 8 g/ℓ. (C) Na-% of the groundwater: (1) above 75; (2) below 75.

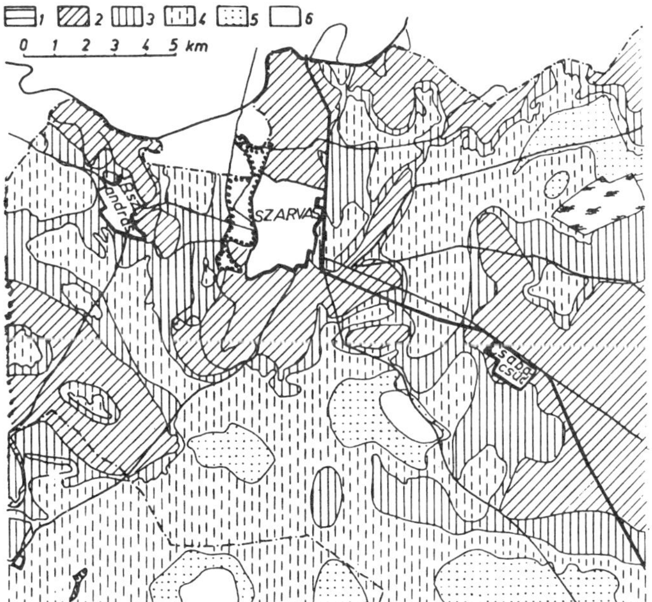

FIGURE 118. Map of the critical depth of the water table, in m.[21] 1. 2.0 m; 2. 2.5 m; 3. 3.0 m; 4. 3.5 m; 5. 4.0 m; 6. 4.5 m

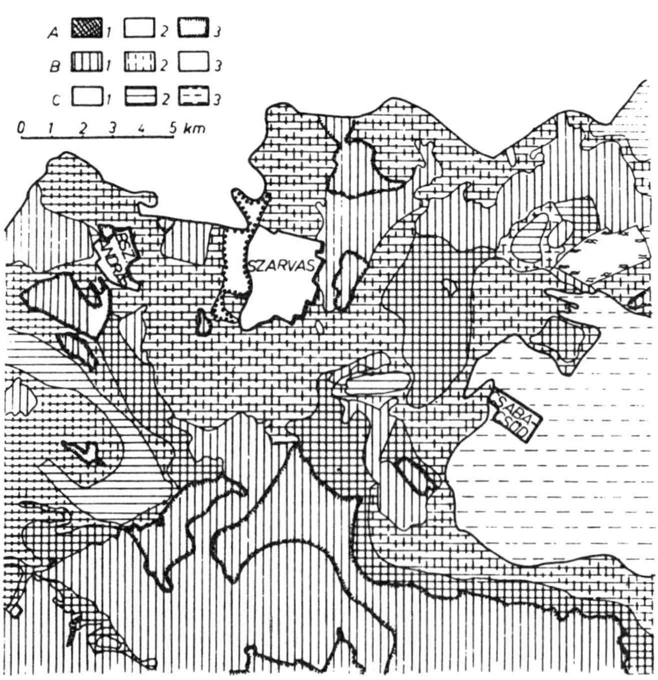

FIGURE 119. Map of possibilities and preconditions of irrigation.[21] (A) The possibilities of effective irrigation from the viewpoint of soil conditions: (1) areas to be irrigated; (2) areas that may be irrigated conditionally; (3) areas not to be irrigated. (B) The necessary groundwater control: (1) lowering of the water table; (2) prevention of a rise in the water table; (3) regular study of the water table. (C) General directives concerning irrigation: (1) frequent irrigation is advisable with low water application rates; (2) moderately frequent irrigation is advisable with medium water application rates; (3) one (or infrequent) irrigation is advisable with a high water application rate.

groundwater control to prevent salinization and alkalization processes due to irrigation. In addition to the above-mentioned information the map (Figure 119) gives some general directives concerning irrigation.

V. CERTAIN ASPECTS OF WATER QUALITY AND IRRIGATION

Irrespective of their sources, all the natural waters of the continents contain dissolved salts. The quality and quantity of these salts depend on the origin and course of the water. Due to the excess of inorganic compounds, the natural waters of the earth may be regarded as an electrolyte solution of mixed salts. The ions prevailing in them are

$$Fe^{2+}, Al^{3+}, Ca^{2+}, Mg^{2+}, Na^{+}, K^{+}, SiO_4^{2-}, CO_3^{2-}, HCO^{-}, SO_4^{2-}, Cl^{-}$$

Boron, fluorine, nitrate, and compounds of heavy metals may also be present in low concentrations or in traces. Among the waters of the continents, rainwater has the lowest salt content. The amount of salts in rainwater varies widely and depends on the distance from the sea and on the eolic deflation.[394]

The salt content of surface and groundwaters is a function of the rocks prevalent at the water source, the kinds of rocks and soils over which the water flows, the climate, and the extent to which the environment has been polluted by human activities. The surface waters can be classified in two groups: (1) Flowing waters (rivers), and (2) Stagnant waters (lakes).

Salt concentration is usually lower in flowing waters than it is in stagnant waters under the same conditions. The average concentration of salts in river waters on the continents appears to vary between 60 and 180 ppm. As regards the mean chemical constituents in

Table 86
VARIATION OF AVERAGE COMPOSITION OF RAIN COLLECTED AT VARIOUS STATIONS IN VICTORIA FROM MARCH 1954 TO DECEMBER 1956

	Distance from ocean							
Station	In S.W. direction (km)	Shortest direction (km)	Annual rainfall (mm)	Na^+ (meq/ℓ)	Ca^{++} (meq/ℓ)	K^+ (meq/ℓ)	Na:K	Na:Ca
Merbein	320	320	240	0.06	0.14	0.007	8	0.4
Walpeup	260	260	310	0.05	0.04	0.004	12	1.2
Seymour	240	160	550	0.03	0.01	0.003	10	3.0
Horsham	190	190	400	0.05	0.03	0.005	10	1.7
Parwan	130	80	490	0.08	0.02	0.004	20	4.0
Coleraine	80	80	590	0.10	0.02	0.005	20	5.0
Cape Bridgewater	1.5	0.80	800	0.63	0.08	0.016	40	8.0

After Hutton, J. T. (cited in Darab[116]).

river waters, calcium and bicarbonate ions are dominant (see Table 27 in Chapter 2). The quantities of sodium salts are low, and their relative amount ranges from 10 to 34%.

The concentration and chemical composition of waters are not static; they are continuously changing and are influenced very much by the climate (precipitation and evaporation in the area).

Table 86 gives data on the average composition of rain collected at various stations in Victoria.

The salt concentration and the chemistry of the salts in the groundwater are determined by the geochemical properties of the source of the water and the course over which it flows. They are also influenced by the climate and the level of the groundwater table.[404]

Irrigation waters can be evaluated by:

1. The salt regime and salt movement in irrigated soils and the soils surrounding the irrigated area
2. The sodium hazard[405]
3. The kinds and quantities of compounds that contain magnesium
4. The amount of compounds capable of alkaline hydrolysis
5. The amount of compounds toxic to plants

The harmful effects of saline water may appear in two ways:

Directly — Highly salinized water can have an adverse effect on plants. However, the damaging concentration of salts in irrigation water is high, and it takes a solution with a total salt concentration of 5 to 10 g/ℓ to inhibit plant growth.

Indirectly — Mineralized water may be damaging due to the accumulative enrichment of soluble salts in the soil after a large number of irrigations where drainage and leaching are not sufficient.

This is why the maximum permissible salt concentration of irrigation waters varies very widely and depends on several factors.[392] The most important factors determining the maximum permissible concentration of salts are

1. The texture of the soil. Usually the permissible concentration of salts in irrigation waters increases as the texture of the soils becomes lighter.

Table 87
CLASSIFICATION OF IRRIGATION WATER SALINITY IN THE U.S.S.R.

Salt concentration (g/ℓ)	Evaluation
0.2—0.5	Water of best quality
1.0—2.0	Water causing salinity hazard
3.0—7.0	Water may be used only with leaching and perfect drainage

2. The degree and chemistry of soil salinity. If the soil is nonsaline, even slightly salinized irrigation water can promote the accumulation of salts in the soil and result in an increase in the salt concentration of the soil solution. On the other hand, when irrigating a saline soil with mineralized water, the water may dilute the concentration of salts in the soil solution. Some of the salts may be leached from the soil with the percolating water under good drainage conditions.[394] The chemistry of water and soil salinity also influences the direct and indirect effects of irrigation on the soil properties. The toxic effects of carbonates, chlorides and sulfates on the crop are different.[406]
3. The natural and artificial drainage conditions of irrigated areas. Among the natural drainage factors, mention must be made of physical soil properties such as field water capacity, the moisture content at wilting point, the rate of infiltration and water conductivity, and the lack or presence of a layer impermeable to water and its depth.
4. The salt tolerance of the plants grown and the intensity of plant production.
5. The methods of irrigation, the quantity of irrigation water in a given growth season, and the frequency of irrigation.

As these factors influence the level of maximum permissible salt concentration in the irrigation water, it is not possible to present a uniform evaluation for utilization on all sites and under all conditions.

The evaluation of irrigation water applied in the U.S.S.R., the standard set up by the U.S. Salinity Laboratory, and the limit values applied in Hungary are given in Tables 87, 88, and 89.

In Table 87 a simple but practical irrigation water grouping system from the U.S.S.R. is demonstrated. In this system, the total salt concentration of the irrigation water is indicated as the basis for its applicability. Due to the fact that in the U.S.S.R. irrigation waters with different chemical compositions exist, more detailed classifications have been elaborated locally.

In Table 88 the grouping of irrigation water according to the U.S. Salinity Laboratory is given; here the electrical conductivity is the basic measurement for determining the quality of irrigation water.

A detailed irrigation water quality system is shown in Table 89, which has proved to be of practical use in Hungary. The three main properties for determining the quality of the irrigation water are the total salt content, the sodium percentage, and the alkalinity. When evaluating the suitability of water for irrigation, in addition to the degree of salinity, the type of salinity and the sodium hazard have to be taken into account.

The chemistry of salinity is characterized by the anion composition of the salts in the irrigation water (Table 90). The sodium hazard of irrigation water is evaluated by the Sodium Adsorption Ratio (SAR) in the system, elaborated by the staff of the U.S. Salinity Laboratory, based on the Gapon equation:

Table 88
GROUPING OF IRRIGATION WATER BY THE U.S. SALINITY LABORATORY[2]

Classification of water	Electrical conductivity (μmno/cm at 25°C)	Salt conc. in g/ℓ (approx.)
C1, Low salinity water — can be used for irrigation with most crops on most soils, with little likelihood that a salinity problem will develop. Some leaching is required, but this occurs under normal irrigation practices, except in soils of extremely low permeability.	0 <EC >250	<0.2
C2, Medium salinity water — can be used if a moderate amount of leaching occurs. Plants with moderate salt tolerance can be grown in most instances without special practices for salinity control.	250 <EC >750	0.2—0.5
C3, High salinity water — cannot be used on soils with restricted drainage. Even with adequate drainage, special management for salinity control may be required, and plants with good salt tolerance should be selected.	750 <EC >2250	0.5—1.5
C4, Very high salinity water — is not suitable for irrigation under ordinary conditions but may be used occasionally under very special circumstances. The soils must be permeable, drainage must be adequate, irrigation water must be applied in excess to provide considerable leaching, and very salt-tolerant plants should be selected.	2250 <EC >5000	1.5—3

$$SAR = \frac{Na^+}{\sqrt{\frac{Ca^{2+} + Mg^{2+}}{2}}} \quad (38)$$

Examples of the application of the Gapon equation are shown in Figures 120 and 121. Figure 120 shows the correlation between different alkalinity quotients in the irrigation water and the total salt concentration in the case of different soil ESP values. The same relationships are shown in Figure 121, in the case of waters with different types of chemistry.

Several approaches exist for the characterization of the sodium hazard in irrigation waters. For example, the sodium hazard can be expressed in terms of soluble sodium percentage (SSP):

$$SSP = \frac{[Na^+]}{\sum[\text{cation}]} \cdot 100 \quad (39)$$

where [ ] = the concentration of cations, in meq/ℓ.

A figure of 60 is given as a limit value for SSP by Scofield, Magistad and Christiansen.[287,290] In some places, depending on the chemistry of the water, it may be as high as 60 to 65%, as demonstrated in Figures 120 and 121.

Figure 122 illustrates the relationship between the sodium hazard and the suitability of water for irrigation. When evaluating water on the basis of the SAR, the total salt concentration measured by electrical conductivity and expressed in micro-Siemens should be considered. The relation is again derived from the Gapon equation:

$$\frac{RNa}{RCa + RMg} = K \frac{[Na^+]}{\sqrt{\frac{Ca^{2+} + Mg^{2+}}{2}}} \quad (40)$$

Table 89
QUALITY NORMS FOR IRRIGATION WATER IN HUNGARY

Suitability	Water type	Total salt (mg/ℓ)	Na^+ (%)	Phenolphthalein alkalinity (soda (mg/ℓ))	Soda equivalent
A. Suitable in every case	a. Hydrocarbonate	<500	<35	<10	
	b. Hydrocarbonate-sulfate	<500	<40	<10	
	c. Hydrocarbonate-chloride-sulfate	<500	<45	<10	
B. Suitable for certain soil types	a. Hydrocarbonate	500—650	<35	<10	
	b. Hydrocarbonate-sulfate	500—650	<40	<10	
	c. Hydrocarbonate-chloride-sulfate	500—650	<45	<10	
	a. Hydrocarbonate	650—800	<30	<10	
	b. Hydrocarbonate-sulfate	650—800	<40	<10	
	c. Hydrocarbonate-chloride-sulfate	650—800	<40	<10	
	a. Hydrocarbonate	800—1000	<35	<10	
	b. Hydrocarbonate-sulfate	800—1000	<40	<10	
	c. Hydrocarbonate-chloride-sulfate	800—1000	<45	<10	
C. Suitable for certain salt-affected types if irrigation is not connected with soil amelioration	a. Hydrocarbonate	<800	35—65	10—50	
	b. Hydrocarbonate-sulfate	<800	35—75	10—50	
	c. Hydrocarbonate-chloride-sulfate	<800	40—75	10—50	
	a. Hydrocarbonate	<1000	<35	<50	
	b. Hydrocarbonate-sulfate	<1000	<40	<50	
	c. Hydrocarbonate-chloride-sulfate	<1000	<45	<50	
D. Suitable in every case after chemical amelioration	a. Hydrocarbonate	<300	<40	50—100	2.0—3
	b. Hydrocarbonate-sulfate	<300	<45	50—100	2.0—3
	c. Hydrocarbonate-chloride-sulfate	<300	<50	50—100	2.0—3
E. Suitable for certain soil types after chemical amelioration	a. Hydrocarbonate	300—500	<40	100—200	3—6
	b. Hydrocarbonate-sulfate	300—500	<45	100—200	3—6
	c. Hydrocarbonate-chloride-sulfate	300—500	<50	100—200	3—6
F. Suitable in every case if improved by dilution	a. Hydrocarbonate	500—1000	<60	10—30	
	b. Hydrocarbonate-sulfate	500—1000	<60	10—30	
	c. Hydrocarbonate-chloride-sulfate	500—1000	<60	10—30	
G. Suitable for certain soil types if improved by dilution	a. Hydrocarbonate	1000—2000	<70	10—50	
	b. Hydrocarbonate-sulfate	1000—2000	<70	10—50	
	c. Hydrocarbonate-chloride-sulfate	1000—2000	<70	10—50	

Table 90
CLASSIFICATION OF NATURAL WATER ACCORDING TO THE ANION COMPOSITION OF DISSOLVED SALTS

Chemistry of the water	$\frac{[CO_3^{2-}] + [HCO_3^-]}{\Sigma \text{ anion}}$	$\frac{[SO_4^{2-}]}{\Sigma \text{ anion}}$	$\frac{[Cl^-]}{\Sigma \text{ anion}}$
Hydrocarbonatic	0.5—1	<0.25	<0.25
Hydrocarbonatic-sulfatic	0.5—1	0.25—0.5	<0.25
Sulfatic-hydrocarbonatic	0.25—0.5	0.5—1	<0.25
Sulfatic	<0.25	0.5—1	<0.25
Hydrocarbonatic-chloridic	0.5—1	<0.25	0.25—0.50
Chloridic-hydrocarbonatic	0.25—0.5	<0.25	0.50—1.00
Sulfatic-chloridic	<0.25	0.5—1	0.25—0.5
Chloridic-sulfatic	<0.25	0.25—0.5	0.5—1.00
Chloridic	<0.25	<0.25	0.5—1.00
Hydrocarbonatic-chloridic-sulfatic	0.25—0.5	<0.25	0.25—0.5

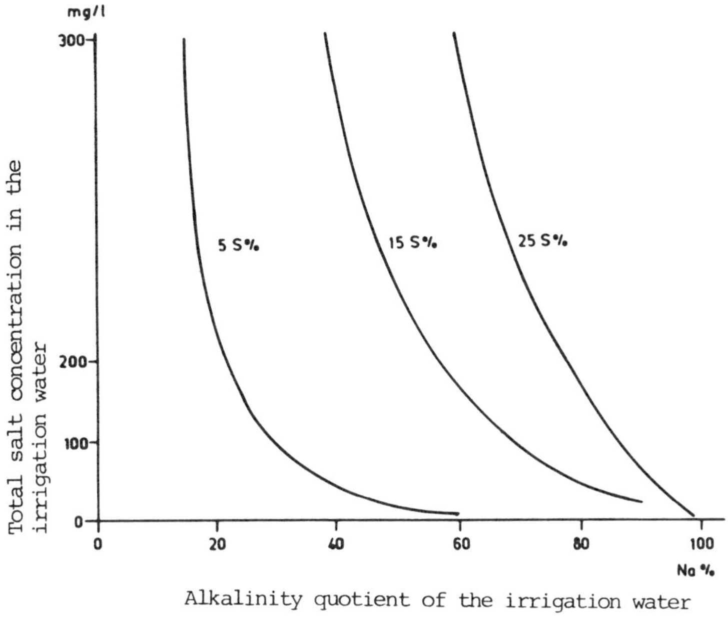

FIGURE 120. Graphical form of the Gapon equation.[368]

where RNa, RCa, RMG = the amount of exchangeable Na, Ca, Mg ions, in meq/100 g; [ ] = the concentration of cations, in meq/ℓ; and K = exchange constant varying between 0.01 and 0.015.

The carbonate and bicarbonate hazard of irrigation water depends mainly on the occurrence and quantity of sodium carbonate and bicarbonate. Calcium and magnesium bicarbonates precipitate in the form of calcium carbonate and magnesium carbonate, and the soil solution becomes alkaline. In order to take account of this reaction when evaluating the applicability of irrigation waters, Eaton introduced the term Residual Sodium Carbonate (RSC). RSC is calculated by subtracting the sum of calcium and magnesium ion concentrations in the water from its total alkalinity:[405]

$$RSC = (CO_3^{--} + HCO_3^-) - (Ca^{++} + Mg^{++}) \tag{41}$$

The limit values introduced by Eaton[290] can be accepted. In this case, if irrigation water has an RSC:

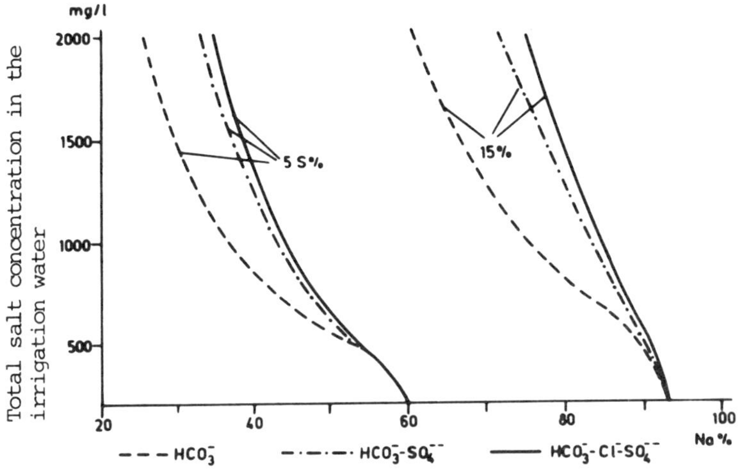

FIGURE 121. Application of the Gapon equation to different types of water.

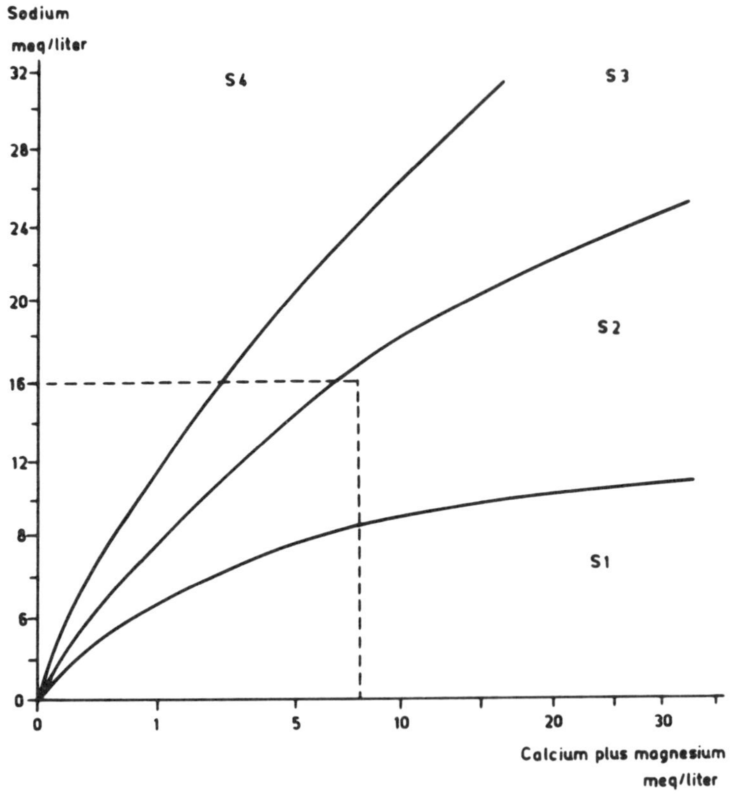

FIGURE 122. Sodium diagram. (S1) Low sodium water; (S2) medium sodium water; (S3) high sodium water; (S4) very high sodium water. (After Wilcox; cited in Darab.[368])

- Higher than 2.5 it is not suitable for irrigation
- Between 1.25 and 2.5 it is marginal
- Less than 1.25 it is applicable for irrigation

The salt content of irrigation water may have a directly toxic effect on plants. Another effect of the salinity and alkalinity of irrigation water results in changes in the soil, as described above (see Figure 123).

Chapter 6

PRINCIPLES OF UTILIZATION AND RECLAMATION OF SALT-AFFECTED SOILS

The great diversity of salt-affected soils explains why different approaches are necessary for their utilization and reclamation under different conditions. The methods to be used for the improvement of various groups of salt-affected soils are determined not only by the physical, chemical, and biological properties of the soils, but also by economic conditions influencing the possibilities for utilizing salt-affected soils. In many cases, especially in developing countries, the shortage of technical and financial means limits the utilization and reclamation of salt-affected soils, but even if these means are available, reclamation projects often seem irrational on the basis of cost-benefit analyses.[407] This is why salt-affected areas are often not reclaimed, but left in their original state, either as part of National Parks or in use in extensive agriculture as grazing land. Figure 123 represents a salt-affected area of this type on the Hungarian Plain. The salt-affected spot itself can only be used for grazing sheep, though it is surrounded by arable land.

In different salt-affected areas of the World, under different natural and economic conditions, countless approaches and methods exist for the utilization and reclamation of salt-affected soils. To give even a short review on these would require at least a separate volume; therefore, only the main principles will be discussed briefly below.

Whenever the improvement, reclamation, and agricultural utilization of salt-affected soils are concerned, the principal task is to diminish or weaken the factors reducing fertility. High salinity and/or alkalinity acts adversely on the physical, chemical, and biological properties of these soils. One or both of these effects disturbs the nutrient and water regimes and hinders the support and maintenance of plants.[408,409]

In saline soils the high electrolyte concentration increases the osmotic pressure in the soil solution and hinders the uptake of ions by the plant root system. According to numerous authors, this phenomenon is one of the major causes of reduced fertility in saline soils. The natural vegetation and various crops have different degrees of tolerance to soil salinity.[410]

The natural vegetation of arid and semi-arid soils, consisting of xerophytes and halophytes, has a high tolerance, but as a rule it has no agricultural value. In literature on plant physiology, soil science, irrigation, agronomy, and other fields, the tolerance of different field, orchard, and other crops was characterized on the basis of various types of experimental data.

In Table 91 saline soils are classified according to the effects of their salt contents on field crops with medium tolerance.[19] Of course, the values given in Table 91 are only approximate. Tolerance obviously varies with soil conditions, farming conditions, and with the different crops and their varieties. Moreover, considering the varying climatic conditions and the different methods applied in the experiments, it is not surprising that the results obtained are not always in agreement.

The following is a list of the most widely grown crops, separated into three groups according to their tolerance toward salts:

Highly tolerant — Date palm, sugar beet, cotton, asparagus, alfalfa, spinach.

Moderately tolerant — (a) Fruit trees (pomegranate, fig, olive, grapevine); (b) field crops and vegetables (sorghum, barley, wheat, tomato, oats, rice, corn, flax, potato, carrot, onion, cucumber).

Sensitive — (a) Fruit trees (pear, apple, orange, plum, almond, apricot, peach); (b) other crops (red clover, peas, beans, sugar-cane).

Within each group of fruit trees or other crops, the species are listed as far as possible in order of decreasing tolerance.

The group of highly tolerant crops can withstand salinity up to an electrical conductivity

FIGURE 123. Salt-affected area not under reclamation.

(EC) of about 16-8 mmho/cm of salt concentraion in a saturated extract of the soil, if the soluble salts are mainly chlorides and sulfates. The moderately tolerant crops can withstand an EC of about 8-4; the limit of the sensitive group is about 4-2. Expressed as the percentage of salts in a dry soil with average texture, the limiting figures for the three groups are about 1.0 to 0.5%, 0.5 to 0.2%, and <0.2%, respectively.

An example of the relative tolerance of certain fruit trees, taking into account the quality of the salts, is presented in Table 92. It is clear that the order of magnitude of tolerance is about the same in this case for chloride salinity and sulfate salinity. The trees included in this table appear in a somewhat different order than they do in a listing on the basis of total salts, but this may well have to do with differences in varieties, rootstocks, and climate. In both systems, all the fruit trees in Table 92 belong to the sensitive group.[19]

For several crops, sensitivity to salts is known to differ markedly according to the salt type. In such cases the order of tolerance will depend on the type of salinity. But much work has yet to be done before a complete picture of plant tolerance to chlorides, sulfates, or carbonates can be given.

The sensitivity of plants to salt also varies according to the phase of development. At the germination and sprouting stages, *Zea mays* is only slightly affected by soil salinity, whereas later on it is extremely sensitive to salinity in the initial phases of its growth and development. In the case of *Oryza* it was established that sensitivity to salts is greater during tillering than during earing. Similar findings are described throughout the literature.

In addition to the total salt concentration, the pH of the soil also has a great influence on the values indicated above. With pH values below 7, there will be a marked increase in the solubility of Fe, Al, and Mn, to the detriment of the vital activity of plants. It may be assumed that there is a direct relationship between the pH of the soil solution, the salt concentration and the growth of the plants. It should be mentioned that the pH of the soil solution fluctuates considerably depending on the soil moisture status and temperature, especially in cases of carbonate salinity.

It is significant that besides causing high osmotic pressure, the salts capable of alkaline hydrolysis may affect plants and crops through their alkalinity, because of varying tolerance of alkali conditions.

In certain areas the direct toxic effects of several salts and ions (S, Mn, Fe, etc.) may be dominant and should be suspected wherever salts have an adverse effect on soils. In California

Table 91
CLASSIFICATION OF SALINE SOILS BY DEGREE AND TYPE OF SALINITY IN RELATION TO FIELD CROPS[19]

Condition of agricultural crops with medium salt resistance	Degree of soil salinity	Type of salts dominating in soils						
		Soda	Cloridic soda and soda chloridic	Sulfatic soda and soda sulfatic	Choridic	Sulfatic-chloridic	Chlorodic-sulfatic	Sulfatic
Good growth and development (no bare patches, crop normal)	Practically non-saline (or only very slightly saline)	<0.10	<0.15	<0.15	<0.15	<0.20	<0.25	<0.30
Slight withering (bare patches and decrease of crop by 10 — 20%)	Slightly saline	0.10 — 0.20	0.15 — 0.25	0.15 — 0.30	0.15 — 0.30	0.20 — 0.30	0.25 — 0.40	0.30 — 0.60
Medium withering (bare patches and decrease of crop by 20-50%)	Medium saline	0.20 — 0.30	0.25 — 0.40	0.30 — 0.50	0.30 — 0.50	0.30 — 0.60	0.40 — 0.70	0.60 — 1.0
Marked withering (bare patches and decrease of crop by 50-80%)	Strongly saline	0.30 — 0.50	0.40 — 0.60	0.50 — 0.70	0.50 — 0.80	0.60 — 1.0	0.70 — 1.20	1.0 — 2.0
A few scattered plants survive (virtually no crop)	Solonchaks	>0.50	>0.60	>0.70	>0.80	>1.0	>1.20	>2

Table 92
MAXIMUM TOLERABLE SALT CONTENT[19]

Species	Total chlorides of Na, Mg, and Ca	Total of harmful sulfates (Mg and Na)
Quince pear	1.5—2	4.5—5
Cherry-plum	1.5—2	4—5
Pear	1.2	4
Apricot and sour cherry	0.9	3
Apple	0.8	3
Peach and sweet cherry	0.5—0.8	2—2.5

Note: Values are meq/100 g dry soil.

and other areas of the U.S., firm requirements limiting the boron content of soils and waters are enforced.

Soil salinity and alkalinity also have adverse effects on the water economy of soils and plants. In Figure 124 the main factors limiting the water supply of plants in salt-affected soils are shown.[1,411]

In saline soils, as a rule, in addition to direct toxic effects, the high osmotic pressure caused by high electrolyte concentration is mainly responsible for low fertility, while in alkali soils the high pH value and the physical effect of the high percentage of exchangeable Na lead to adverse trends in the fertility of these soils.

In Figure 125 the water-physical properties of heavy- and light-textured soils are indicated, as affected by different ESP values.

In Figure 126 the hydraulic conductivity of a chernozem soil saturated with sodium at different levels of ESP is demonstrated. The figure shows that increasing ESP values sharply decrease the hydraulic conductivity of a given soil.

It should be noted that these figures reflect the results of analyses of solonetz soils in Europe.[412,413] Therefore, the relationship between the ESP and the water-physical properties in alkali soils may be different from those in other types of soils.

The foregoing facts lead to the plausible conclusion that if it is salinity which prevents successful agriculture, the salt content must be diminished, in other words, leached out.[414] In irrigated agricultural systems, this is accomplished through proper drainage — a practice which has been associated with irrigation over many centuries of history, and can now be optimally developed, thanks to up-to-date scientific methods. In-depth treatments of irrigation and drainage are available in the literature; e.g., see Richards,[2] and Kovda, van den Berg, and Hagan.[19]

Where alkalinity, rather than salinity, is responsible for the low fertility of a soil, reclamation is sometimes possible without irrigation and drainage. The spectrum of results obtained in reclaiming alkali and solonetz soils and described in the literature is wide. The following is a general summary of these findings.

The discrepancies between the various accounts of experiments on alkali and solonetz soils are due to differences in local climatic conditions, parent materials, levels of agricultural technology, etc. Differences in salt profiles and salt dynamics, particularly arising from differences in the depth and quality of the groundwater, exercise a decisive influence not only on the genetics of alkali, solonetz and solod soils, but on possible methods for improving and utilizing them. Table 93 divides solonetz and solod soils into three principal groups according to their water-soluble salt content and linkage with the groundwater, and summarizes the appropriate methods of amelioration for each group.[415]

In the solonetz and solod soils of Group 1, where the soil profiles — particularly in the

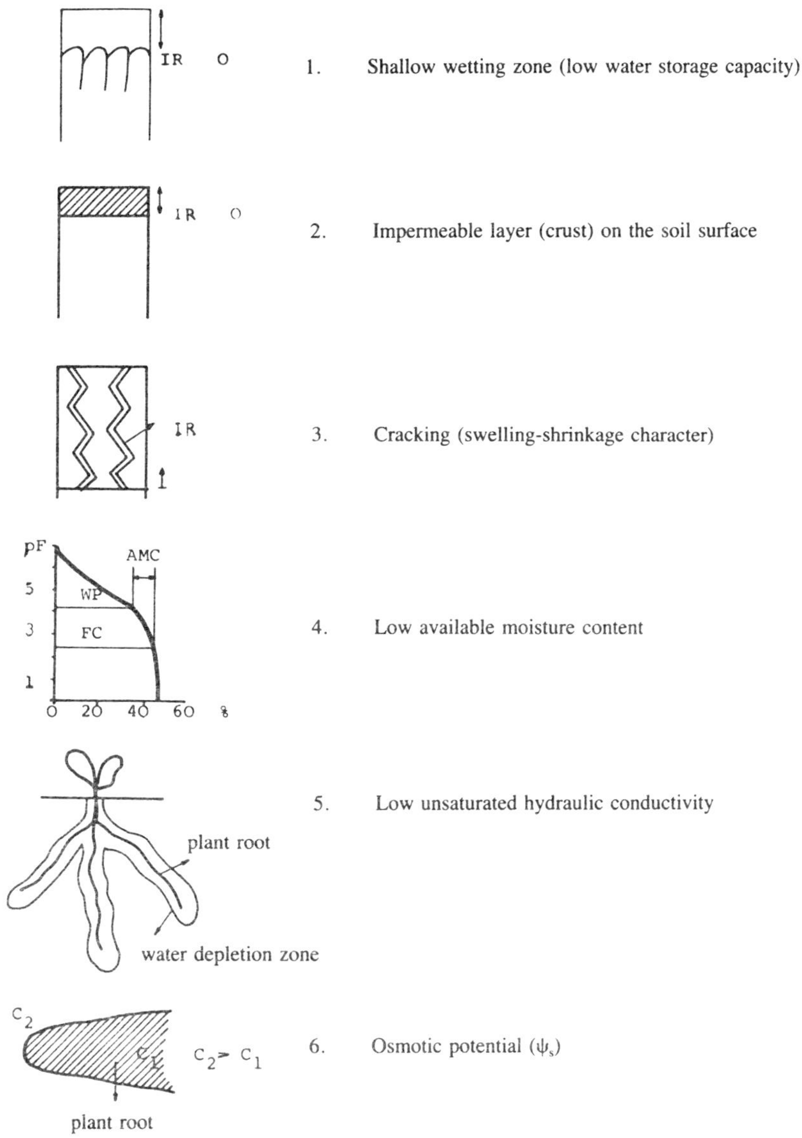

FIGURE 124. Soil factors limiting the water supply of plants. (After Várallyay.[411])

top layers — is capillarily linked with mineralized groundwater, and the A_1, A_2, B_1, and B_2 horizons contain over 0.2% salts at a depth of 40 to 50 cm, the leaching of salts and drainage are absolutely essential. The reclamation of these solonetz and solod soils may be similar to that of solonchak (saline) soils.[416,417] Leaching can be carried out either by applying irrigation water of good quality and providing good drainage or — under more humid climatic conditions, where the annual precipitation is high enough to leach out the salts — by lowering the water table to a depth below the critical level. In many cases chemical amendments should also be applied either concurrently with the leaching and drainage, or afterwards, especially in the case of heavy-textured soils, in order to replace the sodium ions adsorbed in the colloid fraction by calcium ions. Sometimes solonchak-solonetz soils are also reclaimed in this way.[418]

In Group 2, where the profile of a solonetz (often called a meadow solonetz) or a solod soil is only temporarily linked with the groundwater, and the salt content of the A, A_1, and

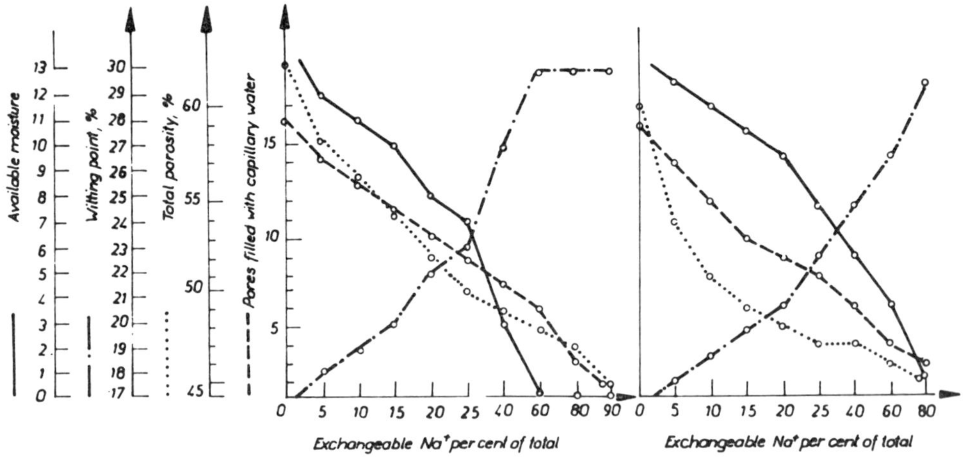

FIGURE 125. Exchangeable cations and water properties of soils depending on the ratio of (Ca^{+} + Mg^{+}) to Na^{+} in sandy and clayey soils.

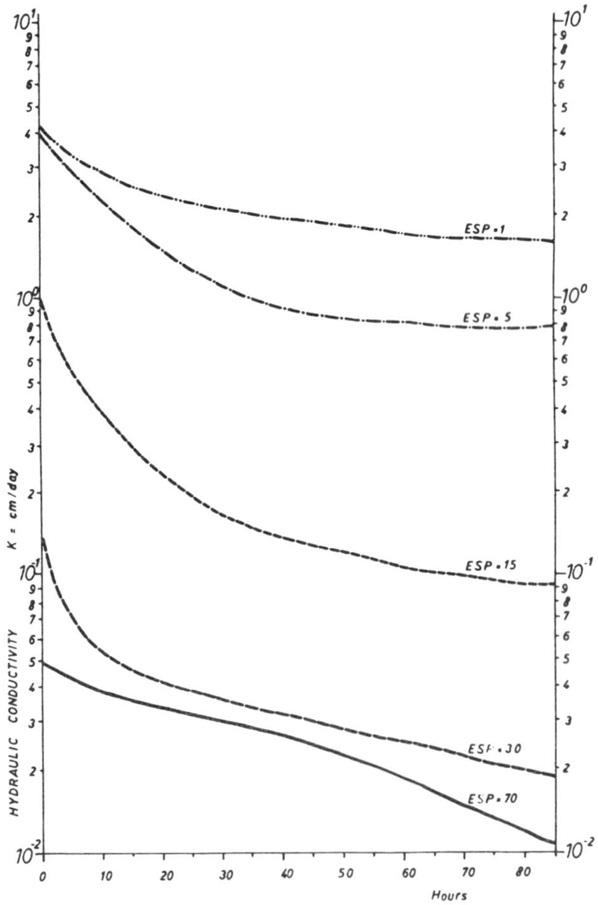

FIGURE 126. Hydraulic conductivity of a chernozem soil saturated with sodium to various degrees.

Table 93
SCHEMATIC GROUPING OF SOLENETZ AND SOLOD SOILS AND SUGGESTED METHODS OF AMELIORATION

Genetic type	Relation with groundwater	Water-soluble salt content in the surface layer	Amelioration[a]
1. Solonchak-solonetz Meadow solonetz Meadow solod (shallow and middle)	Permanently linked	More than 0.2% (about 4 mmho)	Drainage and chemical amendments
2. Meadow solonetz and solod soils turning into steppe formation	Temporarily linked	About 0.2%	Chemical amendments, deep plowing, and drainage if necessary
3. Deep solonetz and solod soils Solonetz-like meadow soils	Not linked	Less than 0.2%	Low amount of chemical amendments, proper agrotechnics (deep plowing and subsoil loosening, etc.), suitable crops (alfalfa, etc.)

[a] The necessity of irrigation depends on local conditions.

B horizons is lower than in soils belonging to the first group, drainage is not always necessary (informative data are presented in Table 92). In these cases, the application of chemical amendments (gypsum and/or others), combined with deep plowing and subsoil loosening, may be sufficient.[419,420] If the quantity of water-soluble sodium salts in the B_2 and C horizons is not high and a considerable amount of gypsum is present, it can be utilized as reclamation material during deep plowing. Depending upon local conditions, the amelioration may be carried out either with or without irrigation. Under irrigated conditions, however, good drainage is more important. These meadow solonetz and solod soils, when rationally treated, can turn into a steppe formation.[421]

In Group 3, where the profile is not linked with the groundwater, the salt content of the upper layers of the profile should be the primary consideration in selecting suitable amendments. When the present processes of natural soil formation assure, to a greater or lesser extent, the leaching out of water-soluble salts, it is sufficient to provide adequate drainage. Frequently, these soils are only moderately solonetzized and/or solodized. Their exchangeable Na^+ content is less than 10 to 15% of the cation exchange capacity, and this value very often occurs in a comparatively deep layer, more than 15 to 20 cm below the surface.[422]

In these soils, which are often referred to as steppe solonetz and solod soils, or solonetz-like meadow soils, it is possible to obtain good results with inexpensive, simple reclamation methods, because the basic aim of reclamation is to facilitate the natural leaching processes. The climatic conditions, the possibility of irrigation, etc. are also decisive factors in choosing the proper methods for salt removal and for the improvement of physical soil properties. Chemical amendments, deep plowing, and subsoil loosening may be used as indicated for Group 2, but the use of proper agrotechnics and the selection of the most suitable crops are very important.

The methods used for the reclamation and utilization of these three types of solonetz and solod soils must always be carefully selected and adjusted to local climatic and farming conditions, which determine whether the reclamation should be carried out with or without irrigation. As we have seen, the chemistry of solonetz soils is also very important, and the occurrence or absence of free soda may have a considerable influence on reclamation methods, as well as on the type and doses of amendments required.

The reclamation of gypsiferous soils is a difficult task because, as was mentioned in previous chapters, these soils are very sensitive to water, and the removal of excess gypsum is only possible by a combination of mechanical, physical and even biological methods of complex reclamation. If the gypsum content is not too high and the calcium sulfate occurs together with a high amount of sodium salts, the leaching of the soils may be very effective, and may even prevent the formation of alkali soils.[419,420]

The reclamation of magnesium soils always needs a practical approach which improves both physical and chemical soil properties. Only in the case of a very high amount of Mg salts can leaching be effective; in other cases dry reclamation methods (e.g., deep plowing and chemical reclamation) are effective.[415]

An efficient system has been elaborated for the reclamation and improvement of acid sulfate soils, and this is widely applied in areas where this group of salt-affected soils is extensive. The main task is to buffer the high acidity, simultaneously diminishing the toxic effect of aluminum and ferric ions. The main method employed is liming, parallel with the application of available fertilizers.[19,29]

The literature contains a much higher number of sources related to the reclamation of salt-affected soils than to any other question regarding their properties, formation, etc.[423,424] The different methods are so dependent on local circumstances that a precise description of soil-forming processes, combined with advanced monitoring and diagnostic systems are always required when reclamation methods are to be adapted for use in places other than the site of origin.

Not only must local conditions be considered during the design of reclamation systems for salt-affected soils, but the economic and social conditions must also be thoroughly analyzed. In order to avoid the disatrous consequences often experienced during the reclamation of salt-affected areas (e.g., secondary salinization of irrigation systems in many countries), up-to-date studies and an integrated multidisciplinary approach should be applied not only during, but also prior to the reclamation of salt-affected soils.

REFERENCES

1. **Szabolcs, I.,** *Review of research on salt-affected soils* (with a bibliography compiled by Várallyay, G.), Natural Resources Research XV, United Nations Educational Scientific and Cultural Organization, Paris, 1979.
2. **Richards, L. A.,** *Diagnosis and Improvement of Saline and Alkaline Soils,* USDA Handbook No. 60., U.S. Department of Agriculture, Washington D.C., 1954.
3. **Kovda, V. A.,** *Origin and Regime of Salt-Affected Soils* (Russian), Vols. 1 and 2, Izd. Ak. Nauk SSSR, Moscow, 1947.
4. **Szabolcs, I.,** Salt-affected soils as a world problem, in *Proc. Int. Symp. Reclamation of Salt-Affected Soils, Jinan, China,* 1, 30, 1985.
5. **Stebutt, A.,** *General Soil Science* (German), Borntraeger, Berlin, 1930.
6. **Dokuchayev, V. V.,** *Selected Papers* (Russian), Sel'khozgiz., Moscow, 1954.
7. **Dregne, H. E.,** *Soils of Arid Regions,* Developments in Soil Science 6, Elsevier, New York, 1976.
8. *Desertification: Its Causes and Consequences,* (Compiled and edited by Secretariat of UN Conference on Desertification, Nairobi, Kenya), Pergamon Press, New York, 1977.
9. **Sokolovsky, A. N.,** *Soil Science* (Russian), Sel'khozgiz, Moscow, 1936.
10. **Hilgard, E. W.,** *Soils: Their Formation, Properties, Composition and Relations to Climate and Plant Growth in the Humid and Arid Regions,* Macmillan, New York, 1906.
11. **Glinka, K. D.,** *Soil Science* (Russian) Sel'khozgiz., Moscow, 1931.
12. **'Sigmond, E.,** *Hungarian Alkali Soils and Methods of their Reclamation* (Hungarian), MTA Kiad., Budapest, 1923.
13. **Ballenegger, R.,** *Salt-Affected Soils and their Amelioration* (Hungarian), Egy. Nyomda, Budapest, 1931.
14. **Treitz, P.,** Alkali soils in Hungary (Hungarian), *Földt. Közl.* 26, 19, 1896.
15. **Treitz, P.,** Saline soils in the Hungarian Plain (Hungarian) *Földt. Közl.,* 38, 38, 1908.
16. **Treitz, P.,** *The Nature of Saline and Alkali Soils* (Hungarian), Pátria, Budapest, 1924.
17. **Vilensky, D. G.,** *Salt-Affected Soils, their Origin, Composition and Methods of Amelioration* (Russian), Novaya Derevnya, Moscow, 1924.
18. **Kelley, W. P.,** *Alkali Soils, their Formation, Properties and Reclamation,* Reinhold, New York, 1951.
19. **Kovda, V. A., van den Berg, C., and Hagan, R. M.,** *International Source-Book on Irrigation and Drainage of Arid Lands in Relation to Salinity and Alkalinity,* FAO/UNESCO, 1967.
20. *Prognosis of Salinity and Alkalinity,* FAO Soil Bulletin No. 31, Food and Agriculture Organization, Rome, 1976.
21. Symposium on the Reclamation of Sodic and Soda-Saline Soils, Yerevan, 1969, *Agrokémia és Talajtan,* 18 (Suppl.), 1969.
22. **Dudal, R.,** Key to Soil Units for the Soil Map of the World, in *Soil Map of the World,* Food and Agriculture Organization, Rome, 1970.
23. USDA, *Soil Taxonomy,* U.S. Department of Agriculture, Washington, D.C., 1976.
24. **Szabolcs, I.,** *Salt-Affected Soils in Europe,* Martinus Nijhoff, The Hague; and Research Institute for Soil Science and Agricultural Chemistry of the Hungarian Academy of Sciences, Budapest, 1974.
25. **Hodge, C. and Dusberg, P. C., Eds.,** *Aridity and Man. The Challenge of the Arid Lands in the United States,* American Association Advancement of Science, Washington, D.C., 1963.
26. **Aubert, G.,** Arid zone soils. A study of their formation characteristics, utilization and conservation, *Arid Zone Res.,* 18, 115, 1962.
27. Proceedings of the Symposium on Sodic Soils, Budapest, 1964. *Agrokémia és Talajtan,* 14 (Suppl.), 1964.
28. Salinity problems in the arid zones, *Arid Zone Res.,* 1961, 14.
29. **Dost, H.,** *Acid Sulphate Soils. 1 and 2,* International Institute for Land Reclamation and Improvement, Wageningen, The Netherlands, 1973.
30. *International Symposium on Salt Affected Soils, February 18 to 21, 1980,* Symposium Papers, Central Soil Salinity Research Institute, Karnal, India, 1980.
31. **Dregne, H. E. and Mojallali, H.,** Salt-fertilizer-specific ion interaction in soil, *Bull. New Mex. Agric. Exp. Sta.,* 1968.
32. **Singh, D. V., Chauhan, R. P. S., and Singh, D. P.,** A note on the effect of saline water on uptake of N, P and K by wheat at different growth stages, *Indian J. Agron.,* 19, 281, 1974.
33. Changes of climate, *Arid Zone Res.,* United Nations Educational Scientific and Cultural Organization, Paris, 1963.
34. **Vink, A. P. A.,** *Land Use in Advancing Agriculture,* Springer Verlag, New York, 1975.
35. **Dando, W. A.,** *The Geography of Famine,* W. H. Winston & Sons, Edward Arnold, 1980.
36. **Bishay, A. and MicGinnies, W. G.,** *Advances in Desert and Arid Land Technology and Development,* Vol. 1, Hardwood Academic Publishers, New York, 1979.
37. **Rode, A. A.,** A biogeocoenological approach to the solution of land reclamation problems, *Agrokémia és Talajtan,* 18, (Suppl.), 278, 1969.

38. **Lieth, H. and Whittaker, R., Eds.,** *Primary Productivity of the Biosphere,* Springer-Verlag, New York, 1975.
39. **Kovda, V. A. and Szabolcs, I., Eds.,** Modelling of soil salinization and alkalinization, *Agrokémia és Talajtan,* 28, (Suppl.), 1979.
40. **Husz, G.,** Systematics of salt-affected soils in the Seewinkel region, Austria (German), *Die Bodenkultur,* 17, 295, 1966.
41. **Raikov, L. and Kavardzhiev, Ya.,** Chemical reclamation and fertilization of sodic meadow solonetzes in Bulgaria (Russian), in *Proc. Int. Symp. Reclam. Soda-saline Soils,* Yerevan, 1969, Trudy Inst. Pochv. Agrokhim (Yerevan) 6, 325, 1971.
42. **Cervenka, L. and Szabolcs, I.,** Genesis and properties of Slovakian salt-affected soils (Hungarian), *Agrokémia és Talajtan,* 17, 269, 1968.
43. **Servant, J.,** The salt profile in soils. Investigational techniques and significance. Application to halomorphic soils in southern France (French), *An. Agron.,* 24, 375, 1973.
44. **Maianu, A.,** *Secondary Soil Salinization in Romania* (Russian), Ed., Akad. RPR, Bucharest, 1964.
45. **Obrejanu, G., Maianu, A., and Aksenova, I.,** Salt accumulation in mineralized groundwaters and saline soils of floodplains in the lower Danube bottomland (Romania), *Pochvovedenie,* 8, 44, 1964.
46. **Sandu, G. et al.,** Study of ground water mineralization and the hydrosaline regime of soils in the Calarasi terrace irrigation systems (Rumania), *Pedol. A. Sect.,* 33, 323, 1965.
47. **Sandu, G.,** Hydrosaline regime of some soils in the Calmatui Valley of the Romanian Plain (French), in *Proc. Int. Symp. on New Dev. in the Field of Salt-Affected Soils, Cairo, December 4-9, 1972,* Ministry of Agriculture, Cairo, 1975.
48. **Villar, E. H. del,** *Soils of the Iberian Peninsula,* Bebidos en Espana, Madrid, 1937.
49. **Grande Covian, R.,** Saline soils in the delta areas of southern Spain. Evolution of saline soils in the Guadalquivir Delta (Es), *Agrokémia és Talajtan,* 17 (Suppl.), 113, 1968.
50. **Grande Covian, R.,** Reclaiming and Guadalquivir marshes, *Fatis Review,* 12, 73, 1965.
51. **Kovda, V. A.,** *Solonchaks and Solonetzes* (Russian), Izd. Ak. Nauk SSSR, Moscow, 1937.
52. **Kovda, V. A.,** Solonetzes, in *Soils of the USSR* (Russian), Vol. 1, Izd. AN SSSR, Moscow, 1939.
53. **Egorov, V. V., Bazilevich, N. I., and Pankova, E. I.,** Principles of the Map of Soil Salinity-Alkalinity Types in the European U.S.S.R. (Russian), in *Proc. Int. Symp. Reclam. Soda-saline Soils,* Yerevan, 1969, Trudy Inst. Pochv. Agrokhim (Yerevan), 6, 601, 1971.
54. **Filipovski, G.,** Results of investigations on the possibilities of reclaiming salt-affected sodic soils in Yugoslavia, *Agrokémia és Talajtan,* 18 (Suppl.), 385, 1969.
55. **Miljkovic, N. and Plamenac, N.,** The improvement of solonetz soils in Yugoslavia, *Agrokémia és Talajtan,* 18 (Suppl.), 377, 1969.
56. **Neugebauer, V. and Miljkovic, N.,** General Review of Salt-Affected Soils of Yugoslavia, in *Salt-Affected Soils Subcommission Meeting, Sevilla, May, 1971,* IRYDA, 1971, 191.
57. **Zivkovic, B.,** Salinization and comparative characteristics of normal soils, soils in the process of salinization and solonchaks in Vojvodina (Yugoslavia), *Savremena poljoprivreda,* 5, 1, 1965.
58. **Cairns, R. R.,** Canadian Solonetz Soils and their Reclamation, *Agrokémia és Talajtan,* 18 (Suppl.), 233, 1969.
59. **Clayton, J. S., Ehrlich, W. A., Cann, D. B., Day, J. H., and Marshall, I. B.,** *Soils of Canada, I-II,* Canada Department of Agriculture, Ottawa, 1977.
60. **Kelley, W. P.,** *The Reclamation of Alkali Soils,* Reinhold, New York, 1937.
61. **Bonnet, J. A.,** Edafologia de los suelos salinos y sodicos, University of Puerto Rico, 1960.
62. **Molina, J. S. and Sauberan, C.,** Reclamation of sodic soils by biological methods, *Agrokémia és Talajtan,* 14 (Suppl.), 411, 1965.
63. **Pla, S. I.,** Salt and water balances in irrigated soils under tropical conditions, in *Proc. Int. Symp. on New Dev. in the Field of Salt-Affected Soils, Cairo, December 4-9, 1972,* Ministry of Agriculture, Cairo, 1975, 117.
64. **Zavaleta, A.,** Saline and alkali soils in Peru, *Agrokémia és Talajtan,* 18 (Suppl.), 169, 1969.
65. **Fournier, F.,** Halomorphic soils of the African Continent (French), *Agrokémia és Talajtan,* 14 (Suppl.), 243, 1965.
66. **Massoud, F. I.,** Factors to be considered for prognosis in soil management and agronomic practices, *FAO Soils Bull.,* 31, 111, 1976.
67. **Zonn, I. S.,** Saline soils of central Sahara oases and their reclamation. In *Proc. Int. Symp. on New Dev. in the Field of Salt-Affected Soils, Cairo, December 4-9, 1972,* Ministry of Agriculture, Cairo, 1975, 483.
68. **Durand, J. H.,** *Les Sols Rouges et les Croutes en Algerie,* ORSTOM, Alger, 1959.
69. **Frankart R. P. and Herbillon, A. J.,** The presence and genesis of amalcime in sodic soils of the lower Ruzizi (Burundi), (French), *Bull. Grpe fr. Argiles,* 22, 79, 1970.
70. **Frankart, R. P. and Herbillon, A. J.,** *Aspects de la Pedogenese des Sols Halomorphes de la Basse Rusizi (Burundi),* Musée Royal de l'Afrique Centrale, Tervurgen, Belgique, 1971.

71. **Cheverry, C.**, Genesis of saline and alkali soils in the polders of the north-east shore of Lake Chad (French), *Agrokémia és Talajtan,* 18 (Suppl.), 238, 1969.
72. **Cheverry, C.**, Contribution to pedological study of Chad lake polders (French), *Cah. Pedol.*, ORSTOM, 1975.
73. **Segalen, P.**, Soils and geomorphology of Cameroon (French) *Cah. Pedol.*, ORSTOM, 5, 137, 1967.
74. **El-Gabaly, M. M.**, Problems of sampling, analysing and mapping of salt-affected soils, in *Salt-Affected Soils Subcommission Meeting,* IRYDA, 1971, 17-28.
75. **Ruellan, A.**, Saline and alkali soils at the bottom of the Zebra plain of Basse Moulouya, Morocco. Preliminary results of an experiment designed to study their reclamation and their evolution under irrigation (French), *Trans. 8th Int. Congr. Soil Sci.*, 1967, 937.
76. **Durand, J. H.**, An attempt to classify saline soils of the Senegal Delta (French), *Bull. Assoc. Fr. Etude Sol,* 4, 35, 1968.
77. **Buringh, P.**, Differences in salinization between the Mesopotamian Plain and the Nile Delta. *Agrokémia és Talajtan,* 14 (Suppl.), 107, 1965.
78. **Buringh, P.**, Sodic vertisols in central Sudan. *Agrokémia és Talajtan,* 18 (Suppl.), 100, 1969.
79. **Williams, M. A. J.**, Soil salinity in the west central Gezira, Republic of the Sudan, *Soil Sci.*, 105, 451, 1968.
80. **Cooper, G. R.**, A field method for the estimation of sodium status of Rhodesian soils, *Rhodesia Agric. J.*, 69, 126, 1972.
81. **Govinda Rajan, S. V. and Murthy, R. S.**, Physico-chemical properties of some alluvial soils containing high sodium carbonate in the Indo-Gangetic Basin, *Agrokémia és Talajtan,* 18 (Suppl.), 243, 1969.
82. **Kanwar, J. S. and Bhumbla, D. R.**, Physico-chemical characteristics of sodic soils of the Punjab and Haryana and their amelioration by the use of gypsum, *Agrokémia és Talajtan,* 18 (Suppl.), 315, 1969.
83. **Shah, R. K. and Palkhiwala, A. B.**, Saline and alkali soils of savainagar salines near Bhavnagar in Saurashtra, *Indian J. Agric. Res.*, 5, 201, 1971.
84. **Sharma, R. A. et al.**, Characterization of saline and sodic soils of chambal-commanded area of Bhind (Madhya Pradesh), *Indian J. Agric. Res.*, 9, 117, 1975.
85. **Dewan, M. L. and Famouri, J.**, *The Soils of Iran,* Food and Agricultural Organization, Rome, 1964.
86. **Buringh, P.**, *Soils and Soil Conditions in Iraq,* Ministry of Agriculture, Iraq, Baghdad, 1960.
87. **Kadry, L. T.**, A note on the alluvial soils of Iraq, their salinity, sodicity and reclaimability, *Agrokémia és Talajtan,* 18 (Suppl.), 159, 1969.
88. **Zonn, I.S. and Limachev, N. S.**, Characteristics of soil formation in irrigated regions of southern Iraq (Russian), *Pochvovedenie,* 11, 72, 1966.
89. **Dan, I., Yaalon, D. H., Koyumdjisky, H., and Raz, Z.**, *The Soils of Israel,* The Volcani Center, Bet Dagan, 1976.
90. **Choudhury, M. B. and Rafiq, C. M.**, Soils of West Pakistan, *Pakistan J. Soil Sci.*, 3, 1, 1967.
91. **Mulders, M. A.**, *The Arid Soils of the Balikh Basin (Syria),* Bronder, Rotterdam, 1969.
92. **Meester, T. de**, *Soils of the Great Konya Basin, Turkey,* CAPB, Wageningen, The Netherlands, 1970.
93. **Kovda, V. A.**, *Nature and Soils of China,* Publishing House of the Academy of Science, Moscow, 1959.
94. **Chen, E. F. et al.**, The soda-solonchak in the northeast part of China and the experiences in growing rice for the amelioration of soils, *Trans. 8th Int. Congr. Soil Sci.*, Bucharest, Romania, 1964, 1967, 2, 897.
95. **Kang Sek Hjon**, *Geochemical Study of the Paddy Soils,* Kim Il Sung University Press, Pyonyang, North Korea, 1974.
96. **Bespalov, N. D.**, *Soils of the Mongolian People's Republic,* Publishing House of Acad. Sci. U.S.S.R., Moscow, 1951.
97. **Kowalkowski, A. and Borzyszkowski, W.**, Earth hummocks in the dry steppe and in forest-steppe in central Mongolia, *Geomorphologia Carpato-Balcanica, XIX,* Krakow, 1985, 111.
98. **Egorov, V. V., Bazilevich, N. I., and Pankova, E. I.**, The Map of the Salinity Types of the U.S.S.R. Soils, in *Proc. Int. Symp. on New Dev. in the Field of Salt-Affected Soils,* Cairo, December 4-9, 1972, Ministry of Agriculture, Cairo, 1975, 479.
99. **Dames, T. W. G.**, *The Soils of East Central Java,* Pemberitaan, Bogor, Indonesia, 1955.
100. **Northcote, K. H. and Skene, J. K. M.**, *Australian Soils with Saline and Sodic Properties,* Commonwealth Scientific and Industrial Research Organization, Soil Publ. No. 27, 1972.
101. CSIRO, *A Handbook of Australian Soils,* Rellim, Commonwealth Scientific and Industrial Research Organization, South Australia, 1968.
102. CSIRO, *Soils. An Australian Viewpoint,* Academic Press, Commonwealth Scientific and Industrial Research Organization, Melbourne, Australia, 1983.
103. **Bresler, E., McNeal, B. L., and Carter, D. C.**, *Saline and Sodic Soils,* Advanced Series in Agricultural Sciences, 10, Springer-Verlag, New York, 1982.
104. **Vind, A. P. A.**, *Land Use in Advancing Agriculture,* Advanced Series in Agricultural Sciences, 1, Springer-Verlag, New York, 1975.
105. **Egorov, V. V.**, Soil aspects relating to the sodic salinization in the subhumid regions of Europe and Asia (Russian), *Agrokémia és Talajtan,* 18 (Suppl.), 187, 1969.

106. **Vilyams, V. R.**, *Selected Papers* (Russian), Izd. Ak. Nauka SSSR, Moscow, 1950.
107. **Nestroy, O.**, Ökologische Aspekte der Landwirtschaftlichen Bodennutzung in Europa, *Z. für Agrargeographie*, 3, 79, 1985.
108. **Szabolcs, I., Ed.**, *European solonetz soils and their reclamation*, Akadémiai Kiadó, Budapest, 1971.
109. **Hrasko, J.**, Salt sources of alkali soils in South Slovakia, *Agrokémia és Talajtan*, 17 (Suppl.), 105, 1968.
110. **'Sigmond, E.**, The Alkali soils in Hungary and their reclamation, *Soil Sci.*, 17, 379, 1924.
111. **'Sigmond, E.**, *Hungarian Alkali Soils and Methods of their Reclamation*, University of California Press, Berkeley, 1927.
112. **Timkó, I.**, Salt-Affected Soils in the Kiskunság and Jászság Regions (Russian), *Földtani Intézet Évi jelentése*, 1925-1928, 254, 1929.
113. **Szabolcs, I.**, *The Influences of Water Regulation and Irrigation on Soil-formation Processes in the Trans-Tisza Region* (Hungarian), Akadémiai Kiadó, Budapest, 1961.
114. **Arany, S.**, *Salt-Affected Soils and their Amelioration* (Hungarian), Mezögazdasági Kiadó, Budapest, 1956.
115. **Ábrahám, L. and Bocskai, J.**, *Amelioration of Hungarian Salt-Affected Soils* (Hungarian), OMMI Kiadv., (1st series), No. 4, Budapest, 1971.
116. **Darab, K.**, *The Application of Soil Genetic Principles to Irrigation in the Hungarian Plain* (Hungarian), OMMI Kiadv., (1st series), No. 4, Budapest, 1962.
117. **Ábrahám, L.**, Problems of utilization and amelioration of solonchak-solonetz soils in the Hungarian Danube Valley (Hungarian), *Agrokémia és Talajtan*, 17, 283, 1968.
118. **Szabolcs, I.**, Recent problems of investigating and utilizing salt-affected soils, *Beitr. trop. Landw. Vet. Med.*, 4, 5, 1966.
119. **Várallyay, G.**, Salt balances of soils in the region between the rivers Danube and Tisza in Hungary. I. Salt balances under natural, non-irrigated conditions (Hungarian), *Agrokémia és Talajtan*, 15, 423, 1966.
120. **Florea, N. and Munteanu, I.**, Soils with sodic salinization in Romania, *Agrokémia és Talajtan*, 18 (Suppl.), 207, 1969.
121. **Obrejanu, G., Sandu, G., and Rudzic, N.**, Reclamation of solonetzes and solonetzic soils of Europe (Russian), *Pochvovedenie*, 7, 94, 1970.
122. **Oprea, C. V. and Anastasescu, I.**, Saline and alkali soils of the eastern Tisza Plain in Romania (French), *Salt-Affected Soils Subcommission Meeting*, Seville, May, 1971, 77.
123. **Cardoso, J. C.**, Reclamation of saline soils in Portugal, *Agrokémia és Talajtan*, 18 (Suppl.), 329, 1969.
124. **Antipov-Karataev, I. N.**, *Amelioration of Solonetzes in the Soviet Union* (Russian), Izd. Ak. Nauk SSSR, Moscow, 1953.
125. **Kovda, V. A., Rozov, N. N., and Sambur, G. N.**, *Improvement and Utilization of Solonetzes in the Soviet Union* (Russian), Izd. Ak. Nauk SSSR, Moscow, 1950.
126. **Egorov, V. V.**, Origin of soda-salinization in the subhumid regions of Eurasia (Russian), in *Proc. Int. Symp. Reclam. Soda-saline soils*, Trudy Inst. Pochv. Agrokhim. (Yerevan), 6, 365, 1971.
127. **Egorov, V. V. and Bazilevich, N. I.**, Salt-Affected Soils in the European USSR and Trans-Caucasian Soviet Republics (Russian), *Nauchn. Tr. Pochv. Inst. Dokuchaeva*, Moscow, 1973.
128. **Grin, G. S.**, Salt-affected soils in the Ukraine and their origin (Russian), *Tr. Kharkov, Sel'-khoz. Inst.*, 39, 8, 1962.
129. **Mozheyko, A. M.**, Solonetz Complexes and their Chemical Amelioration (Russian), *Tr. Comm. Irr.*, 6, Izd. AN SSSR, Moscow, 1936.
130. **Kovda, V. A.**, *Soils of the Pre-Caspian Bottomland* (Russian), Izd. Ak. Nauk SSSR, Moscow-Leningrad, 1950.
131. **Kozlovsky, F. I.**, Process of Salt Accumulation in Soils, and Methods of its Long-term Prediction, *Bull. Dokuchaev Inst. Soil Sci.*, Moscow, 7, 1972.
132. **Bazilevich, N. I. and Pankova, E. I.**, Classification of soils according to their chemistry and degree of salinization, *Agrokémia és Talajtan*, 18 (Suppl.), 219, 1969.
133. **Petrosyan, G. P. and Avundiyan, Z. S.**, Chemical reclamation effect on physical and water physical properties of soda saline soils in pre-Araks Lowland, In *Proc. Int. Symp. on New Dev. in the Field of Salt-Affected Soils*, Min. of Agric. Cairo, 1975, 513.
134. **Agaev, B. M. and Kulieva, G. S.**, Salt-release capacity of sodic soils of the Karabakh Plain in Azerbaijan, *Agrokémia és Talajtan*, 18 (Suppl.), 345, 1969.
135. **Pankova, E. I.**, Solonetz-solonchak soils of the central Asia sierozem zone with particular reference to their reclamation In *Proc. Int. Symp. on New Dev. in the Field of Salt-Affected Soils*, Cairo, Dec. 4-9, 1972, Ministry of Agriculture, Cairo, 1975, 759.
136. **Petrosyan, G. P. and Chitchyan, A. I.**, Soda-saline soils of the Ararat Plain and methods for their reclamation, *Agrokémia és Talajtan*, 18 (Suppl.), 121, 1969.
137. **Genesis, R.**, Regime and Amelioration of Salt-Affected Soils, Soviet Union (Russian), *Tr. Pochv. Inst. Dokuchaeva*, 54. Izd. Ak. Nauk SSSR, Moscow, 1958.
138. **Agababyan, V. G. and Akhumyan, M. S.**, Genesis of soda in saline soils of the Ararat Valley (Russian), *Pochvovedenie*, 9, 69, 1966.

139. **Blaskovic, P.**, Classification of maritime salt-affected soils in Yugoslavia, *Agrokémia és Talajtan,* 18 (Suppl.), 197, 1969.
140. **Munteanu, I. and Ionescu, M.**, Bog soils with maritime salinization in the Danube Delta (Rumania), *Trans. 8th Int. Cong. Soil Sci.*, 5, 667, 1964.
141. **Miljkovic, N. and Plamenac, N.**, Solonetz soils of Yugoslavia, their properties and possibilities of utilization, in *European Solonetz Soils and their Reclamation,* Szabolcs, I., Ed. 151, Akadémiai Kiadoó, Budapest, 1971.
142. **Zivkovic, B.**, The characteristics of groundwaters in Voyvodina from the standpoint of irrigation and their influence on the soils, in *Proc. Int. Symp. New Development in the Field of Salt-Affected Soils,* Ministry of Agriculture, Cairo, 1975, 149.
143. **Bazilevich, N. I. and Pankova, E. I.**, Tentative classification of soils according to salinity (Russian), *Pochvovedenie,* 11, 3, 1968.
144. *Saline Soils in North Kent,* Soil Survey of England and Wales, Harpenden, England, 1986.
145. *Studies on the Impact of Agricultural Management on the Environment. Summary of Research Results,* U.S.S.R./United Nations Environmental Programme, Project, Moscow, 1982.
146. **Kovda, V. A.**, Problem of Combating Salinization of Irrigated Soils, United Nations Environmental Programme, 1980.
147. **Raychadhuri, S. P. and Govinda-Rajan, S. V.**, *Soils of India,* Indian Council of Agricultural Research (ICAR), New-Delhi, 1971.
148. *Irrigation and drainage,* Paper 13, Food and Agriculture Organization, Rome, 1972.
149. **Boxem, H. W. and Wielmaker, W. M.**, *Soils of the Kucuk Menderes Valley,* PUDOC, Wageningen, The Netherlands, 1972.
150. **Borovsky, V. M.**, Criteria for evaluating the salinity of land for irrigation (Russian), *Vest. Akad. Nauk SSSR,* 9, 85, 1969.
151. **Kovda, V. A. and Egorov, V. V.**, Old and new problems of soil reclamation in an irrigated zone (Russian), *Pochvovedenie,* 4, 3, 1972.
152. **Kovda, V. A. and Lobova, E. V.**, *Geography of Soils in Asia,* Academy of Sciences, U.S.S.R., Moscow, 1965.
153. **Lobova, E. V.**, *Soils of Deserts in the USSR,* Academy of Sciences, U.S.S.R., Moscow, 1960.
154. **Borovsky, V. M.**, *Geochemistry of Soils in Kazakhstan* (Russian), Nauka, Moscow, 1978.
155. **Gorshenin, K. P.**, *Soils of South Siberia* (Russian), Academy of Scicnces, U.S.S.R., Moscow, 1955.
156. **Kovalev, R. V.**, *Soils of Novosibirsk* (Russian), Novosibirsk, 1966.
157. **Borovsky, V. M. and Pogrebinsky, M. A.**, Soil formation and soil amelioration in continental deltas (USSR) (Russian), *Dokl. VIII. Mezhd. Kongr. Pochv,* 173, 1964.
158. **Volobuev, V. R.**, Quantitative criteria for estimation of salt regime of irrigated and ameliorated lands, in *Proc. Int. Symp. New Dev. in the Field of Salt-Affected Soils, Cairo, Dec. 4-9, 1972,* Ministry of Agriculture, Cairo, 1975, 229.
159. **Wang Zunqin**, The processes and regionalization of soil salinization and alkalinization in China, in *Proc. Int. Symp. on Reclamation of Salt-Affected Soils,* Jinan, 1985, 48.
160. **Yu Ren Pei and Xione Yi**, Alkaline Soils in China, in *Proc. Int. Symp. on Reclamation of Salt-Affected Soils, Jinan, Jan. 1-9, 1985,* Jinan, 1985, 74.
161. **Kanwar, J. S.**, Salt-affected soils in India, *Agrokémia és Talajtan,* 18 (Suppl.), 79, 1969.
162. **Nazir Ahmad**, A review of salinity-alkalinity status of irrigated soils of west Pakistan, *Agrokémia és Talajtan,* 14, (Suppl.), 117, 1965.
163. **Hanna, A. B.**, *The nature of the saline (sabakh) soils of Iraq and their desalinization,* M.D. Thesis, University of Arizona, 1956.
164. **Boulaine, J.**, *Etude des Sols des Plaines du Chelif,* Min. de l'Algerie, Clairbois, 1957.
165. **Aubert, G. and Lejeaille, G.**, Observations on ground-water variations in an irrigated sector of Relizane (Algeria) (French), *Ann. Min. Agric.*, 67, 10, 1943.
166. **Aubert, G. and Ollat, C.**, Use of brackish water for irrigation and improvement of saline soils (French), *C. R. Hebd. Séanc. Acad. Agric.*, 55, 244, 1969.
167. **Durand, J. H.**, *Irrigated Soils* (French), Algiers, 1958.
168. **Aubert, G.**, *Sodic Soils in North Africa* (French), INAA, Algiers, 1975.
169. **Aubert, G.**, *Map of salty soils of Africa, Managing Saline Water for Irrigation,* Texas Tech University, Lubbock, Texas, 1966, 598.
170. **Aubert, G.**, Morphology of salt-affected soils (French), *FAO Soils Bull.*, Rome, 31, 187, 1976.
171. *Dryland Saline Seep Control,* Proc. Mting. Subcommission on Salt-Affected Soils, Edmonton, Canada, 1978.
172. *The Yearbook of Agriculture, 1955. Water,* United States Department of Agriculture, Washington, D.C., 1955.
173. **Halvorson, A. D. and Rhoades, J. D.**, Assessing soil salinity and identifying potential saline-seep areas with field soil resistance measurements, *Proc. Soil Sci. Soc. Am.*, 38, 576, 1974.

174. **Beek, C. G. E. M. van and Bolt, G. H.**, The relationship between the composition of the exchange complex and the composition of the soil solution, *Physical Aspects of Soil, Water and Salts in Ecosystems*, 4, 379, 1973.
175. **Janitzky, P. and Whittig, L. D.**, Natural leaching processes in some salt-affected soils of California, *Agrokémia és Talajtan*, 14, Suppl., 251, 1965.
176. **Bower, C. A. and Fireman, M.**, Saline and Alkali Soil, *Yearbook of Agriculture*, United States Department of Agriculture, Washington, D.C., 1957, 282.
177. **Greene, H.**, Using Salty Lands, *FAO Agric. Studies*, Food and Agriculture Organization, Rome, 1948, 3.
178. **Tanji, K. K. et al.**, A computer method for predicting salt concentration in soils at variable moisture contents, *Hilgardia*, 88, 307, 1967.
179. **Marbut, C. F.**, Soils of the United States, *USDA Atlas of American Agriculture*, United States Department of Agriculture, Washington, D.C., 1936.
180. **Bonnet, J. A.**, *The Soil Science of Saline and Alkali Soils* (Spanish), University of Puerto Rico, Rio Piedras, Est. Exp. Agric., 1962.
181. **Bennet y Allison**, *Los Suelos de Cuba*, United Nations Educational Scientific and Cultural Organization, Havana, Cuba, 1962.
182. *Suelos de Cuba*, Editorial Obre, Havana, Cuba, 1975.
183. **Loveday, J.**, Salt affected soils in Australia (Hungarian), *Agrokémia és Talajtan*, 34, 179, 1985.
184. **Blackburn, G.**, Salinity cycles in Australia, with special reference to the Murray Basin, *Trans. 10th Int. Congr. Soil Sci.*, 10, 27, 1974.
185. **Talsma, T. and Philip, J. R.**, *Salinity and Water Use*, Australian Academy of Science, Macmillan, New York, 1971.
186. **Borovsky, V. M.**, Formation of Salt Affected Soils and the Hydrogeology in Kazakstan (Russian), Alma Ata, Publ. House Nauka of Kazakh SSR, 1982.
187. **Kovda, V. A.**, *Geochemistry of Deserts in the Soviet Union*, (Russian), Izd. Ak. Nauk SSSR, Moscow, 1954.
188. **Szabolcs, I.**, Solonetz Soils in Europe. Their Formation and Properties with Particular Regard to Utilization, in *European Solonetz Soils and Their Reclamation*, Szabolcs, I., Ed., Akadémiai Kiadó, Budapest, 1971.
189. *International Symposium on Salt-Affected Soils, February 18-21, 1980*, ISSS-ICAR, Karnal, India, 1980.
190. **Fortescue, J. A.**, *Environmental Geochemistry*, Ecological Studies No. 35, Springer, New York, 1980.
191. **Fersman, A. E.**, *Geochemistry*, Leningrad, 1934.
192. **Vernadsky, V. I.**, *Chemical Composition of the Biosphere* (Russian), Nauka, Moscow, 1965.
193. **Polynov, B. B.**, *Selected Papers* (Russian), Izd. Ak. Nauk SSSR, Moscow, 1956.
194. **Brazilevich, N. I. and Kurachev. V. M.**, Migration of substances with surface and gravitational water in soils of geochemically associated landscapes of Baraba (Russian), *Pochvovedenie*, 11, 3, 1972.
195. **Bear, F. E., Ed.**, *Chemistry of the Soil*, Reinhold, New York, 1964.
196. **Dinauer, R. C.**, *Minerals in Soil Environments*, Soil Science Society of America, Madison, Wisconsin, 1977.
197. **Colman, S. M. and Dethier, D. P.**, *Rates of Chemical Weathering of Rocks and Minerals*, Academic Press Inc., London, 1986.
198. **Beus, A. A.**, *Geochemistry of the Lithosphere*, MIR Publishers, Moscow, 1976.
199. **Degens, E. T.**, *Geochemie der Sedimente*, F. Euke, Stuttgart, 1968.
200. **Berg, I. S.**, *Climate and Life*, Akadémiai Kiadó, Budapest, 1953.
201. **Gorbunov, N. I. et al.**, Mineral composition and physico-chemical properties of irrigated solonetzes and chestnut soils in the Volgograd Trans-Volga region (Russian), *Pochvovedenie*, 5, 77, 1975.
202. **Gorbunov, N. I.**, Sodium potassium and phosphorus reserves in solonetzes in relation to their mineralogical composition and particle size (Russian), *Pochvovedenie*, 5, 67, 1969.
203. **Shopski, N. and Eneva, S.**, The salinization of soils and ground water in the Stara Zagora irrigation system (Bulgaria), *Pochvoznanie i Agrokhimiya*, 10, 50, 1975.
204. **Oster, J. D.**, Measuring soil salinity, *Calif. Citrogr.*, 53, 280, 1968.
205. **Florea, N.**, Migration and accumulation of easily soluble salts in the Romanian Plain, *Agrokémia és Talajtan*, 23 (Suppl.), 154, 1974.
206. **Ianovici, V. and Florea, N.**, The accumulation of salts in the soils of quaternary plains in Rumania, *Studii Pedol.*, Ser. C(14), 5, 1964.
207. **Kabata Pendias, A. and Pendias, H.**, *Trace Elements in Soils and Plants*, CRC Press, Boca Raton, Fla., 1985.
208. **Szabolcs, I.**, Landscape geochemistry of soil salinization and alkalinization, *Agrokémia és Talajtan*, 30 (Suppl.), 47, 1981.
209. **Orlova, M. A.**, *Aeolic Factors in Salt Accumulation* (Russian), Publ. House Nauka of Kazakh SSR, Alma-Ata, 1983.
210. **Kovda, V. A. and Minashina, N. G., Eds.**, *Irrigation and Drainage of Salt-Affected Soils* (Russian), Izd. Nauka, Moscow, 1967.

211. **Kozlovsky, F. I. and Kornblyum, E. A.,** Soil ameliorative conditions in the Volga-Akhtuba bottomland (U.S.S.R.) in connection with its development and evolution, (Russian), *Pochvovedenie*, 7, 73, 1963.
212. **Hyams, E.,** *Soil and Civilisation*, John Murray, London, 1976.
213. **Kovda, V. A.,** *Principles of Soil Science I-II*, (Russian), Nauka, Moscow, 1973.
214. **Bazilevich, N. I.,** *Geochemistry of Soils with Soda Salinity* (Russian), Izd. Nauka, Moscow, 1965.
215. **Jackson, M. L.,** *Soil Chemical Analysis*, Prentice-Hall, Englewood Cliffs, N.J., 1958.
216. **Mohr, E. C. J., Baren, F. A., van, and Schuylenborgh, J. von.,** *Tropical Soils*, Mouton-Ichtiar Baru-Van Hoeve, The Hague, 1972.
217. **Servant, J. and Servat, E.,** Introduction to the study of saline soils on the Languedoc-Roussillon coastline (French), *Ann. Agron.*, 17, 53, 1966.
218. **Filipovski, G. and Ciric, M.,** *Zemljista Jugoslavie*, Beograd, Yugoslavia, 1963.
219. **El-Gabaly, M. M.,** Three types of sodic soils in the United Arab Republic, *Agrokémia és Talajtan*, 18 (Suppl.), 87, 1969.
220. **Egorov, V. V.,** Some unsolved problems of soda salinizaton of soils (Russian), *Pochvovedenia*, 11, 126, 1971.
221. **Darab, K.,** Methods for Investigation of Sodium Compounds and Sodium Hazard of Soils, in *Proc. Int. Symp. on New Dev. in the Field of Salt-Affected Soils*, Ministry of Agriculture, Cairo, 1975, 439.
222. **Chapman, J. V.,** *Salt Marshes and Salt Deserts of the World*, L. Hill, London, 1960.
223. **Zonn, S. V.,** *Tropical and Subtropical Soil Science*, MIR, Moscow, 1986.
224. **Kawaguchi, K. and Kyuma, K.,** *Lowland Rice Soils in Thailand*, Kyoto University, 1969.
225. *Proceedings of the International Symposium on the Reclamation of Salt-Affected Soils, I-II*, Jinan, China, 1985.
226. **Mattson, S.,** *Soil Colloids* (Russian), Selhozgiz, Moscow, 1938.
227. **Rhoades, J. D. and Ingvalson, R. D.,** Laboratory determination of leachable soil boron, *Proc. Soil Sci. Soc. Am.*, 34, 871, 1970.
228. **Poljakoff-Mayber-Gale J.,** *Plants in Saline Environments*, Springer-Verlag, New York, 1975.
229. **Ganev, S.,** Contribution to the chemistry of alkali soils (French), *C. R. Acad. Sci. Agric. Bulg.*, 3, 107, 1970.
230. **Scharpensel, H. W. and Ronzani, C.,** Dating of salt-affected soils with ^{35}Cl (Russian), in *Proc. Int. Symp. Reclam. Soda-Saline Soils*, Trudy Inst. Pochv. Agrokhim., Yerevan, 6, 617, 1971.
231. **Perelman, A. I.,** *Geochemistry of the Landscape* (Russian), Gedizdat, Moscow, 1961.
232. **Borovsky, V. M.,** *Formation of Salt Affected Soils and Halogeochemistry of Kazakhstan*, Alma-Ata, 1982.
233. **El-Gibaly, M. H. et al.,** Critical levels of alkali condition in some soils of Egypt as measured by the chemical and physical characteristics, in *Proc. Int. Symp. on New Dev. in the Field of Salt-Affected Soils, Cairo, December 4-9, 1972*, Ministry of Agriculture, Cairo, 1975, 447.
234. Proceedings of the Hungaro-Indian seminar on salt affected soils, Budapest, 1981, *Agrokémia és Talajtan*, 30 (Suppl.), 1981.
235. **Bekhar, A., Khubenov, G. R., and Marel, H. W. van.,** Highly interstratified clay minerals in salt-affected soils, *Contributions to Mineralogy and Petrology*, 34, 229, 1972.
236. **Kizilova, A. A.,** Movement of easily soluble salts in solonchak soils under leaching, *Arid Zone Res.*, 14, 227, 1961.
237. **Gedroitz, K. K.,** Solonetz soils, their origin, properties and amelioration (Russian), *Leningrad Izd. No-sovskaya opyt st.*, 46, 3, 1928.
238. **Gedroitz, K. K.,** *Selected Papers in Soil Science* (Russian), Vol. I., Soil Colloids and Exchange Properties of Soils; Vol. II. Chemical Analyses of Soils; Vol. III. Use of Fertilizers. Soil Amelioration and Field Experiments, Sel'-khozgiz, Moscow, 1955.
239. **Darab, K.,** Modern aspects of sampling and sodium balance studies in salt-affected areas, *Agrokémia és Talajtan*, 23 (Suppl.), 45, 1974.
240. **Szabolcs, I. and Darab, K.,** Dynamics of soluble salts in irrigated soils (Hungary), *Agrokémia és Talajtan*, 4, 251, 1955.
241. **Bhumbla, D. R. and Poonia, S. R.,** Effect of exchangeable sodium percentage on availability of calcium from gypsum and $CaCO_3$ applied to barley, dhaincha *(Sesbania cannabina)*, (Retz Pers.), and maize., *Indian J. Agric. Sci.*, 43, 1032, 1973.
242. **Gerei, L.,** Mineralogical composition of the <0.001 mm fraction of saline soils in the region of Lake Konyár (Russian), *Pochvovedenie*, 6, 109, 1967.
243. **Gerei, L.,** Transformation and destruction of clay minerals in alkali soils as affected by soil-forming processes, *Agrokémia és Talajtan*, 17 (Suppl.), 119, 1968.
244. **Gerei, L.,** Role of clay minerals, formation and transformation in sodic processes of the Danube Valley in Hungary, *Trans. 10th Int. Congr. Soil Sci.*, 7, 52, 1974.
245. **Szabolcs, I. and Leszták, J.,** The movements of different salt solutions in soil profiles, *IASH/UNESCO Symp.*, Wageningen, The Netherlands, 83, 611, 1968.

246. **Orlovsky, N. V.,** *Salt-Affected Soils in Western Siberia and their Amelioration* (Russian), Novosibirsk, 1941.
247. **Hissink, D. J.,** Base exchange in soils, *Trans. Faraday Soc.*, 20, 1925.
248. **'Sigmond, E.,** New theories on the formation of salt-affected soils and recommendations for their amelioration (Hungarian), *Földt. Közl.*, 67, 182, 1937.
249. **Kelley, W. P.,** *Cation Exchange in Soils*, Reinhold, New York, 1948.
250. **Cairns, R. R. and Szabolcs, I.,** Effects of sodium sulphate and sodium carbonate solutions on chemical and physical properties of chernozem soils, *Can. J. Soil Sci.*, 53, 399, 1973.
251. **Szabolcs, I.,** Sodium balance in alkali soils, *Trans. 10th Int. Congr. Soil Sci.*, 10, 49, 1974.
252. **Chaudhry, G. H. and Warkentin, B. P.,** Studies on exchange of sodium from soils leaching with calcium sulphate, *Soil Sci.*, 105, 190, 1968.
253. **Szendrei, G.,** Micromorphological Study of Alkali, or Szik Soils of the Kiskunság (Hungarian), *Agrokémia és Talajtan*, 19, 231, 1970.
254. **Raychaudhuri, S. P.,** Transformation of clay minerals under saline and alkali soil conditions, in *Proc. Int. Symp. on New Dev. in the Field of Salt-Affected Soils, Cairo, December 4-9, 1972*, Ministry of Agriculture, Cairo, 1975, 803.
255. **Abdel—Aal, S. I. and Bergseth, H.,** Ion exchange removal of carbonates from saline calcareous soils prior to determination of cation exchange capacity using $^{89}Sr^{2+}$, *Soil Sci.*, 120, 349, 1975.
256. **Wiklander, Ch.,** Cation and anion exchange phenomena, in *Chemistry of the Soil*, Bear, F. E. Ed., Reinhold, New York, 1964, 163.
257. **Gedroitz, K. K.,** *Investigation of the Cation Exchange Properties of Soils* (Russian), Izd. N. K. Z., Moscow-Leningrad, 1922.
258. **Gedroitz, K. K.,** Exchangeable cations and the physical properties of soils (Russian), *Zhurn. Prikhl. Him.*, 2, 180, 1925.
259. **Thomas, H. C.,** Toward a connection between ionic equilibrium and ionic migration in clay, in *Plant and Nutrient Supply and Movement*, International Atomic Energy Agency, Tech. Report No. 48, 4, Vienna, 1965.
260. **Basta, N. T. and Tabatabai, M. A.,** Determination of exchangeable bases in soils by ion chromatography, *Soil Sci. Soc. Am. J.*, 49, 84, 1985.
261. **Schainberg, I. and Letey, J.,** Response of soils to sodic and saline conditions, *Hilgardia*, 52, 2, 1984.
262. **Sparks, D. L.,** *Soil Physical Chemistry*, CRC Press, Boca Raton, Fla., 1986.
263. **Szabolcs, I. and Darab, K.,** New trends in ion-exchange studies on saline and alkali soils, in *Plant Nutrient and Supply and Movement*, International Atomic Energy Agency, Tech. Report No. 48, 24, Vienna, 1965.
264. **Sposito, G.,** *The Thermodynamics of Soil Solutions*, Oxford University Press, New York, 1981.
265. **Babcock, K. L.,** Theory of the chemical properties of soil colloidal systems at equilibrium, *Hilgardia*, 34, 417, 1963.
266. **Bolt, G. H. and Page, A. L.,** Ion exchange equation based on double layer theory, *Soil Sci.*, 99, 357, 1965.
267. **Krisnamoorthy, C. and Overstreet, R.,** An experimental evaluation of ion exchange relationships, *Soil Sci.*, 69, 41, 1950.
268. **Langerwerff, J. V. and Bolt, G. H.,** Theoretical and experimental analysis of Gapon's equation for ion exchange, *Soil Sci.*, 87, 217, 1959.
269. **Landelout, H. and Bladel, R. van.,** Interpretation of thermodynamic and thermochemical measurement of ion exchange reactions in clay, in *Limiting Steps in Ion Uptake*, International Atomic Energy Agency, 8, Vienna, 1965.
270. **Laudelout, H.,** Mathematical modelling of salt movement, *FAO Soils Bull.*, 31, 213, 1976.
271. **Gilbert, M.,** Thermodynamics of the ion exchange reaction between Na^+ and Mg^{2+}, *Agrokémia és Talajtan*, 18 (Suppl.), 251, 1969.
272. **Gapon, E. N.,** Ion exchange reactions of soils (Russian), *Pochvovedenie*, 2, 190, 1934.
273. **Darab, K. and Rédly, L.,** Investigations on the absorption of exchangeable calcium and sodium, with special reference to the properties of alkali (Szik) soils. I. Studies with equilibrium systems of bentonite-sodium chloride and bentonite—sodium sulphate (Hungarian) *Agrokémia és Talajtan*, 16, 575, 1967.
274. **Darab, K. and Rédly, M.,** Chemical and physico-chemical aspects of soil salinization and alkalization processes, *Agrokémia és Talajtan*, 28 (Suppl.), 83, 1979.
275. **Darab, K.,** Ion Exchange in the Formation and Amelioration of Salt-Affected Soils, in *Proc. Int. Symp. on Reclamation of Salt-Affected Soils*, Jinan, China, 1985, 309.
276. **Davis, C. W.,** *Ion Association*, Butterworths, London, 1962.
277. **Minkin, M. B., Endovitsky, A. P. and Levchenko, V. M.,** Ion association in soil solutions (Russian), *Pochvovedenie*, 2, 49, 1977.
278. **Rao, T. W., Page, A. Z., and Coleman, N. T.,** The influence of ionic strength and ion-pair formation between alkaline earth metals and sulphate on Na-Divalent cation exchange equilibria, *Proc. Soil Sci. Soc. Am.*, 32, 639, 1968.

279. **Tanji, K. K.**, Solubility of gypsum in aqueous electrolytes, *Environ. Sci. Technol.*, 3, 656, 1969.
280. **Adams, F.**, Ionic concentrations and activities in soil solutions, *Proc. Soil Sci. Soc. Am.*, 35, 420, 1971.
281. **Bennett, A. C. and Adams, F.**, Solubility and solubility product of gypsum in soil solution and other aqueous solutions, *Proc. Soil Sci. Soc. Am.*, 36, 288, 1972.
282. **Bennett, A. C. and Adams, F.**, Solubility and solubility product of dicalcium phosphate dihydrate in aqueous solutions and soil solutions, *Proc. Soil Sci. Soc. Am.*, 40, 39, 1976.
283. **Nakayama, F. S.**, Calcium activity complex and ion-pair in saturated $CaCO_3$ solutions, *Soil Sci.*, 106, 429, 1968.
284. **Darab, K., Csillag, J., and Pintér I.**, Studies on the ion composition of salt solutions and saturation extracts of salt-affected soils, *Geoderma*, 23, 95, 1980.
285. **Babcock, K. L.**, Effect of anions on the sodium-calcium exchange in soils, *Proc. Soil Sci. Soc. Am.*, 27, 630, 1963.
286. **Szabolcs, I. and Darab, K.**, The effect of the kind of anions on the calcium-sodium ion exchange in soils, *Agrochimica*, 10, 344, 1966.
287. **Szabolcs, I. and Darab, K.**, Investigation on the effect of anions on Na-Ca ion exchange, *Agrokémia és Talajtan*, 14 (Suppl.), 21, 1968.
288. **Sposito, G. and Mattigod, S. V.**, On the chemical foundation of the sodium adsorption ratio, *Soil Sci. Soc. Am. J.*, 41, 323, 1977.
289. **Vanselow, A. V.**, Equilibria of the base-exchange reactions of bentonites, permutites, soil colloids and zeolites, *Soil Sci.*, 1, 371, 1932.
290. **Darab, K.**, Studies on Ca-Na exchange in model systems, *Trans. 10th Int. Congr. Soil Sci.*, 2, 392, 1974.
291. **Boyko, H., Ed.**, Salinity and aridity, new approaches to old problems, *Monographiae Biologicae* (The Hague), 16, 408, 1966.
292. **Kovda, V. A.**, Arid land irrigation and soil fertility. Problems of salinity, alkalinity and compaction, in *Arid Land Irrigation in Developing Countries*, Pergamon Press, New York 1977, 211.
293. **Dudal, R. Tavernier, R., and Osmond, D.**, *Soil Map of Europe, 1:2,500,000*, Food and Agriculture Organization, Rome, 1966.
294. **Ruellan, A.**, Deep saline and alkali soils of Zebra Plain (Morocco) (French), *Proc. 8th Congr. ISSS*, 6, 937, 1964.
295. **Bols'hakov, A. F.**, Changes of soil-formation processes in solonchakous solonetzes due to radical disturbance of their profiles (Russian), *Pochvovedenie*, 10, 86, 1975.
296. **Pagel, H.**, Relationshps between salt content and other properties in the Guinea mangrove soils (German), in *Proc. Int. Symp. on New Dev. in the Field of Salt-Affected Soils, Cairo, December 4-9, 1972*, Ministry of Agriculture, Cairo, 1975, 62.
297. **Paliwal, K. V. and Anjaneyulu, B. S.**, Water-soluble boron in some saline-alkaline soils of Delhi, *J. Indian Soc. Soil Sci.*, 15, 103, 1967.
298. **Peck, A. J.**, Thomas, J. F. and Williamson, D. R., *Salinity Issues*, A.G.P.S., Canberra, 1983.
299. **Szabolcs, I. and Latkovics, I.**, The effect of various nitrogen fertilizers on the mineral nutrition of winter wheat studied on two different types of salt-affected soils, *Symp. Use Isot. Pl. Nutr. Physiol. Stud.*, International Atomic Energy Agency, Vienna, 1967, 3.
300. **Abrol, I. P. and Bhumbla, D. R.**, Field studies on salt leaching in a highly saline sodic soil, *Soil Sci.*, 115, 429, 1973.
301. **Malcolm, C. V.**, Wheatbelt Salinity, A review of the Salt Land Problem in South Western Australia, *Western Australian Department of Agriculture Tech. Bull.*, 52, 1982.
302. **Szabolcs, I. and Darab, K.**, Accumulation and dynamism of silicic acid in irrigated alkali soils, *Acta Agr. Acad. Sci. Hung.*, 8, 213, 1958.
303. **Arany, S.**, *Salt Affected Soils of Hortobágy* (Hungarian), Budapest, 1934.
304. **Pringle, W. L. et al.**, Salt status of some soils of the Slave River lowlands in Canada's northwest territories, *Can. J. Soil Sci.*, 55, 399, 1975.
305. **Pak, K. P.**, Solonetzes of the European Part of the USSR and their Reclamation, in *European Solonetz Soils and Their Reclamation*, Szabolcs, I., Ed., Akadémiai Kiadó, Budapest, 1971, 139.
306. **Kouhestani, E.**, Contribution to the study of alkali and earthy-alkali ions and some physical properties of soils (French), *Cah. Pedol.*, ORSTOM, 1964.
307. **Borovsky, V. M.**, Change of soils under the impact of irrigation (Russian), *Trans. 10th Int. Congr. Soil Sci.*, 10, 35, 1974.
308. **Rhoades, J. D.**, Measuring, mapping and monitoring field salinity and water-table depths with soil resistance measurement, *FAO Soils Bull.*, 31, 159, 1976.
309. **Ollat, C. et al.**, Salinization and alkalization of a saline calcareous soil irrigated with saline water in Tunisia (French), *Agrokémia és Talajtan*, 18 (Suppl.), 113, 1969.
310. **Grinchenko, A. M., Ed.**, Origin and amelioration of solonetz soils in the Ukraine (Russian), *Tr. Kharkov, Sel'khoz. Inst.*, 39, 1, 1962.
311. **Darab, K. and Ferencz, K.**, *Soil Mapping and Control of Irrigated Areas* (Hungarian), OMMI Kiadv., Ser. 1, No. 10, Budapest, 1969.

312. **Darab, K.**, Laboratory analyses of soils related to the prognosis and monitoring of salinity and alkalinity, *FAO Soils Bull.*, 31, 147, 1976.
313. **Antipov-Karataev, I. N.**, *Solonetz Amelioration in the USSR* (Russian), Academy of Science, Moscow, 1953.
314. **Viljams, V. R.**, *Soil Science* (Russian), Selhozgiz, Moscow, 1950.
315. **Szabolcs, I.**, *The Influences of Drainage and Irrigation on Soil Forming Processes in the Trans-Tisza Region* (Hungarian) Akadémiai Kiadó, Budapest, 1961.
316. **Gerei, L. and Szendrei, G.**, Micromorphological investigaion of some representative meadow and solonetz soils of the alluvial plain of the Zagyva River in Hungary, *Trans. 10th Int. Congr. Soil Sci*, 7, 264, 1974.
317. **Bower, C. A. and Rhoades, J. D.**, Fractionation of cation-exchange capacity for assessing soil and water sodicity, *Proc. Soil Sci. Soc. Am.*, 36, 174, 1972.
318. **El-Gabaly, M. M.**, Reclamation and Management of Salt-Affected Soils in *Proc. Int. Symp. on New Dev. in the Field of Salt-Affected Soils, Cairo, December 4-9, 1972*, Ministry of Agriculture, Cairo, 1975, 401.
319. **Rozanov, A. N.**, System of scientific investigations in projects for the irrigation and reclamation of saline soils, *Arid Zone Res.*, 14, 223, 1961.
320. **Artjushin, A. M. et al.**, Reclamation of solonetz soils in the USSR and principles for planning reclamation work, *Agrokémia és Talajtan*, 18 (Suppl), 307, 1969.
321. **Bower, C. A.**, Chemical Amendments for Improving Sodium Soils, *USDA Agric. Inform. Bull.*, 195, 1958.
322. **Muraközy, K.**, A talajról (Hungarian), *Természettud. Közl.*, 33, 5, 1902.
323. **Tyrurin, I. V.**, Biological silicium accumulation in soils (Russian), *Problems of Soviet Soil Science*, 4, 1934.
324. **Brinkman, R.**, *Ferrolysis, a Soil Forming Process in Hydromorphic Conditions*, Centre for Agriculture Publications and Documents, Wageningen, The Netherlands, 1979.
325. **Driessen, P. M.**, Soil Salinity and Alkalinity in the Great Konya Basin, Turkey *PUDOC Bull.*, Wageningen, The Netherlands, 743, 99, 1970.
326. **Naga, M. and Ahmed, S. H.**, A study on clay mineralogy of ancient saline-alkali soils in upper Egypt, in *Proc. Int. Symp. on New Dev. in the Field of Salt-Affected Soils, Cairo, December 4-9, 1972*, Ministry of Agriculture, Cairo, 1975, 827.
327. **Szabolcs, I. and Szendrei, G.**, Different forms of silicon compounds and their distribution in solod and solodized soils, *Agrokémia és Talajtan*, 29, 167, 1980.
328. **Szabolcs, I., Rédly, M., Pártay, G., and Szendrei, G.**, Properties of accumulation of calcium sulphate in gypsiferous soils from Iraq (Hungarian), *Agrokémia és Talajtan*, 29, 135, 1980.
329. **Dieleman, P. J., Ed.**, *Reclamation of Salt-Affected Soils in Iraq*, International Institute for Land Reclamation and Improvement, Wageningen, The Netherlands, Publ. No. 11, 1963.
330. **Schilfgaard, J. van et al.**, Irrigation management for salt control, *J. Irrig. Drain. Div. Am. Soc. Civ. Eng.*, 100, 321, 1974.
331. **Antipov-Karataev, I. N. and Filipova, V. N.**, *The Effect of Long-Term Irrigation on Soils* (Russian) Academy Science, U.S.S.R., Moscow, 1955.
332. **Arany, S.**, Behaviour of magnesium in soils (Hungarian), *Mezögazd. Kut.*, 4, 439, 1931.
333. **Mondal, R. C.**, Effect of high Mg Waters on a bentonite clay complex, *Indian J. Agric. Sci.*, 42, 1095, 1972.
334. **Polynov, B. B.**, *New Trends in the Study of Weathering* (Russian), Selected Works, Academy of Science, U.S.S.R., Moscow, 1956.
335. **Joffe, J. I.**, *Pedoloy*, New Brunswick, 1949.
336. **Es, F. W. J. van and Schuylenborgh, J. van.**, Contribution to the knowledge of a solonetzic, magnesium-rich alluvial silty clay in the Maro-Koembe Plain, *Neth. J. Agric. Sci.*, 15, 11, 1967.
337. **Oucharenko, M. N.**, On the transformation of primary and secondary minerals in solonetz profile (Russian), *Trans. Congr. Soil Sci.*, Moscow, 7, 146, 1974.
338. **Emerson, W. W. and Smith, B. H.**, Magnesium, organic matter and soil structure, *Nature*, 228, 453, 1970.
339. **Máté, F.**, Magnesium ion accumulation in the adsorption complex of meadow soil of the Tisza Valley (Hungarian), *Agrokémia és Talajtan*, 4, 291, 1955.
340. **Antipov-Karataev, I. N. and Mamaeva, L. J.**, The role of magnesium in the formation of solonetz properties of soils (Hungarian), *Agrokémia és Talajtan*, 4, 1, 1955.
341. **Black, C. A.**, Methods of soil analysis, *Am. Soc. Agric.*, 9, 1, 1965.
342. **Greenlands, D. J. and Hayes, M. H.**, *The Chemistry of Soil Constituents*, Wiley, London, 1978.
343. **Darab, K. and Reményi, M.**, Mineralogical composition of Mg soils in Hungary (Russian), *Trans. 10th Congr. Soil Sci.*, Moscow, 7, 125, 1974.
344. **Szendrei, G.**, Distribution of calcite and dolomite in soils of Kiskunság (Hungarian) *Agrokémia és Talajtan*, 26, 1, 1977.
345. **Darab, K.**, Magnesium in Solonetz Soils in *Int. Symp. of Salt-Affected Soils*, Karnal, India, 1980, 92.

346. **Kreybig, L.**, Vorkommen, Eigenschaften und Wirtschaftlicher Wert von Magnesia und Kaliböden in Ungarn, *Trans. 3rd Int. Congr. Soil Sci.*, 1, 353, 1935.
347. **Janson, W. K. and Moss, H. C.**, Exchangeable cations in solodized solonetz soils of Saskatchewan, *J. Soil Sci.*, 7, 203, 1956.
348. **Yakovleva, M. E.**, Chemical properties and fertility of solonetzes of high exchangeable-magnesium content, *Agrokémia és Talajtan*, 18 (Suppl.), 290, 1969.
349. **Rickard, D. T.**, Sedimentary iron sulphide formation, in *Acid Sulphate Soils* I, 3, IILRI Publ. No. 18, Wageningen, The Netherlands, 1973.
350. **Breemen, N. van.**, Soil forming processes, in *Acid Sulphate Soils* I, 66, IILRI Publ. 18, Wageningen, The Netherlands, 1973.
351. **Szabolcs, I.**, Accumulation of water-soluble salts in certain soils of Finland (Hungarian), *Agrokémia és Talajtan*, 11, 308, 1962.
352. **Kevie, W. van der.**, Physiography, classification and mapping of acid sulphate soils, in *Acid Sulphate Soils*, I, 204, IILRI Publ. 18, Wageningen, The Netherlands, 1973.
353. **Szabolcs, I.**, Agronomical and ecological impact of irrigation on soil and water salinity, *Advances in Soil Science*, 4, 189, 1986.
354. **Worthington, E. B.**, *Arid Land Irrigation*, Pergamon Press, Elmsford, New York, ICSU, 1977.
355. **Boyko, H., Ed.**, *Saline Irrigation for Agriculture and Forestry*, Dr. W. Junk N. V. Publishers, The Hague, 1968.
356. **Thorne, D. W. and Peterson, H. B.**, *Irrigated Soils*, Blakston Co., New York 1954.
357. **Talsma, T. and Philip, J. R., Ed.**, *Salinity and Water Use*, Wiley Interscience, New York, 1971.
358. **Hobbs, E. H. and Russell, G. C.**, Salt accumulation as influenced by irrigation method, *Can. J. Soil Sci.*, 47, 73, 1967.
359. **Alekseevsky, E. E.**, *Irrigation and Drainage of the World* (Russian), Kolos, Moscow, 1971.
360. **Cox, G. W. and Atkins, M. D.**, *Agricultural Ecology: an Analysis of World Food Production Systems*, W. H. Freeman, San Francisco, 1979.
361. **Novikova, A. V.**, Prognosis of secondary salinization and alkalization of soils due to irrigation (USSR) (Russian), *Agrokhim. Poch.* 5, 3, 1967.
362. **Dregne, H. E.**, *Managing Saline Water for Irrigation*, Technology University, Lubbock, Texas, 1967.
363. **Balba, A. M.**, Predicting soil salinization, alkalization and waterlogging, *FAO Soils Bull.*, 31, 241, 1976.
364. **Szabolcs, I., Darab, K., and Várallyay, G.**, Prediction and prevention of secondary salinization and alkalinization processes due to irrigation in the Hungarian Plain (Russian), *Pochvovedenie*, 1, 115, 1972.
365. **Szabolcs, I., Darab, K., and Várallyay, G.**, The Tisza irrigation systems and soil fertility in the Hungarian lowland. II. The critical depth of the water table in the area belonging to the irrigation system of Kisköre (Hungarian), *Agrokémia és Talajtan*, 18, 211, 1969.
366. **Szabolcs, I., Darab, K., and Várallyay, G.**, The Tisza irrigation system and soil fertility in the Hungarian lowland. III. Methods of preparation of 1:25,000 scale maps indicating the possibilities and the conditions of irrigation (Hungarian), *Agrokémia és Talajtan*, 18, 221, 1969.
367. **Jenkin, J. J.**, Terrain, groundwater and secondary salinity in Victoria, Australia, *Agric. Water Manage.*, 4, 143, 1981.
368. **Darab, K.**, Salt Balance and Salt Regime of Irrigated Soil in Hungary (Hungarian), *Agrokémia és Talajtan*, 10, 305, 1961.
369. **Maianu, A.**, Secondary Alkalization of Danube River Floodplains in Romania (Rumanian) in *Proc. Int. Symp. Reclam. Soda-Saline Soils, Yerevan, 1969*, Trudy Inst. Pochv. Agrokhim. 299, 1971, 6.
370. **El-Swaify, S. A. and Swindale, L. D.**, Effects of saline water on the chemical properties of some tropical soils, *Proc. Soil Sci. Soc. Am.*, 34, 207, 1970.
371. **Shainberg, I. and Shalhevet, J.**, Soil salinity under irrigation, *Ecological Studies*, Analysis and synthesis, Vol. 51. Springer-Verlag, New York, 1984.
372. **Osman, A. M. and El-Swaify, S. A.**, Effect of Irrigation Management Practices on Salt Accumulation in a Tropical Oxisol, in *Proc. Int. Symp. on New Dev. in the Field of Salt-Affected Soils, Cairo, December 4-9, 1972*, Ministry of Agriculture, Cairo, 1975, 74.
373. **Molen, W. H. van der and Boumans, J. H.**, Drainage requirements of irrigated soils in relation to salinity, *Trans. 8th Int. Congr. Soil Sci.*, 2, 847, 1964.
374. **Szabolcs, I., Darab, K., and Várallyay, G.**, Methods of predicting salinization and alkalinization processes due to irrigation on the Hungarian Plain, *Agrokémia és Talajtan*, 18, (Suppl.), 351, 1969.
375. **Raheja, P. C.**, Saline soil problems with particular reference to irrigation with saline water in India, in *Saline Irrigation for Agriculture and Forestry*, Boyko, H., Ed. Junk N. V., The Hague, 1968, 217.
376. **Den, G. D.**, Secondary salinization of soils and salt accumulation in the irrigated region of Lokwei (Chinese), *Acta Pedol. Sin.*, 13, 59, 1965.
377. **Nazir Ahmad,** A review of salinity-alkalinity status of irrigated soils of West Pakistan, *Agrokémia és Talajtan*, 14, (Suppl.), 117, 1965.

378. **Ontanon, J. M.,** *In situ* study of salinization in the new irrigation system at La Agra by conductivity measurements (Spanish), *Salt-Affected Soil Subcommission Meeting, Seville, May, 1971,* IRYDA Inform., 1971, 57.
379. **Pekatoros, L. G.,** Secondary salinization of soils in the Dneper left bank bottomlands and Danube delta, in Ukraine (Russian), *Pochvovedenie,* 2, 26, 1962.
380. **Bohoci, L., Ionescu, C., and Covaliov, T.,** Soil and groundwater salinization in the dammed enclosure from Calmatui-Gropeni, *Studii Hidroamelior,* 4, 263, 1968.
381. **Szabolcs, I., Darab, K., and Várallyay, G.,** Salt balance of irrigated soils (Hungary), *Beitr. trop. Landw. Vet. Med.,* 4, 123, 1966.
382. **Novikova, A. V.,** Prognosis of secondary salinization and alkalization of soils due to irrigation (Russian), *Agrokh. Pochv.,* 5, 3, 1967.
383. **Dregne, H. E.,** Predicting soil salinity from irrigation-water salinity, *Trans. 10th Int. Congr. Soil Sci.,* 10, 198, 1974.
384. **Kolpakov, V. V.,** Effect of a water table rise on evaporation and the water soluble salt regime of soils in the Presivash Region (USSR) (Russian), *Izv. Timiryazev Sel'khoz Acad.,* 1, 132, 1961.
385. **Abrol, I. P., Bhumbla, D. R., and Bhargava, G. P.,** Some Observations on the Effect of Ground-Water Tables on Soil Salinization, in *Proc. Int. Symp. on New Dev. in the Field of Salt-Affected Soils, Cairo, December 4-9, 1972,* Ministry of Agriculture, Cairo, 1975, 40.
386. **Szabolcs, I., Várallyay, G. and Darab, K.,** Soil and hydrologic survey for the prognosis and monitoring of salinity and alkalinity, *FAO Soils Bull.,* 31, 119, 1976.
387. **Nazir Ahmad,** *Tubewell,* Lahore, Pakistan, 1979.
388. **Szabolcs, I.,** Current and potential salinity of soils, in Assessing soil degradation, *FAO Soils Bull.,* 34, 10, 1977.
389. **Pels, S. and Stannard, M. E.,** Environmental changes due to irrigation development in semi-arid parts of New South Wales, Australia, in *Arid Land Irrigation in Developing Countries, Environmental Problems and Effects,* Worthington, E. B., Ed., Pergamon Press, Elmsford, New York, ICSU, 1977, 171.
390. **Bresler, E. and Yaron, D.,** Soil-water regime in economic evaluation of salinity in irrigation, *Water Resour. Res.,* 8, 791, 1972.
391. **McIntyre, D. S., Loveday, J., and Watson, C. L.,** Field studies of water and salt movement in an irrigated swelling clay soil. III. Salt movement during ponding, *Aust. J. Soil Res.,* 20, 101, 1982.
392. **Szabolcs, I. and Darab, K.,** Salt balance and salt transport processes in irrigated soils (Hungary), *9th Int. Congr. Soil Sci.,* 1, 491, 1968.
393. **Yaron, B. and Thomas, G. V.,** Soil hydraulic conductivity as affected by sodic water, *Water Resour. Res.,* 4, 545, 1968.
394. **Szabolcs, I. and Darab, K.,** Irrigation water quality and problems of soil salinity, *Acta Agric. Sci. Hung.,* 31, 173, 1982.
395. **Eckholm, E. P.,** *Losing Ground,* Norton Co., New York, 1976.
396. **Talsma, T.,** Control of saline ground-water, *Meded. Landbouwhogeschool* (Wageningen) 63, 1, 1963.
397. **Várallyay, G.,** Soil/Water Problems Related to Salinity and Alkalinity of Irrigated Lands in *Arid Land Irrigation in Developing Countries,* Pergamon Press, New York, 241.
398. **Anayan, A. K.,** *Drainage for the Amelioration of Soda Solonchaks (USSR)* (Russian), Izd. Kolos, Moscow, 1971.
399. **Boumans, J. H. and Molen, W. H. van der.,** Drainage requirements of irrigated soils in relation to their salinity, *Landb. Tijdschr.,* 76, 880, 1964.
400. **Hoorn, J. W. van.,** Irrigation with salty water in Tunisia, in *Saline Irrigation for Agriculture and Forestry,* Boyko, H., Ed., Junk N. V., The Hague, 1968, 168.
401. **Bower, C. A., Spencer, J. R., and Weeks, L. O.,** Salt and water balance, Coachella Valley, California, *Proc. Am. Soc. Civ. Eng.,* 95, 55, 1969.
402. **Szabolcs, I., Darab, K., and Várallyay, G.,** Salt balance for the prediction and prevention of secondary salinization and alkalinization of irrigated soils (Hungary), *Nemzetk. Mezögazd. Szemle,* 13, 46, 1969.
403. **Yaron, B. J., Shalhevet, J., and Shimshi, D.,** Patterns of salt distribution under trickel irrigation, in *Physical Aspects of Soil Water and Salts in Ecosystems,* 389, 1973.
404. **Kovda, V. A.,** *Biogeochemistry of Soils* (Russian), Nauka, Moscow, 1985.
405. **Rhoades, J. D., Krueger, D. B., and Reed, M. J.,** The effect of soil-mineral weathering on the sodium hazard of irrigation Waters, *Proc. Soil Sci. Soc. Am.,* 32, 643, 1968.
406. **Bresler, E.,** A model for tracing salt distribution in the soil profile and estimating the efficient combination of water quality and quantity under varying conditions, *Soil Sci.,* 104, 227, 1967.
407. **Chou, M., Harmon, D. P., Kahn, H., and Wittwer, S. H.,** *World Food Prospects and Agricultural Potential,* Praeger, New York, 1977.
408. **Zonneveld, I. S.,** Survey methods for performance, monitoring and prognosis of natural vegetation and economic crops with special reference to salt-affected soils, *FAO Soils Bull.,* 31, 195, 1976.
409. *Soil. The 1957 Yearbook of Agriculture,* United States Department of Agriculture, Washington, D.C., 1957.

410. **Kovda, V. A.,** Scientific basis of soil amelioration in the USSR (Russian), *Pochvovedenie,* 11, 14, 1966.
411. **Várallyay G.,** Extreme moisture regime as the main limiting factor of the fertility of salt affected soils, *Agrokémia és Talajtan,* 30 (Suppl.), 73, 1981.
412. **Prettenhoffer, I.,** *Utilization and Amelioration of Hungarian Salt-Affected Soils in the Trans-Tisza Region* (Hungarian), Akadémiai Kiadó, Budapest, 1969.
413. **Dobrzanski, B. et al.,** Effect of chemicals used for snow removal on soil salinization of grassed areas in Warsaw (Polish), *Roczniki Glebozn.,* 22, 59, 1971.
414. **Egorov, V. V. and Minashina, N. G.,** Amelioration of salt-affected soils and meliorative soil science in the USSR (Russian), *Pochvovedenie,* 11, 3, 1967.
415. **Szabolcs, I.,** Reclamation methods developed for solonetz and solod soils with a view to their formation processes, *Acta Agron. Acad. Sci. Hung.,* 18, 339, 1969.
416. **Minashina, N. G.,** Physico-chemical model for calculating rates of water for washing saline soils (Russian), *Pochvovedenie,* 3, 115, 1972.
417. **Molen, W. H. van der,** Drainage principles and applications, theories of field drainage and watershed runoff. Salt balance and leaching requirement, *Int. Inst. Land Reclam. Improv. Publ. No. 16,* 59, 1973.
418. **Anderson, J. U., Bailey, O. F., and Dregne, H. E.,** Short-term effects of irrigation with high-sodium waters, *Soil Sci.,* 113, 358, 1972.
419. **Yadav, J. S. P.,** Improvement of saline alkali soils through biological methods, *Indian Farming,* 101, 385, 1975.
420. **Singh, H. P. and Szabolcs, I.,** The nutrient status and genesis of alkali (szik) soils on the Hungarian Plain. I. Humus and nitrogen status (Hungarian), *Agrokémia és Talajtan,* 18, 351, 1969.
421. **Abrol, I. P., Dahiya, I. S., and Bhumbla, D. R.,** On the method for determining the gypsum requirements of soils, *Soil Sci.,* 120, 30, 1975.
422. **Tóth, B. et al.,** *Afforestation of Salt-Affected Soils* (Hungarian), Akadémiai Kiadó, Budapest, 1972.
423. **Rana, R. S., Ed.,** *Afforestation of Salt-Affected Soils,* Vol. 1-3, CSSRI, Karnal, India, 1987.
424. **Barrett-Lennard, E. G., Malcolm, C. V., Stern, W. R., and Wilkins, S. M., Eds.,** *Forage and Fuel Production from Salt-Affected Wasteland,* Elsevier, New York, 1986.

INDEX

A

D

E

F

G

H

I

J

K

L

M

N

O

P

Q

R

S

T

U